How Wars Are Won and Lost

How Wars Are Won and Lost

Vulnerability and Military Power

JOHN A. GENTRY

Praeger Security International

PRAEGER

AN IMPRINT OF ABC-CLIO, LLC
Santa Barbara, California • Denver, Colorado • Oxford, England

Library of Congress Cataloging-in-Publication Data

Gentry, John A., 1949–
How wars are won and lost : vulnerability and military power / John A. Gentry.
p. cm.
Includes bibliographical references and index.
ISBN 978-0-313-39582-6 (hbk. : alk. paper) — ISBN 978-0-313-39583-3 (ebook)
1. Strategy. 2. Strategic culture—United States. 3. Military art and science—United States. 4. History, Military—United States—20th century. 5. History, Military—United States—21st century.
I. Title. II. Title: Vulnerability and military power.
U162.G348 2012
355.02—dc23 2011030647

ISBN: 978-0-313-39582-6
EISBN: 978-0-313-39583-3

16 15 14 13 12 1 2 3 4 5

This book is also available on the World Wide Web as an eBook.
Visit www.abc-clio.com for details.

Praeger
An Imprint of ABC-CLIO, LLC

ABC-CLIO, LLC
130 Cremona Drive, P.O. Box 1911
Santa Barbara, California 93116-1911

This book is printed on acid-free paper ∞

Manufactured in the United States of America

Contents

Acknowledgments

This book, less chapter 7 on Afghanistan, began as a PhD dissertation submitted to the political science department at the George Washington University in 2008. Chapter 5 draws appreciably from my article "Norms and Military Power: NATO's War Against Yugoslavia," which was published in *Security Studies* 15, no. 2 (April–June 2006), portions of which are adapted and reprinted by permission of the publisher (Taylor & Francis Ltd, http://www.tandf.co.uk/journals). Susan Gentry, Joseph Gordon, Farhana Qazi, Chad Rector, Elizabeth Saunders, Ronald Spector, Erik Voeten, and especially Mike Mochizuki, provided helpful suggestions on drafts of this book.

CHAPTER 1

Introduction

Military power is a key component of state power. Yet for all the attention statesmen, soldiers, and scholars give to military power, it remains a loose concept. A considerable academic effort has catalogued characteristics of great power wars.[1] Scholars know much about individual wars, and they understand general patterns of war.[2] But prominent theories of military power as a determinant of strategic political/military outcomes remain crude.[3]

Many international relations (IR) scholars rely on material-based concepts of military power exemplified by the Correlates of War Project (COW), which postulates six elements of national power in three general areas: military capabilities, defined as military expenditures and personnel; industrial capabilities, defined as energy consumption and iron/steel production; and demographic potential, defined as national and urban populations.[4] Others prefer gross national product as a measure of power.[5]

The term *capabilities* appears often as a synonym for both military power and the stock of military-related material assets a state possesses—independent of actors' abilities to influence military outcomes.[6] For example, Kenneth Waltz holds that states' capabilities consist of unspecified combinations of "size of population and territory, resource endowment, economic capability, military strength, political stability and competence."[7] While recognizing that the outcomes of international interactions are uncertain, he claims power is measurable by estimating state capabilities.[8] Waltz argues that power is homogenous, making it fungible across operational contexts and thereby enabling it consistently to determine political/military outcomes.[9] Some theorists hold that national material wealth is so critical to security that states should refuse to deal with other states unless their "relative gains" from each international interaction exceed those of their "partners."[10] While possessing a much more sophisticated understanding of the nature and complexity of military power, even Klaus Knorr regards material resources as the major determinant of military power.[11]

A body of theoretical work known as "offense-defense theory" holds that advantages to offensive or defensive military operations exist over extended periods of time due primarily to long-lived technologies and geographical factors.[12] Other materialist-oriented scholars argue that while resource endowments largely determine outcomes, the unspecified "skill" of military forces, which evidently means something close to Waltz's similarly vague concept of state "competence," also influences outcomes.[13]

Unfortunately, the empirical record of materialist theories of military power is poor. Seemingly dichotomous power preponderance and power transition theories both have poor records.[14] Offense-defense theory has logical and empirical weaknesses.[15] COW-like measures cannot explain al-Qaeda's worldwide military capacity or the inability of the United States to defeat lightly armed insurgents in Afghanistan. A purported "superpower" since 1945 and "hegemon" since 1991, the United States repeatedly has done poorly in its recent wars.

Wealthier actors not only regularly lose wars, in recent decades they have lost increasing shares of them.[16] Materialist-oriented theorists little address how "skill" or "competence" can account for this trend, although Christopher Layne ponders reasons why allegedly hegemonic U.S. power frequently fails to achieve American political/military goals.[17] Military analysts, who devote much more attention to such issues than do academics, also rarely forecast wars' courses and outcomes accurately.[18] A basic question, therefore, remains unanswered: what is the nature of military power?

The purpose of this book is to present, illustrate, and test an alternative general theory of *military power* defined as *an ability to consistently, favorably influence strategic military outcomes*—that is, how and why actors win and lose wars. I also aim to use the theory to help explain why the United States has done very well in some of its military operations but very poorly in others, sometimes virtually simultaneously.

THEORY

I offer here a theory that accounts for the deficiencies of materialist theories of power and incorporates and builds upon a set of existing theories to explain how and why actors win and lose wars. I call it *vulnerability theory.*

I argue that a political/military actor (a state or nonstate group like al-Qaeda or Hezbollah that conducts organized military operations for political purposes) has military-related assets if it chooses to generate, produces, and effectively uses the assets in six broad, interrelated but discrete ideational and material arenas, or what I call dimensions, each of which is divisible: (1) national will, (2) resource mobilization, (3) resource conversion, (4) force generation, (5) leadership, and (6) operational execution. Some activity in all of the dimensions is required to produce military power. However, an actor can defeat an enemy by disrupting the adversary's production of military power at any point in its process. While power production processes are continuous, I focus on the dimensions as the key stations or segments along an "assembly line" on which actors produce capacities to influence strategic military outcomes.

An actor has military power if the actor: (1) accurately identifies exploitable vulnerabilities in a target adversary in any dimension *and* (2) successfully exploits one or more of the target's critical vulnerabilities, leading to physical military defeat or disruption of power production processes, leading to a decision to forego or to stop fighting, which concedes the point of contention. I distinguish *critical* vulnerabilities from other vulnerabilities as ones that if exploited significantly alter strategic outcomes.

The first condition requires institutional powers of observation and abilities to learn about enemies and operational environments, broadly defined. The second requires building and maintaining operationally effective organizations capable of military-related action against critical adversary vulnerabilities through good preparations before wars and effective adaptations during conflicts. The importance of identifying and exploiting enemy vulnerabilities, given the interactive nature of military power, means a third independent variable—an ability to identify and ameliorate one's own exploitable vulnerabilities—also contributes importantly to the determination of strategic outcomes. Successful vulnerability amelioration may disguise vulnerabilities, preventing their identification by enemies, or alter physical or institutional characteristics in ways that make vulnerabilities less exploitable.

The specific characteristics and relative importance of the three independent variables depend upon the dynamic interaction of belligerents. To possess military power, actors must identify opportunities and make in advance adequate political/military forces or adapt their forces during wars in the context of ongoing learning and adaptation by their adversaries. In long wars, especially, much learning and effective, adaptive adjustments, often made in the context of considerable uncertainty, may be necessary to achieve success.

These variables embed, reflect, and influence many normal functions of government, which largely determine whether, and how, actors identify, exploit, and ameliorate vulnerabilities. These factors amount to actors' collective institutional skills, providing a theoretical link between vulnerability theory and the *skill* and *competence* parts of various materialist theories. They also appear regularly in military operations, including the case studies below. While details vary by case, general categories of activities actors need to perform to successfully identify, exploit, and ameliorate vulnerabilities include the following:

1. To *identify* adversary vulnerabilities, successful actors accurately observe their adversaries' actions, assess their intentions, and understand operational environments, meaning they need good intelligence collection and analytical capabilities, which typically are provided by troops in the field, intelligence agencies, and foreign ministries. Successful senior national leaders accept accurate intelligence warnings and assessments of the opportunities and threats that foreign situations present, which help but do not ensure effective policy making. Leaders prevent relevant and accurate intelligence messages from being killed or warped by cognitive blinders or analytical errors. They keep unwelcome assessments from being quashed or slanted to serve parochial institutional interests.
2. To *exploit* adversary vulnerabilities, successful actors convert learning from their operational successes and failures, from intelligence assessments, and from learning vicariously about the experiences of other belligerents into effective new or revised policies, strategies, doctrines, operational plans, and target packages. They field competent military and complementary civilian forces that work well together and effectively implement policies. Military organizations promptly and effectively adapt internally to changing operational environments, and when directed by political leaders, they change to serve collective needs even if such actions damage their parochial institutional interests. That is, senior leaders make good policies, build competent and cooperative subordinate organizations, and supervise effective and timely policy implementation.[19]
3. To *ameliorate* their own vulnerabilities, successful actors convert perceptions of threatening actors' technical capabilities and intentions, and judgments about their own internal weaknesses, into recognition of exploitable vulnerabilities in their own governmental institutions, electorates, economies, physical infrastructures, and military forces. They identify effective remedial action and implement it despite political and economic constraints they recognize. They accurately estimate the financial costs and time required to ameliorate critical vulnerabilities.

The specific means by which actors achieve these institutional performance characteristics vary within states and across time and conflicts, sometimes significantly.

Vulnerability is not the same as weakness, although material weaknesses sometimes produce vulnerabilities. It is a condition in one actor that another actor may use to its advantage. To use an economic example for illustrative purposes, the specialization that makes modern economies highly productive also creates vulnerabilities, including dependence on electricity. In a hypothetical conflict in which the reliability of the supply of electricity of each belligerent is a critical part of the resource mobilization dimension, belligerents may *identify* vulnerabilities in enemy electricity generation stations, transmission lines, fuel supplies, or transformers. To *exploit* vulnerabilities, they may develop forces of electrical engineers who understand the physics of electricity, computer hackers who can command networks to overload or shut down, commandos able to down transmission towers, or air forces able to destroy enemy generating stations. And, to *ameliorate* their own vulnerabilities, they may harden electric motors against electromagnetic pulse attacks, equip industrial facilities with backup generators to dampen the effects of electric power supply disruptions, train more repair crews, or convert electrically powered facilities to alternative energy sources. This simple example suggests that many permutations of options are available to actors in even a narrowly limited "war"; real conflicts are much more complex.

Military power exists independently of belligerents' absolute and relative material asset holdings and technological prowess. While it does not determine military power, some material wealth always is needed, and large material wealth holdings may confer advantages by, for example, enabling actors to produce specialized assets that help identify and exploit others' vulnerabilities. But more assets may also contribute to the growth of bureaucracies whose size and complex structures generate organizational rigidities that inhibit collective institutional learning and adaptation, thereby hindering military operational effectiveness and efficiency.

Military power is relational. It exists only for a given set of adversary vulnerabilities, exploitation capacities, and actions. It depends on what I call the "direction" and type of the interaction.[20] Put simply, for any adversary pair, or *dyad*, an actor may be able to win an offensive fight but not a defensive one, or vice versa, or effectively conduct some kinds of offensive actions but not others. As Bernard Finel similarly reasons, war is like the children's rock-paper-scissors game: a player who chooses paper wins if his opponent chooses rock but loses if his adversary picks scissors.[21] By more plausibly real example, al-Qaeda is a potential threat to detonate a radiological bomb near the White House but not to conquer Kansas. There is no consistent ordinal or cardinal measure of power that ranks any set of actors' military power. Indeed, there is no such thing as generic military power.[22]

I argue that the number of operationally relevant subdimensions has risen over time, increasing opportunities for vulnerability identification and exploitation, and therefore the complexity and importance of vulnerability amelioration. The scope of conflict has expanded due to technological change, which added many new types of warfare—like air, space, nuclear, and cyberwarfare—in the 20th century alone. In addition, the democratization of polities and evolving norms reflecting aversion to casualties in war—particularly civilian casualties since the end of World War II and, in the U.S. case, military casualties since the 1980s—have created new opportunities for actors to use tactical military actions that generate casualties, independent of traditional measures of military success like taking ground or smashing enemy combat units, to achieve strategic political objectives. Technologies that enable precision air and missile strikes from afar now help militaries ameliorate vulnerabilities to the political costs of violating both casualty aversion norms. Communications technologies like television and the Internet enable propagandists to deliver politically relevant sights and sounds of war to target audiences. Weak actors also have new tactical military techniques, like suicide bombings, not in the inventories of conventional state military forces.

The expansion in recent decades of the types of possible political/military operations has repeatedly enabled effective warfare by materially weak states like North Vietnam and nonstate groups like al-Qaeda. Dimensions in which critical operational decisions occur similarly have evolved over time—prominently shifting from material to ideational dimensions, especially national will and leadership.[23] These factors largely account for the poor military performance of wealthy states like the United States since 1945.

Time plays an important role in vulnerability theory in two ways. First, the speed of effective learning and adaptation during conflicts is a major determinant of strategic outcomes. Successful actors typically learn and adapt faster than their opponents. Second, because actors learn from the historical record about strategies, tactics, and operational techniques that worked (and did not) in actual conflicts, time between conflicts enables adaptation that generates evolution in the conduct of warfare. Capable actors learn to fight their last wars better but also generate new approaches that reflect peacetime assessments of potential adversaries' ongoing adaptations and changing circumstances.

COW-like materialist concepts of military power apply only when actors cannot identify or asymmetrically exploit opponents' vulnerabilities, leading to symmetrical conflicts—like infantrymen attacking similarly armed infantrymen in parallel lines across open fields—in which the residual vulnerability is relative material weakness in the form, in this example, of fewer soldiers. Because asymmetric attack opportunities now are abundant, the default condition of attritional warfare that IR theory focuses upon is increasingly rare, making COW-like measures progressively less meaningful for contemporary political-military analyses.

Actors' vulnerabilities, their abilities to exploit others' vulnerabilities, and the six dimensions are variables, not constants, although some characteristics of them evolve so slowly that at any moment they may seem fixed. Each variable is shaped primarily by actors' internal policies and institutions before wars, but belligerents frequently try both to exacerbate existing adversary vulnerabilities and actualize latent ones by using propaganda, mobilizing sympathetic forces within enemy states, and conducting military operations to influence political events like national elections, for example. Some actors also try to identify and ameliorate their own vulnerabilities before wars but often fail to do so—including the United States before September 2001.[24]

Like most theorists of military power, I do not address why actors choose to use their military assets.[25] Motivations for using force vary considerably, and a full discussion of this topic lies beyond the scope of this work. I therefore do not offer a theory of war in general, its causes, or the reasons people fight.[26] Nevertheless, motives for, and concerns about, using force in any given circumstance affect the national will and leadership dimensions of military power, especially, and create or limit exploitable vulnerabilities. Hence, actors' goals and constraints are essential parts of the analysis of the military power present in all conflicts.

Vulnerability theory addresses the nature of actors' abilities to achieve favorable strategic military outcomes and offers influence mechanisms that materialist theories of military power typically lack.[27] The dimensions include actions in all three levels of war—tactical, operational, and strategic—reflecting actors' integration of various types of military and military-related political and economic activities. The scope of vulnerability theory, therefore, is much broader than many of the strategies that appear in IR theories—like "blitzkrieg," "limited war," "deterrence," and "attrition" strategies that all are vaguely defined, sometimes are closer to tactics than strategy, and often are more narrowly military than the dimensions I posit here.[28]

Military power is best characterized by the extent to which actors' strategic political-military goals, including, but not restricted to, battlefield victory, are achieved. In general, standards of assessment of degrees of success and failure should be the goals that actors postulated when entering conflicts, not ones modified later to deflect domestic criticisms of failure, for example.

The six dimensions are intermediate variables in the sense that actors operate in them to influence strategic outcomes. All are deeply embedded in military historical and IR literatures. The dimensions are the "places" where actors make and use military power, but the specifics of power making and war fighting vary considerably from conflict to conflict.

I briefly define below the dimensions, place them in theoretical context, and outline means of identifying dimensional vulnerabilities, their exploitation, and their relationships to strategic outcomes. All are identifiable independently of outcomes. Aspects of the dimensions are identifiable in many ways that do not feature common measurement standards. The dimensions are listed in the order they produce capacities to influence military outcomes. A deficiency in any one may preclude effective action in later stages; however, during long wars, especially, feedback loops mean a lack of success in a later stage (like battlefield fighting) may affect an earlier one (like popular desire to continue waging a losing war). No one dimension is inherently more or less important than any other, although for political or technological reasons one or more may be relatively more important for any one actor at a specific time or generally for most actors in an era.

National Will

Political-military actors must desire to produce military capacities and to use them. If will is absent, no amount of material wealth can produce military power.[29] The concept of collective national will (independent of the desires of absolute monarchs) goes back at least to revolutionary France, is embedded in the people portion of Carl von Clausewitz's trinity of actors (people, the military, and the state), and appears in the work of many academics and military practitioners.[30] I use the term *national will* even though some nonstate actors like al-Qaeda are driven by variants of collective will that are not national in nature. Will exists (or not) to conduct specific types of operations against specific opponents; the will of a polity to conduct each of a variety of operations it is capable of conducting may vary considerably vis-à-vis different potential adversaries for many reasons, as the democratic peace literature has partly documented.[31]

While aspects of national will are identifiable in many ways, for purposes of linking it to military power, national will often is best evidenced qualitatively by the influence of popular opinion on legislative decisions to enter, end, or modify the conduct of wars and in the orders of civilian leaders, responding to perceived national will, to their military commanders. In the U.S. case, this also means presidential orders to heads of security-related civilian agencies like the Department of State and the Central Intelligence Agency (CIA).[32]

This dimension has become a prominent arena of the modern wars of Western, democratic states, and national will has prominently influenced the terms of settlement of several recent American military operations.[33] Statesmen variously seek to influence national will by appealing to motives of patriotism, material gain, revenge, and altruism. Sometimes, as in 1898 against Spain and after the surprise attacks of December 7, 1941, and September 11, 2001, popular opinion helps drive U.S. politicians to instigate war. But two concerns recurrently in the last half century have led Americans to demand limits on the conduct of, or ends to, ongoing U.S. wars—the level and trend of U.S. military casualties and civilian casualties caused by U.S. military actions, especially when military operations were not going well. Actors commonly recognize such concerns and often try to influence adversaries' national will by using information operations, especially, to delegitimize wars by publicizing civilian casualties, and by both literally increasing and propagandistically emphasizing the human costs of casualties. Expansion of the scope of conflict in recent decades into arenas not traditionally considered warfare—prominently including versions of "terrorism"—offer even materially very weak actors ways of using targeted violence to attack materially strong actors' strategically important political vulnerabilities in the national will dimension.[34]

National will influences military operations through its impact on leaders' operational goals, the *rules of engagement* (ROE) that guide military units' use of force, choices about the composition of forces (or *force structure*) employed in operations, and targeting decisions. National will–driven, casualty aversion–motivated restrictions on Western militaries often produce highly restrictive ROE and force protection imperatives. These create vulnerabilities that numerous enemies of Western countries in general, and of the United States in particular, have in recent decades exploited. Together, many kinds of efforts to exploit and ameliorate norm-based vulnerabilities generate significant ideational influences on the conduct of military operations and strategic outcomes.

National will is an ideational dimension. It has no consistent material causes and is little recognized in materialist theories of military power, although some materialist-oriented theorists note the resource-mobilizing importance of nationalism—a related concept.[35] Evidence of national will is mainly qualitative in nature, except when it affects decisions about topics that are quantitatively measurable—like budget appropriations. Societies must desire to produce military power before they decide to tax themselves or borrow; otherwise, they will not mobilize the resources necessary to produce military forces.

Resource Mobilization

A dimension long recognized in the IR literature, resource mobilization refers to actors' acquisition of tangible resources for military purposes. Aspects of this dimension are measurable in monetary terms like tax revenues and in physical quantities like numbers of men conscripted and tons of iron mined, which are not directly comparable in physical units or in terms of their political/military significance. Operations in this dimension include attacks on transportation infrastructures, naval blockades, economic sanctions, and many others. Actors must mobilize enough financial, material, and human resources to enable production of military-related goods and services; otherwise, resource conversion will be inadequate.

Resource Conversion

Another traditional IR dimension, resource conversion refers to actors' ability to convert raw resources into usable political-military assets—like weapons.[36] Activities in this dimension are variously measurable in quantities of industrial output (tons of bombs, numbers of tanks and aircraft, and so on) and the financial value of such production, which again are usually not comparable in physical terms or in their impact on outcomes.

Modern forms of attack in this dimension include bombing factories that produce important military-related products. Consequences of such attacks are variously measurable by their effects on levels of output of key products, military logistics, and civilian supply networks, which affect outcomes through mechanisms that logisticians, economists, and military historians understand fairly well. To be successful, actors must produce adequate numbers and types of military-related goods and services; otherwise, they will not be able to form and support adequate numbers of field units.

Force Generation

In this dimension, actors combine material, human, organizational, and ideational assets to produce militarily relevant field forces. Material assets like weapons and soldiers are individually measurable in quantitative physical and monetary terms. Intangible institutional factors like military training, doctrine, and organizational cultures produce operational strengths, limitations, and vulnerabilities that often are not quantifiable in any meaningful way.[37] Military force structures are describable in qualitative terms whose meanings are sometimes quite imprecise. For example, the military unit known as *battalion* is virtually always bigger in terms of numbers of assigned troops than a *company* and smaller than a *division,* but its personnel strength, equipment holdings, and operational specialization can vary radically across and even within military establishments, sometimes making assessment of the operational significance of a battalion difficult even for skilled military analysts. Actors must produce enough capable field units; otherwise, military and civilian leaders will not be able to make and implement effective political-military plans.

Leadership

This primarily intangible dimension, very important to military practitioners and military historians, appears infrequently in IR theory.[38] It includes civilian and military commanders' objectives, their orders and actions, and their ability to inspire their troops. The directives of civilian leaders of modern democracies, especially, to their generals often reflect goals and constraints produced originally in the national will dimension. Military leaders' actions and orders reflect military culture and doctrine and are channeled operationally by force structures produced in the force generation dimension. Because leaders act mainly through subordinate organizations, the effectiveness of leadership depends sometimes heavily on the ability and desire of subordinate organizations to implement policies. This means that parochial interests, organizational cultures, and bureaucratic politics affect leadership, sometimes significantly.

Good (and bad) leadership traits are often discernable only after leaders are severely tested. Military organizations work hard to identify and groom promising young officers for senior leadership, but they often err because traits that make good small-unit leaders differ from those of good senior commanders, and promotion boards cannot consistently forecast how young officers will mature or react to the stresses of combat and senior command. History also is full of cases in which senior civilian leaders surprised observers by unexpectedly good and bad wartime performances. Hence, I offer no explanation of what makes good leaders or suggestion about how to identify them before crises—only discussion of how leadership affects military outcomes.

Modern militaries view leadership as also including the staff activities that support leaders' decision-making and the command-and-control infrastructures that link commanders and their troops, including headquarters facilities and communications networks.[39] Senior civilian and military leaders must be competent and in communication with their deployed units; otherwise, even capable units will not perform well in the field.

Operational Execution

A traditional IR dimension, the operational-execution dimension is where uniformed field forces and related civilian agencies conduct political/military operations and where outcomes defined as battlefield victory or defeat usually are apparent in measurable terms like square kilometers of ground taken or lost, numbers of casualties inflicted or suffered, prisoners taken, tons of materiel captured or destroyed, and the presence (or not) of formal surrender. Materialist-oriented IR theories of military power typically presume that material *capabilities* determine strategic outcomes in this dimension. Battlefield results may not determine outcomes, however, if failure occurs first in another dimension.

THEORETICAL CONTEXT OF VULNERABILITY THEORY

The idea of dimensional vulnerability is not developed systematically in IR theory, but partial recognition of the idea appears repeatedly in the academic literature and in practitioners' ideas about war. Together, they provide much of the foundation of vulnerability theory. The concept of vulnerability exploitation and the six dimensions all appear in varying ways in the literature.

Soldiers long have recognized the importance of asymmetrically exploiting enemy vulnerabilities, and some integrate material and ideational dimensions of military power. Some 2,500 years ago, Sun Tzu urged warriors to know their enemies and themselves, and to attack enemy weak points while avoiding enemy strengths.[40] He emphasizes deception in war (implicitly nonviolent attacks in the leadership dimension) and argues that the pinnacle of generalship is manipulating enemy will to produce favorable outcomes without fighting. Clausewitz, a participant in the Napoleonic Wars, observes that war is a "collision between two living forces," emphasizing the dynamically interactive nature of warfare—not a mechanistic clash of materials or technologies.[41] Basil Liddell Hart, a veteran of World War I, advocates the "indirect approach" to avoid enemy strengths and to exploit weaknesses by attacking paths of least enemy resistance.[42] Other military analysts make similar points.[43]

Academic theorists typically discuss implications of dimensional strengths and vulnerabilities in narrow slices of the spectrum of military operations.[44] Four analysts particularly well discuss one or two dimensions of political/military conflict: Andrew Mack, Gil Merom, Jonathan Shimshoni, and Stephen Biddle.[45] None of their theories alone can consistently explain wars' outcomes, however, because they are narrowly focused, and strategic outcomes, as the case studies demonstrate, typically are determined in several dimensions that vary from war to war. But together these theories provide key constituent arguments of vulnerability theory.

Mack and Merom show how materially superior powers lose wars when their wills crumble. Basing his argument on lessons of the U.S. war in Vietnam, Mack argues that strong actors fighting far from home have limited interests that are dwarfed by those of weak actors determined to achieve important objectives like independence or national unification. For such weak actors, wars are total. For their big state opponents, limited wars often produce bitter domestic disputes over war goals, tactics, and morality that lead to defeat through popular demands for political decisions to quit fighting; this amounts to activity in what I call the *national will* and *leadership* dimensions. Mack argues that the effect on the metropolis of "small wars"—or insurgency wars—depends upon the "structure" of interest asymmetry of each conflict, which links actors' interests and political vulnerabilities.[46]

Merom focuses on regime type, arguing that Western democratic states since 1945 have developed political characteristics that make them vulnerable to defeat in small wars. These prominently include unwillingness to accept the human costs of waging wars effectively.[47] Discussing cases in which stronger actors quit counterinsurgency wars they were winning on the battlefield—France (in Algeria) and Israel (in southern Lebanon)—Merom identifies characteristics of bigger powers that led to their defeats. He assesses domestic influence mechanisms within the *national will* and *leadership* dimensions that affect strategic military outcomes.

Shimshoni argues that what military leaders do with their material assets matters more than the physical attributes of the assets—stressing the importance of the *leadership* dimension. Good leaders analyze their situations and make relational advantages by manipulating interactions of technology, doctrine, and war plans. Because people manage differently, warfare is a "never-ending competitive enterprise" of identifying opportunities and exploiting them in dynamic interaction; "military entrepreneurship," the creative use of resources, is the mechanism by which belligerents exploit adversaries' vulnerabilities.[48] While geography and political culture are constraints, all aspects of warfare can be manipulated to some extent, and there is no systemic role for technology, geography, or other variables.[49] Shimshoni emphasizes military leadership because commanders must recognize enemy vulnerabilities before they can effectively direct their troops to exploit them. An Israel Defense Forces officer when he wrote, Shimshoni focuses on operations consistent with the conventional wars Israel has fought, but his points apply more generally.

Biddle, concentrating on conventional land warfare, argues that *force employment*—the use of firepower against opponents and mobility to avoid enemy firepower—is used by armies organized to employ the *modern system* of warfare, which he judges to have begun during World War I. Good modern armies increase their abilities to attack enemy tactical vulnerabilities and minimize their own by assembling combined arms teams.[50] Biddle says, "combined arms integration reduces net vulnerability by teaming together weapon types with contrasting strengths, weaknesses, and limitations."[51] In terms of vulnerability theory, commanders tailor their units for asymmetrical use against enemy forces with both enemy vulnerability exploitation and friendly vulnerability amelioration in mind. Biddle argues that the modern system is difficult to master and that many armies have not done so—accounting for the lopsided outcomes of some wars in which belligerents' material asset holdings were closely matched. Biddle primarily emphasizes battles and what I call the *force generation* and *operational execution* dimensions.[52]

The cases below show that mastery of Biddle's modern system applies in arenas other than ground combat (like air warfare) but does not guarantee success in small wars. Against an American military that demonstrated in 1944–1945, 1991, and March–April 2003 that it has mastered conventional land warfare, insurgents repeatedly in Vietnam, Iraq, and Afghanistan operated in ways the United States was ill prepared to counter, and that Biddle's modern system does not address. These conflicts reflect a newer, broader, more ideationally oriented type of war, which some military theorists call fourth-generation warfare.[53] Biddle's modern system reflects what these analysts call second- and third-generation warfare—wars characterized by mass and maneuver like the world wars of 1914–1918 and 1939–1945.[54]

Two contemporary, competing IR paradigms each address some of the six dimensions. In general, realists focus on the material dimensions of resource mobilization, resource conversion, force generation, and operational execution. Constructivists offer insights about the ideational dimensions of national will, force generation, and leadership. Both paradigms address force generation because it has material and ideational subdimensions; putting well-equipped troops in the field requires material wealth, but making units competent also requires ideational inputs of effective training, doctrine, operating procedures, and functional military cultures. A major theme of this book is that the ability to influence military outcomes requires simultaneous operations in material and ideational arenas, meaning that for analysis of modern wars, especially, realist and constructivist perspectives are complementary. Alone, each is incomplete.

Constructivists, while little focusing on military power directly, nevertheless insightfully discuss aspects of some of its dimensions.[55] Literatures on logics of appropriateness and strategic culture address influences in the *national will* and *leadership* dimensions. Constructivists more directly analyze aspects of *force generation*—including Elizabeth Kier on doctrine formation, Stephen Rosen and Kenneth Pollack on effects of military culture on operational effectiveness, and Nina Tannenwald and Richard Price on weapons "taboos" that influence actors' choices about the force structures they build and use.[56] Jeffrey Legro discusses norms' impact on submarine warfare during World War II, developing a cultural explanation of operational behavior but not considering implications for the outcome of the Battle of the Atlantic or the war.[57] Theo Farrell considers military power from a sociological institutionalist perspective, arguing incorrectly, in my view, that norms affect military operations uniformly and globally.[58] More persuasively, Ward Thomas addresses insightfully the impact of ideas on military operations—hence on military power as I define it.[59] Liberal institutionalists also little discuss military power; when they do, they typically adopt materialist perspectives. Therefore, neither the constructivist nor liberal institutionalist paradigms offer coherent theories of military power against which vulnerability theory can be tested.

Materialist-oriented realist theorists, in contrast, regularly and directly address the nature of military power. They also discuss vulnerabilities, albeit typically differently than in vulnerability theory. For example, realist scholars commonly use *vulnerability* to refer to disparities in quantities of material military assets that states hold. Stephen Van Evera avers that offense-dominance causes war when "windows of vulnerability" exist.[60] States can close such windows by restoring economic growth or building standing military forces, effectively operating in what I call the *resource mobilization, resource conversion,* and *force generation* dimensions.[61] Barry Posen argues that inability to distinguish offensive from defensive capabilities creates vulnerabilities—effectively talking about the *force generation* and *operational execution* dimensions.[62] Waltz argues that state vulnerabilities stem from interdependent provision of economically valuable goods and services—referring to the *resource mobilization* dimension.[63] Offense-defense theorists' argument that systemic, technology-determined, offensive and defensive advantages help determine both the occurrence of war and military outcomes amounts to an assertion that operational vulnerabilities and means of vulnerability exploitation are produced in the *force generation* dimension and appear in the *operational execution* dimension. The skill and competence of actors, regularly mentioned but very weakly analyzed in the realist literature, implicitly help identify, exploit, and ameliorate vulnerabilities in all dimensions.

Given the dominance of materialist perspectives in the IR literature and materialist-oriented theorists' claims to be able to explain military outcomes, the materialist perspective is an important theoretical approach against which to test vulnerability theory. Because COW-like measures of power clearly cannot well explain strategic outcomes, I do not dwell on reconfirming the deficiencies of the most basic materialist notion that literal mass of material wealth directly determines outcomes. Instead, I focus on *why* and *how* aspects of relative material balances modified by technological and *skill* characteristics of belligerents, as well as real or perceived offensive or defensive advantages, influence outcomes. In this way, I link military material assets to the conduct of operations in the various dimensions and help establish causal mechanisms.

While many theorists privilege material preponderance as the source of military victory despite the increasingly apparent empirical deficiencies of this view, a few theorists who believe that COW-like measures are important also recognize that other factors influence military outcomes. Most prominently, Ivan Arreguín-Toft offers a theory of how weak actors win small wars.[64] In analyzing 202 "asymmetric" wars from the COW database featuring belligerents with ratios of material assets holdings of at least 10:1, he emphasizes the importance of belligerents' decisions about how to fight each other—a process he calls "strategic interaction," which modifies the influence of material holdings.[65] He emphasizes the COW's population and armed forces factors as important determinants of power.[66] Arreguín-Toft's use of the term *asymmetric* to reflect COW-defined power imbalances differs sharply from that of some scholars and most contemporary military practitioners, who use the term to refer to adversaries' differential *conduct* of warfare.[67] I use the latter definition herein.

Arreguín-Toft posits that materially strong actors, assumed to be aggressors operating on the territory of materially weak actors, either attack conventionally (that is, by using conventional forces) or use "barbarism"—defined as "the deliberate or systematic harming of noncombatants (e.g., rape, murder, and torture) in pursuit of a military or political objective."[68] Weak actors defend conventionally or employ "guerrilla warfare strategy" (GWS), whose characteristics closely resemble those of "people's wars" as explicated by Mao Zedong.[69]

Conventional attack and conventional defense are "direct" options that attack actors' "physical capacity" to fight. Barbarism and GWS are "indirect" options that attack enemy will.[70] GWS imposes costs on opponents without direct confrontation and is particularly effective against opponents with "political vulnerabilities"—a notion he borrows from

Andrew Mack.[71] Strong actors tend to win, Arreguín-Toft contends, when their interactions with weak opponents are what he calls "same-approach" (conventional attack–conventional defense or barbarism-GWS), and they tend to lose "opposite-approach" (conventional attack–GWS or barbarism-conventional defense) contests.[72] The opposite occurs for weak actors. Arreguín-Toft's analytical framework, a 2 x 2 matrix, offers only four interaction possibilities, although his case studies also feature an ad hoc "conciliation" option that actors use to end wars rather than to win them.[73] His five case studies all are colonial or insurgency wars in which GWS is a plausible option.

Arreguín-Toft's model captures only part of the operational variation that normally occurs in wars as belligerents interact, but he claims it captures their essence.[74] He does not much discuss the dynamic aspects of interaction or actor learning and adaptation, meaning his *strategic interaction* is fairly static in nature. He deemphasizes his own statistical analyses because he recognizes data inadequacies that suggest correlations but cannot demonstrate the causality he theorizes.[75] His summary coding of whole wars as same or opposite approach based on the four strategies is simplistic given the many strategies and tactics belligerents typically use in various stages of fighting within conflicts, especially long ones, as he recognizes in case studies that all feature more than one strategic interaction.[76] Despite such limitations, which I consider to be significant, I use Arreguín-Toft's theory as the second of two competitors against which to test vulnerability theory because I believe it is the best revision of materialist power theory now widely recognized by academic IR specialists.[77]

RESEARCH DESIGN

I offer a general, inductively derived theory of military power that incorporates insights of military practitioners and scholars, especially Mack, Merom, Biddle, and Shimshoni. I then test vulnerability theory against two theoretical competitors: (1) the proposition of COW-like theories that material preponderance generates military power and, especially, its technology, offense-defense, and skill modifications; and (2) Arreguín-Toft's strategic interaction theory. In case studies, I evaluate whether and, if so, how each actor identified and tried to exploit its enemy's (and in some cases friends') vulnerabilities. I discuss whether and, if so, how actors identified and sought to ameliorate their own vulnerabilities. I track the independent variables in each case through, typically, several relevant dimensions to show how they produced strategic outcomes. While specific events vary substantially from case to case, I focus on how the facts relate to the independent variables, the six dimensions, and outcomes.

To illustrate and test vulnerability theory against its competitors, I assess a sample of U.S. wars since 1899, focusing on how and why the United States produced and deployed types of military forces, how it used its forces in operations, how always weaker U.S. opponents sought to defeat the United States despite their material inferiority, and how key dyadic interactions affected strategic outcomes. The United States is a major power that repeatedly fought wars and smaller violent conflicts over the past century. It won some of them and lost some. It learned with varying degrees of insight and wisdom from earlier wars and fought adversaries who also derived lessons from past and ongoing U.S. wars. Examination of several types of U.S. wars with different outcomes enables me to discern behavioral patterns, including consistency in, and evolution of, the operational styles of U.S. forces and their enemies. I test the following propositions:

1. *Materialist theory*. Conventional COW-type theories are supported if the materially stronger United States consistently mobilized its superior material wealth for military purposes, bludgeoned opponents with massed military assets, fully and effectively used its technological advantages, rationally organized and skillfully used its forces, and won all of its wars.
2. *Arreguín-Toft's strategic interaction theory*. Strategic interaction theory is supported if the materially stronger United States always initiated its wars, relied only on conventional warfare or barbaric attacks on noncombatant civilians, and won (lost) if it fought same-approach (opposite-approach) wars against its enemies. The theory is supported if U.S. enemies fought in response only conventionally or with guerrilla warfare used strategically.
3. *Vulnerability theory*. Vulnerability theory is supported if actors consciously analyzed how best to defeat their enemies by identifying their vulnerabilities and implemented plans based on assessments of enemy strengths and weaknesses; the United States, despite its material dominance, looked for vulnerabilities in its weaker opponents in order to fight with smaller masses of material resources or to win more easily; the United States worried about its own vulnerabilities despite its material dominance; American and adversary strategies and tactics evolved within and across wars based on their experiences; wars' outcomes varied based on the dynamic interactions of belligerents' differentially effective vulnerability identification, exploitation, and amelioration efforts; and reflecting actors' learning and adaptation, critical dimensions varied by war and over time.

Because the theories make divergent points about how belligerents perceive their enemies and choose to act, I focus on identifying belligerents' perceptions of their adversaries and their situations, and their resultant plans and actions. This approach simplifies data collection and establishes the causal logics of belligerent strategies that permit process-tracing analyses that in turn lead to conclusions about the causes of strategic outcomes. Military historians and investigative journalists (for recent wars) have done much of this work—albeit in different theoretical contexts and sometimes using somewhat different terminology.

I illustrate and explain vulnerability theory and test it qualitatively against its competitors but do not present formal models or test hypotheses statistically. While some of its competitors' premises lead to straightforward and consistent forecasts about the determinants of outcomes, vulnerability theory holds that dyad-specific vulnerabilities—and whether they are exploited—determine strategic outcomes and therefore the presence of military power. Because belligerents are variously motivated and endowed with resources, and because they variously learn from history and in interaction with each other, the derivation of asymmetric advantage is always contingent upon individual actors' circumstances that evolve over time. This view is consistent with aspects of chaos theory and the longer military analytical tradition that wars are nonlinear events full of surprise, fog, and friction.[78]

Therefore, I theorize that there is no consistent set of military *capabilities* or influence mechanisms—and correspondingly no single general factor or metric—that consistently explains strategic outcomes. As noted above, even measures that quantify specific military-related assets typically are narrowly applicable within subdimensions and are not readily comparable to other measures. They do not alone reflect outcome-determining operational performance. Instead, the learning feature of vulnerability theory predicts an evolving pattern of strategies and tactics. A finding that a single measure consistently explains outcomes would tend to refute vulnerability theory.

This means that I provide no simple measure of the success of theoretical propositions, which many scholars like.[79] Instead, vulnerability theory offers a framework for analysis of material, ideational, and institutional factors that affect military power in complex and evolving ways. Consistent with many methodological perspectives, I take the position that parsimony is relative. I offer theory at the minimum complexity necessary to produce meaningful explanations and forecasting accuracy. I see little to be gained by a stronger commitment to parsimony that generates simplistic theory that reflects simple-mindedness or has no convincing explanatory or predictive power. Indeed, I think the basic materialist concept that military power is a direct function of wealth and Arreguín-Toft's strategic interaction theory both are too parsimonious to convincingly explain military power.

The IR literature on military power cited herein is generally formatted in terms of qualitative propositions of the sort I use. My approach also is consistent with most of Arreguín-Toft's book.

CASE SELECTION

I test the various theoretical approaches in six case studies. All focus on significant U.S. wars: the Philippine War of 1899–1902; the U.S. strategic bombing campaign against Germany during World War II; Vietnam; the North Atlantic Treaty Organization's (NATO) war against Yugoslavia in 1999; the war in Iraq that began in 2003; and Afghanistan.

The cases collectively have several desirable features: (1) the cases are numerous enough to limit dangers of selection bias; (2) available data are adequate to indicate how all major participants sought to defeat their enemies and how their strategic choices affected outcomes; (3) the wars lasted long enough to feature at least one shift in strategy by opposing sides using similar force structures or changes in force structures in similar operating environments, enabling assessment of whether and, if so, how and why changes in operational dimensions affected outcomes; (4) cases are spread over time, enabling some trend analysis; (5) two contemporary cases offer policy relevance; (6) cases include a range of types of conflicts, including large and small wars, conventional and unconventional wars, and wars in which relative dyadic material imbalances varied considerably in magnitude; and (7) the always materially dominant United States both won and lost wars.

I focus in the case studies on testing vulnerability theory, which produces analyses that also enable testing of the competing theoretical perspectives. I use analyses of how actors operated in the operational execution dimension as the primary basis for judging whether technology and skills played critical roles and whether Arreguín-Toft's four interaction possibilities occurred in the ways he theorizes.

Because outcomes in each case (except the ongoing war in Afghanistan) are clear, little additional work needs to be done to test whether the materially dominant power won or not. My case selection criterion that the United States both won and lost wars ensures in advance that the basic materialist theoretical claim that preponderant mass of resources directly

Table 1.1
Arreguín-Toft's Expectations about Five Wars

War Name	Strong Actor	Strategic Interaction
Philippine ("Second Philippine")	wins	same approach
Vietnam	loses/tie	opposite approach
NATO/Yugoslavia ("Kosovo II")	wins	same approach
Iraq ("Gulf War II")	wins	same approach
Afghanistan ("Anti-Taliban")	wins	same approach

Notes: War names in parentheses are Arreguín-Toft's names for the wars of this study. Arreguín-Toft's "Gulf War II" occurred in 2003 only, indicating that it is roughly equivalent to what I call the conventional phase of the Iraq war. Arreguín-Toft does not address any aspect of World War II, accounting for the lack of reference to my bombing campaign case.

Source: Arreguín-Toft, *How the Weak Win Small Wars*, 228–32.

determines victory is refuted. I aim instead to assess materialist theories' more diffuse but interesting ideas about technology and skills, which do not feature such simple explanations of, or predictions about, strategic outcomes, and to identify and assess *ways* that material wealth affects military operations and outcomes. In my cases, belligerents' perceptions of military technologies, the uses and effects of technologies, and actor skills varied considerably, with demonstrable effects on belligerents' ability to find and exploit enemy vulnerabilities and on the creation and amelioration of their own. I assess actor skills with special attention to whether and, if so, how they affected the learning and adaptation features of vulnerability identification, exploitation, and amelioration, thereby testing aspects of both materialist and vulnerability theories.

Arreguín-Toft offers a detailed case study of one of my cases, Vietnam, and he offers general assessments of four of the other five of my cases as whole wars in an appendix. Table 1.1 summarizes his findings.

OUTLINE OF THE BOOK

This analysis covers eight chapters. After the introductory chapter and six case study chapters, I conclude with a chapter that assesses the lessons of the analysis for vulnerability and other theories, policy makers, and practitioners. I address the major causes of U.S. political/military operational performance—an issue that in the last five case studies features a striking inability of U.S. military forces to consistently conduct operations efficiently even when they prevailed. The case studies feature similar analytical techniques, but discussions differ in detail because relevant operations and dimensions varied appreciably varied across wars. In the case study chapters, I stick generally to following format: (1) an introductory section that discusses the political context of the war; (2) an overview of the war's major military events; (3) analysis of the war from the standpoint of materialist theory, focusing on the role of technology and actors' salient skills; (4) analysis of the case through the lens of Arreguín-Toft's strategic interaction theory; (5) a dimensional analysis that links the conduct of the war to strategic outcomes from the perspective of vulnerability theory; and (6) concluding comments, including assessments of adversaries' learning and adaptation characteristics and the relevance of the case for the theories of Mack, Merom, Shimshoni, and Biddle.

Chapter 2 discusses the U.S. war in the Philippines in 1899–1902, in which the U.S. Army and Navy conducted a brief and successful, but strategically inconclusive, conventional campaign followed by longer but still relatively quickly successful counterinsurgency operations. U.S. forces effectively targeted insurgent vulnerabilities in several dimensions while Filipino nationalists both poorly responded to American attacks and failed to effectively exploit any U.S. vulnerabilities, including a political vulnerability they understood in general terms.

Chapter 3 assesses the U.S. strategic bombing campaign against Germany in 1943–1945. I analyze part of a major war because American airmen argued that by identifying and attacking key German vulnerabilities, strategic bombers alone could defeat Germany. In loose cooperation with the Royal Air Force, U.S. Army Air Forces strategists targeted perceived German vulnerabilities in three dimensions (national will, resource mobilization, and resource conversion) they poorly understood, and bomber units poorly exploited the few German vulnerabilities Allied intelligence officers accurately identified. The failure of bomber forces to find and exploit vulnerabilities that airpower theorists postulated before the war compelled Allied bomber organizations to use their residual superiority—mass of material resources—to gain battlefield victory. The United States and Britain won a costly air war of attrition in the operational execution dimension.

Chapter 4 evaluates the U.S. war in Vietnam during 1961–1975. In this case, the United States failed to find an effective strategy to counter enemies in North Vietnam and South Vietnam but faced a coherent and generally well-executed North Vietnam–directed political-military strategy that defeated the United States in the national will and leadership dimensions. Because one of Arreguín-Toft's cases also is America's war in Vietnam, I assess in detail the four strategic interactions Arreguín-Toft identifies.

Chapter 5 assesses the dominant U.S. role in NATO's war against Yugoslavia in 1999. It discusses how strong casualty aversion norms interacted and dominated U.S. war goals, influenced creation of the operationally relevant U.S./NATO force structure, affected NATO target lists, and influenced operations through restrictive rules of engagement—that is, by influencing the national will, force generation, and leadership dimensions. In a three-sided fight, the Yugoslav government and the Kosovo Liberation Army conducted tactical operations designed to exploit different U.S. political and military vulnerabilities; they achieved different levels of success at influencing the strategic outcome. Despite weak U.S. and NATO policy making and military operational execution, NATO won the war.

Chapter 6 addresses the U.S. war in Iraq (2003–2011). It assesses how the United States and its adversaries conducted different types of operations with very different operational results in the war's two phases—impressive U.S. success during a short conventional phase (March–April 2003) and significant U.S. problems in the unconventional phase thereafter. After April 2003, Iraqi insurgents and al-Qaeda targeted U.S. casualty aversion vulnerabilities in different ways and with greater success than did Saddam Hussein's regime. This chapter evaluates how and why U.S. performance deteriorated when insurgents shifted from conventional to unconventional tactics designed to move the critical dimension from operational execution to national will. Improved U.S. learning and adaptation after 2007, and some luck, led to a costly but modestly successful outcome thereafter.

Chapter 7 discusses the ongoing and more problematic U.S. war in Afghanistan. It assesses the effective conventional campaign against the Taliban in 2001 and much less effective operations thereafter. It notes similarities and differences with the roughly contemporaneous war in Iraq and explains why the United States made a different set of strategically important mistakes.

Chapter 8 summarizes the analysis, presents conclusions, and assesses implications for vulnerability theory and its competitors. It critiques theories embedding materialist concepts of military power, strategic interaction theory, and some IR theories not directly focused on the issue of military power. Finally, it discusses implications of the analysis for American policy makers and military practitioners.

To summarize findings, I conclude that materialist theory's mass, technology, and offense-defense theoretical arguments fail badly, but that *skill* is very important. Several prominent materialist theories are refuted and the study has both positive and negative implications for several other theories. I identify in the cases studies 16 strategic interactions consistent with my understanding of Arreguín-Toft's theory; only seven of the interactions match his expected outcomes, and the facts of the cases are often inconsistent with structural features of his model. I therefore conclude that Arreguín-Toft's theory is flawed in important respects. I also conclude that the recurrently poorly conducted U.S. military operations stemmed mainly from poor institutional learning and adaptation capabilities—not resource shortages, information deficiencies, or cognitive challenges—which chronically limited American abilities to identify, exploit, and ameliorate vulnerabilities.

The cases show consistent actor efforts to identify, exploit, and ameliorate vulnerabilities. Successful actors used a wide variety of strategies, tactics, and operational techniques to achieve their goals—consistent with vulnerability theory's central prediction that successful political/military actors learn and adapt to use approaches and operational techniques that vary within wars and across wars over time. These generate the evolving pattern of critical dimensions that vulnerability theory also predicts.

CHAPTER 2

The Philippine War of 1899–1902

During February 1899–July 1902, the United States fought and decisively defeated the Army of Liberation of the Philippines, the army of the short-lived Philippine Republic led by its president and generalissimo Emilio Aguinaldo.[1] The Philippine War was an unintended consequence of America's war with Spain, which Congress authorized on April 19, 1898. President William McKinley ordered the U.S. Navy, then the U.S. Army, to the Philippines as a subordinate theater of the Spanish-American War. McKinley and his advisors worried that the Spanish fleet based at Cavite, south of the city of Manila on Manila Bay, posed a potential threat to the ships the United States was assembling to transport an invasion force to Cuba. Consistent with its contingency plans for war with Spain, the Navy Department on April 25, 1898, ordered Commodore George Dewey's Asiatic Squadron, based in Hong Kong, to destroy Spanish warships in the Philippines.[2]

On May 1, Dewey's six warships engaged the Spanish fleet off Cavite, sinking several ships and inducing the Spaniards to scuttle several others—eliminating the naval threat; Dewey's sailors then seized the port of Cavite.[3] Dewey's success presented McKinley with a new policy question: what did the United States want to do with the Philippines? While there is debate about McKinley's private aspirations, within days of Dewey's victory, and at Dewey's request, McKinley ordered army units to the Philippines without detailed instructions but made clear his generally benevolent attitude toward Filipinos.[4] Army troops, including professional (regular) regiments and newly formed "volunteer" regiments comprised of state militiamen and military neophytes attracted to war with Spain, quickly assembled in California; the first contingent landed at Cavite on June 30, 1898.[5] By July 25, some 11,000 U.S. soldiers were in the Philippines. They faced two potential enemies—Spanish troops and the Army of Liberation, a predominantly light infantry force that had intermittently since 1896 fought against Spain using guerrilla tactics, which was besieging the Spaniards in Manila.[6]

Despite the nominal friendship conferred by having a mutual enemy—Dewey arranged for Aguinaldo's return to the Philippines from exile in Hong Kong on May 19 and facilitated shipment to the Philippines of arms Aguinaldo's agents bought abroad—relations between Aguinaldo and the Americans cooled as Aguinaldo grew increasingly suspicious of American intentions toward the Philippines and Washington ordered newly promoted Rear Admiral Dewey and army commanders not to cooperate with Aguinaldo.[7] U.S. troops on August 13, 1898, attacked Manila after the Americans agreed to Spanish desires for a short fight designed to preserve Spanish honor and denial of entry of Army of Liberation troops into the city; consistent with their agreement, the Spanish garrison quickly surrendered, and U.S. troops occupied Manila without coordination with Aguinaldo or any involvement by Army of Liberation troops.[8]

With the Spanish-American War effectively but not yet formally ended by a U.S.-Spanish agreement on August 12, the rest of 1898 featured a tense but peaceful standoff between U.S. troops in Manila and Army of Liberation surrounding the city.[9] Aguinaldo, who on May 23, 1898, declared himself dictator for the duration of the crisis, declared Philippine independence on June 12, and was elected president in August. He used the last months of 1898 to establish the capital of the Philippine Republic at Malolos, north of Manila; build the Army of Liberation; establish government structures in parts of the country; and capture isolated Spanish garrisons throughout the archipelago along with their weapons and supplies.[10] Spain sold the Philippines to the United States for $20 million as part of the Treaty of Paris of December 10, 1898, that ended the Spanish-American War.

Finally, on December 21, 1898, after a long internal U.S. government debate about U.S. goals in the Philippines, McKinley ordered the army, supported by the navy, to occupy the Philippines for purposes of "benevolent assimilation" of the islands and their people.[11] U.S. officers discussed terms with Aguinaldo, but America's goals were by then incompatible with Aguinaldo's desires for immediate sovereignty and negotiations led nowhere.[12]

On February 4, 1899, apparently unintentionally from the standpoint of senior commanders on both sides, the outbreak of small-scale skirmishing around Manila escalated into general warfare, which the United States won decisively.[13] Aguinaldo and most other senior Army of Liberation leaders were captured, killed, or surrendered by early 1901; mopping-up operations continued to May 1902; and the United States declared the war won on July 4, 1902.

OVERVIEW OF THE WAR

The war can be divided roughly into two phases—an initial period of conventional fighting followed by a longer unconventional phase. First, the adversaries fought a conventional war in the vicinity of Manila and northward on Luzon Island in 1899. The Army of Liberation's Antonio Luna, first as "general in chief of operations" and then as the army's field commander under Aguinaldo's overall military and political leadership, during 1898–1899 built the guerrilla bands that fought Spanish troops unconventionally and Filipino deserters from the Spanish army into a conventionally organized force and chose to fight the U.S. Army conventionally.[14] Despite possessing some 15,000–40,000 soldiers in the Manila area, compared to only about 20,000 American troops in the entire archipelago in early 1899, the Army of Liberation consistently lost the conventional fights.[15] In the initial battle for Manila, for example, the U.S. Army suffered 59 fatalities but killed several thousand Filipino combatants.[16] Many other early battles were nearly as lopsided. As a result, in April 1899 Filipino leaders asked the U.S. Army commander in the Philippines, Major General Elwell Otis, for a ceasefire.[17] Seeing the overture as a sign of weakness and a ploy to gain time to regroup, Otis refused.[18]

Initial military failures, General Luna's personal obstreperousness, and Luna's commitment to terrorism, generated much dissension among the nationalists and apparently led Aguinaldo to order Luna's murder; whether directed by Aguinaldo or not, Army of Liberation personnel loyal to Aguinaldo killed Luna on June 5, 1899.[19] Insurgent military operational performance did not improve, however, and the U.S. Army continued to drive nationalist forces northward into central and then northern Luzon. On November 13, 1899, after American pressure caused the Filipinos to abandon Tarlac in central Luzon, the republic's last capital, Aguinaldo ordered his army to disperse and to shift to guerrilla tactics; he also moved his headquarters into the mountains of northern Luzon.[20]

Meanwhile, at McKinley's insistence and with the enthusiastic support of U.S. military commanders, General Otis issued General Orders (G.O.) 100—"Instructions for the Government of Armies of the United States in the Field"—which imposed strict limits on the behavior of U.S. forces.[21] G.O. 100 emphasized U.S. troops' responsibilities to restore order, protect property, and to treat civilians with justice and dignity.[22] The army was to be the vanguard of the effort to create a stable American colonial government.[23] The policy emphasized reciprocity; U.S. forces would well treat Filipinos who cooperated with them while simultaneously attacking armed resistance and rooting out insurgent sympathizers. The Americans were relatively gentle at first—even releasing some captured Army of

Liberation fighters after they were disarmed and little punishing Aguinaldo's noncombatant sympathizers. U.S. civilian administrators, including in 1901–1904 future President William Howard Taft as governor, cooperated closely with the military in undertaking civic action programs designed by order of Secretary of War Elihu Root to promote the "happiness, peace, and prosperity" of the people of the Philippines and to prepare the islands for independence.[24]

The U.S. Army performed many of the civic action tasks. For example, army personnel improved local water and sewer infrastructure; established primary schools and taught English and other courses; established hospitals for lepers and vaccinated many people against smallpox; revised the Spanish legal system then in use in the Philippines and established a supreme court; and, after July 1899, formed local governments run by Filipinos under the aegis of U.S. military advisers.[25] These programs were very popular with the Filipinos and generated much consternation among the revolutionaries.[26] In more modern terminology, Otis ordered tight rules of engagement that were designed to protect civilians, and he enthusiastically committed his force to what we now call "nation building." This meant U.S. troops actively participated in political as well as military aspects of the war.[27]

By the end of 1899, senior American officers thought the war was nearly won because the insurgents seemed discouraged and pacification programs seemed to be working.[28] But the Americans misread Aguinaldo's change in approach. The Army of Liberation regained its composure after reverting to unconventional warfare, launched periodically successful small-unit attacks, and showed signs of being able to fight a long war of attrition. Aguinaldo's strategy called for bleeding American forces by using guerrilla tactics in protracted warfare, hoping time and U.S. casualties would lead Americans to demand withdrawal of U.S. troops from the Philippines.[29] U.S. troops soon experienced higher casualties than they had during conventional fighting and were frustrated by their enemies' tactics.[30] The guerrillas also exploited the U.S. policy of benevolence to find and maintain supporters and to extort funds and supplies from less sympathetic Filipinos.

By mid-1900, it was obvious to American officials that their relatively mild policies toward enemy belligerents and their supporters—which led even some friendly Filipinos to question the U.S. commitment to the war effort—were no longer working.[31] Another approach was needed. Under pressure from Washington to end the war quickly, U.S. officers in the Philippines debated how to modify policy.[32] Some argued that because civic action programs and ROE consistent with G.O. 100 were inadequate, much harsher measures were needed.[33]

After much internal discussion, on December 20, 1900, Major General Arthur MacArthur, who replaced Otis as commanding general, ordered a new aggressiveness against guerrillas and their sympathizers along with continuation of the policy of benevolence—which the nationalists called America's "policy of attraction"—toward friendly and neutral Filipinos.[34] Captured guerrillas were shipped to prisons on Guam, not released. By increasing recruitment of Filipino troops (known as "scouts"), establishing police forces run by local personnel, more aggressively pursuing guerrilla bands, and providing more security for Filipino civilians in towns, the Americans with few but well-publicized exceptions balanced increased aggressiveness against guerrillas and continued benevolence toward civilians.[35]

Virtually simultaneously, in December 1900, soon after Republican President McKinley trounced Democratic presidential candidate William Jennings Bryan and the Republicans increased their majorities in both houses of Congress in the elections of November 1900, eliminating chances that they would be in danger if Bryan had won and withdrew U.S. troops, Filipinos sympathetic to the United States formed the Federal Party, which recognized U.S. sovereignty over the islands and advocated working with American occupation personnel for purposes of eventually establishing representative government in the Philippines.[36] Aguinaldo's guerrillas immediately attacked Federal Party members and similarly minded Filipinos in an accelerated campaign of terrorism, further alienating many people and pushing them toward the Americans.[37] By July 1, 1901, membership in the Federal Party had grown to about 150,000.[38]

Also in early 1901, more aggressive American small-unit, counterguerrilla field operations, aided by Filipino scouts and police and resultant improvements in tactical intelligence, killed or captured many key guerrilla leaders.[39] Aguinaldo was captured in a daring raid on his headquarters on March 23, 1901. Others surrendered. Under terms of the Americans' amnesty program, some former guerrillas became officials in the U.S. colonial administration.[40] Aguinaldo signed an oath of allegiance to the United States on April 1, 1901, and urged his followers to lay down their arms three weeks later.[41] By mid-1901, the insurgency was largely over in most of the archipelago.

Mopping-up operations continued until mid-1902, punctuated by occasional guerrilla successes like the surprise attack on a U.S. Army company base camp at Balangiga on Samar Island on September 28, 1901, which killed 48 soldiers. The "massacre," as some Americans called it, prompted retaliation that in turn generated much contemporary press coverage and a prominent place in most histories of the war.[42] Major General Adna Chaffee, who replaced MacArthur in July 1901, ordered Brigadier General Jacob Smith, commander of the 6th Separate Brigade, to pacify Samar; Smith in turn told at least one officer to turn Samar into a "howling wilderness."[43] Smith's troops, by now tired of the war and interested in retribution for the Balangiga defeat, used harsh measures against the guerrillas and many local residents. Americans widely saw these actions as excessive if ultimately effective. They prompted U.S. Senate hearings on army behavior in the Philippines and led to courts-martial for Smith and five other officers.[44] The last two major local campaigns, on Samar and in Batangas province in southern Luzon, were over by May 1902.[45] A wide-ranging amnesty offer on July 4, 1902, marked the end of America's war with Aguinaldo's nationalists.

THE PHILIPPINE WAR AND MATERIALIST THEORY

The materially dominant United States won the war—consistent with basic materialist theory. But how did the Americans win? Did they rely on material dominance or advanced technologies or the advantages of offensive operations to bludgeon the Army of Liberation, the guerrillas, and their supporters into submission? They did not.

On the eve of the Spanish-American War, the United States had a moderate-sized but fairly modern navy and a small and poorly equipped army organized for use on the frontier and for harbor defense.[46] While the navy contributed appreciably, the comparatively weaker U.S. Army mainly fought the Philippine War on the ground.[47] The volunteer regiments initially sent to Manila were poorly equipped and trained; they were mobilized for use only against with Spain. When that war ended, many volunteers clamored for discharge. With a new war on in the Philippines, Congress on March 2, 1999, increased regular army strength from 28,000 to 65,000 troops and authorized another personnel category of 35,000 volunteer troops for service in the Philippines only.[48] The army was able to pick some of the best soldiers of the first volunteer regiments for the new regiments, a considerable advantage. But even the much expanded army did not satisfy commanders' requests for troops, and the rotations of the new and old volunteer regiments negatively affected U.S. operations in the Philippines in 1899. The modest quantities of resources available to the army, and to the military as a whole, little reflected the substantial military potential of America's large population and its dynamic and rapidly growing economy, which by this time was already the biggest in the world.

The U.S. Navy contributed blue-water warships and coastal gunboats for blockade purposes, thereby precluding significant external support for Aguinaldo's forces. The navy's maritime combat capabilities and the U.S. government's ability to charter sea transportation services far exceeded those of the Filipinos and enabled uncontested U.S. troop deployments to the Philippines and movement among the islands of the archipelago, which gave the Americans fairly good logistical and amphibious assault capabilities. U.S. forces' superior means of command and control, aided materially by telegraphy in parts of Luzon and control of the sea, were still very weak by modern standards, however. Aircraft and electronic communication means other than the telegraph had not yet entered military service.

The U.S. Army expeditionary force consisted primarily of infantry regiments backed by a few artillery and cavalry units and some freshwater (lake and river) gunboats variously bought and captured from Spain.[49] Army strength remained modest even after the regiments authorized in 1899 reached the Philippines. U.S. troop strength averaged 40,000 during the war and peaked at about 70,000 in December 1900—far fewer than the Army of Liberation's original estimated force of 80,000–100,000 regulars and many thousands of part-time fighters and auxiliaries.[50] The United States therefore had an initial force ratio of about 1:2—a very unfavorable force ratio by the standards of post–World War II counterinsurgency wars in which winning governments often need 10:1 personnel strength advantages and many years to win.[51]

The United States had, on balance, a considerable but not decisive advantage in the technological component of its military materiel. The 1898 volunteer regiments, which initially outnumbered the regular regiments in the Philippines, at first were equipped with Civil War–vintage, single-shot Springfield rifles whose cartridges used smoky powder—a significant tactical disadvantage for troops fighting enemies using modern rifles firing ammunition with smokeless powder.[52] Only later did the 1899 volunteers get the modern Krag-Jørgensen rifles the regulars had from the outset. In contrast, the Army of Liberation was mainly equipped with modern German-made Mauser rifles it took from the Spaniards and other Mausers Aguinaldo's people procured commercially with American help in 1898 before relations soured. American troops had better ammunition supplies. American soldiers had more and better artillery, some Maxim machine guns, and could at times get fire support from navy ships and gunboats. By comparison, the Army of Liberation used artillery pieces made of wood.[53] These American advantages applied mainly to the conventional fighting, not the small-unit infantry fights, intelligence collection, and local security work in the longer unconventional phase of the war in which the strategic outcome was determined; most such weapons were too heavy for infantrymen to carry over rough terrain. In addition, the all-weather roads the army built and largely controlled facilitated U.S. troop movements and logistics on Luzon during the rainy season, especially. No technology offered distinctive offensive or defensive advantages to either side.

The biggest American advantages were not quantitatively material or technological; rather, they prominently involved *skill*. American political and military leadership, cultural traits, and organizational and command-and-control institutions were substantially superior to those of the Filipinos.[54] The volunteers, despite poor formal military training, showed considerable initiative and tactical prowess that often led to operational successes. Some historians believe the enlightened guidance and high quality of senior commanders and unusually good individual American soldiers, especially the 1899 volunteers, deserve much of the credit for the U.S. victory.[55]

Technical means of command and control were limited, despite some telegraphy and U.S. control of the sea, meaning most army garrisons operated largely independently and most American troops were posted in one locale for the duration of their service in the Philippines. Soldiers often learned local languages sufficiently well to communicate passably with local people, and they learned about their locales. Such knowledge enabled local U.S. commanders to adapt general policy guidance to local conditions and skillfully run somewhat different but locally effective counterinsurgency operations.[56] American success in 1901 also stemmed from ideational and institutional factors like better aggressiveness in the field, better intelligence collection, use of a relatively small number of capable Filipino scouts and police, sound policy that stimulated assistance from Federal Party members, and effective U.S. information operations. These measures were largely consistent with the principles of counterinsurgency warfare widely publicized much later by practitioners like David Galula and Robert Thompson.[57]

Why did the U.S Army promulgate and for the most part well execute a sophisticated counterinsurgency campaign? Historians cite two important institutional reasons in addition to the apparently strong personal commitments of senior military officers, including Generals Otis and MacArthur, and most of the senior field commanders who conducted key local campaigns. First, the senior leaders were veterans of the Civil War who understood the importance of pacification to successful war termination.[58] G.O. 100, their major pacification policy document, was the U.S. Army order of 1863 of the same name that guided federal troops' actions in occupied Confederate territory.[59]

Second, most of the regulars were veterans of the Indian wars, in which pacification was often as important as fighting.[60] Senior commanders in the Philippines prohibited brutality against locals and banned looting because they knew such activities sew animosity and are therefore operationally counterproductive.[61] The result was a U.S. Army force that embraced nation building and usually eschewed anything close to barbarism. During the unconventional phase of the war, soldiers' frustration with guerrilla tactics, guerrilla atrocities, and combat losses of friends led to frustrations and occasional episodes of excessive use of force. But senior civilian leaders in Washington and senior military commanders in the Philippines never condoned brutality, and even late in the war, when the worst lapses of U.S. troop discipline occurred, many local U.S. commanders conducted model counterinsurgency campaigns in terms of both exemplary troop behavior and effective operational performance.[62]

The lessons of the Civil War and the Indian wars were institutionalized in ways conducive to effective counterinsurgency operations in the Philippines. Immediately after the war broke out, General Otis and the War Department reached for G.O. 100 as a doctrinal guide, drawing on what we now might call counterinsurgency doctrine that was consistent with existing army culture. Troops were comfortable with nation building as a legitimate military task.

These capacities developed despite some other, major constraints and overt problems. For example, the terms of the mobilization law of April 1898 that helped prepare the army for war with Spain forced release of the 1898 volunteer regiments in early 1899, immediately after the Treaty of Paris of December 1898 ended the Spanish-American War and just as the Philippine War was beginning, hampering U.S. operations in the Philippines in much of 1899. Interservice rivalry between the army and the navy hindered cooperation somewhat, especially early in the war while the obstreperous Admiral Dewey was the local navy commander.[63] And, the rapid mobilization of 1898 strained War Department managerial resources, leading to scandals over poor-quality supplies and medical care for U.S. troops even within the United States.[64]

THE PHILIPPINE WAR AND STRATEGIC INTERACTION THEORY

Ivan Arreguín-Toft's strategic interaction theory, to recapitulate, holds that belligerents are materially strong or materially weak and have four strategic options. Strong aggressors can attack conventionally or employ barbarism against enemy civilians while weak defenders can defend conventionally or employ GWS. Arreguín-Toft theorizes that strong actors tend to win same-approach interactions and tend to lose opposite-approach wars. The opposite applies for weak actors.

Arreguín-Toft says the materially dominant United States won a same-approach interaction in the Philippines, consistent with his theoretical expectations.[65] Using Arreguín-Toft's analytical approach as exemplified by his case studies, I identify two strategic interactions in the war—the consecutive conventional and unconventional phases of the war.[66] In the first interaction, to use Arreguín-Toft's terminology, the United States generally attacked conventionally, and the Army of Liberation generally defended conventionally. The United States won the same-approach interaction of the conventional fighting in Luzon in 1899, leading to Aguinaldo's decision to shift to unconventional warfare. The result of this interaction is consistent with Arreguín-Toft's expectations.

Aguinaldo drove the second interaction by shifting in late 1899 from conventional to irregular operations, or roughly what Arreguín-Toft calls GWS. According to Arreguín-Toft, the United States then had two choices—to continue to fight conventionally or to resort to barbarism. But American forces did neither. MacArthur's December 1900 policy called for more aggressive counterguerrilla operations and continuation of the political, economic, and social aspects of the counterinsurgency program—a policy carefully balanced to kill insurgents and separate the guerrillas from their supporters while at the same time convincingly reiterating U.S.

desire, and showing U.S. ability, to protect Filipinos against increasing terrorist attacks and to help average Filipinos tangibly through civic action projects.[67]

The unconventional phase produced frustration among American troops, who occasionally retaliated against insurgents with brutality and who increasingly used the infamous "water cure" as an interrogation technique, but U.S. policy against brutality against civilians never wavered.[68] Senior U.S. commanders repeatedly reissued orders against abuse of Filipinos—both combatants and civilians—as the war got tougher in 1900 and 1901. Recent historical analysis indicates that U.S. troop behavior, despite some notable exceptions, generally conformed to the restrictive ROE imposed throughout the war even though anti-imperialist and anti-U.S. military propaganda about atrocities continues to color histories of American conduct of the war.[69] Historian Brian McAllister Linn, a prominent American expert on the war, argues that no senior local commander, with the possible exception of General Smith, commander of the Samar campaign of late 1901 and early 1902, conducted his counterinsurgency operations based on a local policy of extralegal repression.[70] Arreguín-Toft recognizes that isolated atrocities occur often in war and makes policy choice a key criterion for identifying the presence of barbarism.[71] By this standard, U.S. forces in the Philippines did not employ barbarism. The generally good relations Filipinos and Americans had during the war, and for a century thereafter, suggest in another way that American atrocities committed in the Philippines were behavioral exceptions.[72]

In contrast, the Army of Liberation's General Luna from the beginning of the war directed his troops to use what Arreguín-Toft calls barbarism against Filipino civilians and foreigners.[73] Nationalist fighters throughout the war killed large numbers of Filipinos who worked with and for the Americans, regularly perpetrated acts of common criminality against civilians in non-Tagalog areas, especially, and burned Filipinos' houses and other property in scorched-earth operations as they lost territory to U.S. troops; the terror campaign accelerated in early 1901 after MacArthur's policy shift of late 1900 and the Federal Party attracted many Filipinos.[74] Terror generated intimidation that produced resources for the insurgents in 1899 and 1900, but by 1901 it backfired by increasingly driving Filipinos to seek protection from U.S. soldiers and leading some Filipinos to fight guerrilla bands—creating low-level civil wars in some areas.[75]

Luna at the beginning of the war urged his troops to "exterminate" all foreigners.[76] The Army of Liberation periodically tortured, otherwise abused, and executed U.S. soldiers it took prisoner. Army of Liberation troops and guerrillas repeatedly mutilated the corpses of dead American soldiers, angering American troops and precipitating some retaliation. They did so despite Aguinaldo's warnings that such acts might alienate American public opinion.[77] Nationalist fighters also killed some of the deeply unpopular Spanish priests in the islands.[78] All of these actions were war crimes by modern standards. In short, Army of Liberation conventional soldiers and nationalist guerrillas, not U.S troops, in policy and in practice chronically resorted to barbarism during both strategic interactions.

Brutality toward civilians is not an inherent part of the guerrilla tactics organized military forces use against each other and is not part of Arreguín-Toft's concept of GWS.[79] Arreguín-Toft does not offer barbarism as a strategic option that weak actors can choose. Therefore, the facts of the second interaction, where the strategic outcome was determined, are substantially inconsistent with the structure of Arreguín-Toft's theory, meaning he mistakenly asserts that the United States won a same-approach interaction.[80]

DIMENSIONAL ANALYSIS

The United States successfully fought a counterinsurgency war in the Philippines. Given the holistic nature of such wars, the Americans operated in all six dimensions. Filipino nationalists poorly identified and exploited a smaller range of U.S. vulnerabilities.

The United States

The U.S. military and American civilian administrators found and exploited appreciable Army of Liberation and nationalist vulnerabilities in all six dimensions—*national will, resource mobilization, resource conversion, force generation, leadership,* and *operational execution*. First, Aguinaldo had a weak political base. Organizationally, he could rely only somewhat on the infrastructure of the Katipunan secret society he led, which was dedicated to independence for the Philippines. Katipunan started and led the rebellion against Spain that began in 1896 and was temporarily settled in late 1897 largely on Spanish terms, leading Aguinaldo to go into short-lived exile. Personal animosities and regional tensions were so great within Katipunan that leaders meeting in war councils often drew weapons on each other; in one such encounter in 1897, Aguinaldo supporters arrested and later executed the founder and leader of the movement, Andres Bonifacio, enabling Aguinaldo to assume command of the organization.[81] The murder of General Luna, an ethnic Ilocano, by Tagalog speakers, further exacerbated tensions within the movement.[82]

With only a few months between his return from exile and the outbreak of war with the United States, Aguinaldo built only modest political and government structures in parts of the Philippines.[83] He relied extensively on support from his own Tagalog-speaking region of central Luzon. Filipinos of various ethnicities who had been united politically only by authoritarian Spanish colonial rule did not yet have widespread or deep feelings of Filipino nationalism, and Aguinaldo did not offer attractive alternative political, economic, or social programs like land reform, for example, which might have appealed to peasants.[84] Aguinaldo therefore did not generate widespread popular support through a coherent nationalist or ideological program like those that characterized later "people's wars" in China and Vietnam.[85] In David Galula's language, Aguinaldo's organization failed to produce an attractive "cause."[86] There was no unified revolutionary movement; revolutionaries operated independently of Aguinaldo's organization from the start on the islands of Panay and Mindanao, for example.[87] And, the sham elections of September 15, 1898, that restricted suffrage mainly to Tagalogs made a mockery of Aguinaldo's claims to a popular national mandate.[88]

Moreover, Aguinaldo was a poor military and revolutionary political leader and never established widespread credibility as national leader.[89] Born in 1869, he was not yet 30 years of age when he was elected president. An accomplished coalition builder within the revolutionary movement, he had no broad political or military experience except chasing down small numbers of bandits.

In short, Aguinaldo had little strength in the *national will* dimension necessary to build and support his field forces and to sustain them during hard times. He had meager defenses against what proved to an ideationally and materially attractive American policy of benevolence. Moreover, his fractured organizational leadership and narrow ethnic and socioeconomic base of support contributed to vulnerabilities in other dimensions.

Recognizing the weakness of Filipino nationalists' organizations and appeal, the Americans made clear through practical implementation that their "policy of attraction" genuinely had Filipinos' best interests at heart. The American promise of eventual independence appealed to many Filipinos, including senior members of Aguinaldo's movement, generating more internal dissension.[90] U.S. leaders practiced a carefully calibrated campaign of carrot and stick, chastisement and attraction, which appealed to many other Filipinos, leading eventually to large numbers of collaborators, prominently including Federal Party members.[91] After becoming governor in July 1901, William Howard Taft subsidized the Federal Party's newspaper and gave the party exclusive control over appointments to the colonial administrative jobs reserved for Filipinos, a substantial power of patronage that reflected American confidence in the party's leaders.[92]

The Americans exploited the regional and ethnic vulnerabilities of Aguinaldo's government and the Army of Liberation, whose leaders were mostly upper class Tagalog speakers who alienated other ethnic groups. U.S. troops provided security for civilians in non-Tagalog areas, where the mainly Tagalog-manned Army of Liberation was particularly oppressive.[93] On July 1, 1901, the United States manned 491 garrisons throughout the islands with nearly 50,000 troops.[94] The Americans also recruited scouts from ethnic groups that disliked Tagalogs—most notably the operationally very effective Macabebe Scouts from the town of Macabebe on Luzon, who were descendants of Mexican Indians the Spaniards brought to the Philippines.[95] The Americans' campaign of attraction and their ability to exploit ethnic tensions helped keep Aguinaldo from generating much enthusiasm for his independence campaign. This lack of support hampered efforts of covert local governments loyal to the Philippine Republic and local warlords to mobilize resources via taxation and kept guerrilla recruitment low—weaknesses other American actions exacerbated.

American counterinsurgency efforts were designed to deny the guerrillas abilities to *mobilize resources* through voluntary contributions from their supporters and by extortion and theft from less sympathetic Filipinos. A prominent part of the MacArthur program of December 1900 involved resettling people from the countryside into towns, where they and the islands' limited food stores could be better protected from guerrilla predations.[96] The result, given the large number of islands in the archipelago, the limited agricultural potential of some of them, plantation farming of nonfood cash crops like hemp that reduced food crops and correspondingly led to relatively small food stocks in much of the countryside, and U.S. control of lines of communication (via army control of roads and ports and navy dominance of the sea), was chronic guerrilla shortages of essential supplies, including food.[97] The geography of the archipelago's more than 7,000 islands, many of them small, covering 116,000 square miles but spread over some 500,000 square miles of land and sea, combined with American control of the sea, also helped preclude establishment of sanctuaries where guerrillas could rest, train, and reequip.[98]

The guerrillas had minimal ability to *convert* their meager *resources* into militarily significant weapons, equipment, and other supplies. There was no significant military-industrial base in the Philippines. Good American naval intelligence, the naval blockade, and Aguinaldo's inability to generate international aid—both the Katipunan and the Philippine Republic mused internally about appealing to Japan and Germany for recognition and support, but Filipino diplomacy achieved neither—sharply limited the ability of the Army of Liberation to convert external sympathy into tangible military materiel.[99] The Army of Liberation (and later the nationalist guerrillas) therefore did not attract the external support and sanctuaries that often are critical for insurgent success; they were forced to rely heavily on the weapons and ammunition they captured from the Spanish army in 1898 and what Filipino deserters from the Spanish army brought with them. As a result, nationalist fighters throughout the war were chronically short of basic military equipment, including rifles and ammunition.[100] The Americans exacerbated the problem by paying guerrillas to turn in weapons and even empty shell casings to keep them from being refilled; late in the war, they also exchanged captured guerrillas for weapons—a policy that particularly appealed to prisoners' families.[101]

The Army of Liberation never developed a coherent training program, had poor equipment, and had minimal doctrine—meaning its *force generation* capacities were weak.[102] While some commanders were good and many individual Filipino troops were brave and capable fighters, the Army of Liberation consistently fielded weak units that fought poorly.[103] Aguinaldo wrote after the war that most of his senior officers were incapable of leading large formations due to lack of military training and experience.[104] Filipino troops, many of whom were illiterate, received little formal military training, and combat units did not train or exercise in the field, meaning they were poorly prepared to engage their enemy; some imagined that U.S. troops would flee the presence of Filipino bolo (long) knives, leading to charges by soldiers armed only with bolos against rifle-armed Americans that unsurprisingly led to heavy insurgent casualties.[105] American interdiction of supplies, internal controls of resources like food, and the amnesty program contributed to the Army of Liberation's inability to produce and maintain effective combat units. The inducements of the U.S. amnesty program were fairly effective throughout the war but generated a flood of turncoats early in 1901—more than 800 guerrillas with nearly 23,000 weapons surrendered in January alone—crippling the nationalists' fighting capabilities.[106] U.S. actions effectively exploited both material and ideational vulnerabilities in the force generation dimension.

The nationalists' political and military *leadership* as a whole was weak. Policy disagreements between factions favoring accommodation with the Americans and fighting to the end sometimes paralyzed decision making.[107] Aguinaldo gave commissions and senior positions in the Army of Liberation to unqualified cronies.[108] The lack of coherent training programs afflicted the officer corps as well as soldiers. Chronically bad decisions by Aguinaldo and Luna, who Aguinaldo considered to be his best field commander, led to numerous tactical defeats and some missed strategic opportunities.[109] For example, historian Linn argues that Aguinaldo made a critical strategic error by not attacking and occupying Manila in June 1898, after Dewey destroyed the Spanish fleet and demoralized the Spanish troops defending Manila but before American soldiers arrived to reinforce Dewey's small force of sailors.[110] Had the numerically superior Army of Liberation taken and occupied even part of the city, the Americans probably would have had to make some accommodation with Aguinaldo for sharing power in the city, at least. Instead, the Army of Liberation fairly passively manned trench lines around Manila while capturing small Spanish garrisons elsewhere. Similarly, the decision to use terror widely was a strategic political blunder.[111]

Insurgent command and control was poor. With Aguinaldo holed up in remote northern Luzon from November 1899 to March 1901, he controlled little more than north-central Luzon and made little effort to resurrect his moribund government.[112] Hence, there was no central control of the movement during most of the guerrilla phase of the war, and local commanders operated largely autonomously, precluding coherent and coordinated archipelago-wide policies and campaigns.[113] U.S. Navy control of the sea inhibited insurgent leaders' communications with guerrilla bands operating on the small central islands, especially. And, personal and ethnic rivalries chronically divided the leadership. Not surprisingly given these unfavorable characteristics, the nationalist movement was riddled with factions and lacked confidence in its senior leadership, making the American amnesty program that offered even senior guerrilla officers full amnesty and posts in the colonial administration more effective.[114]

Finally, American units were superior in both conventional and unconventional aspects of the *operational execution* dimension. The Americans routed the Army of Liberation in most conventional fights in 1899. And, the conventionally organized Americans adapted well to the Army of Liberation's shift to guerrilla tactics by the end of 1900. With the help of Filipino scouts, they consistently outfought the guerrillas in small, light infantry engagements from late 1900 through mid-1902. The nationalists in turn could not adapt sufficiently to ameliorate the chronic weaknesses their equipment shortages, poor training, and weak leadership caused—which demoralized many fighters.[115] Without sanctuaries, from early 1901 the Americans chased them constantly, killed many, and so exhausted many others that they surrendered. And, very importantly, American military personnel well executed many of the nonviolent activities critical to the success of the pacification program, winning the support of many Filipinos and illustrating the importance of multidimensional counterinsurgency operations carefully balanced for local conditions.

The Filipino Nationalists

On the other side, Filipino nationalists simultaneously fought a political struggle for the support of Filipinos and a political/military war against the United States. They failed in both for related reasons. Focusing here on the war with the United States, Aguinaldo and the nationalist movement as a whole targeted a narrow range of perceived American vulnerabilities in the *national will, leadership,* and *operational execution* dimensions. Their inability to find and exploit American vulnerabilities stem largely from the leadership and force generation weaknesses noted above.

Initially, Aguinaldo and the Army of Liberation sought a conventional military battlefield victory in the vicinity of Manila in the *operational execution* dimension. But they found no critical U.S. military weaknesses they could exploit. Even the Army of Liberation's numerical superiority in manpower could not produce victory by attrition. Instead, the Americans consistently captured fields of battle and inflicted far more casualties than they suffered. Finally recognizing his army's weaknesses and vulnerabilities in conventional fighting, Aguinaldo in late 1899 shifted to unconventional operations to try to exploit American forces' conventional orientation. The shift initially produced some tactical successes and frustrated the Americans in 1900, but MacArthur's adaptive adjustments of

December 1900 ameliorated most U.S. weaknesses and gave the Americans operational techniques that enabled them to attack guerrilla vulnerabilities more effectively soon thereafter. In short, Aguinaldo and his military forces could neither find nor effectively exploit enduring U.S. military tactical- or operational-level vulnerabilities. They also did not ameliorate their own military vulnerabilities, which the Americans both identified and, after late 1900, consistently and effectively exploited.

As part of the shift to guerrilla operations in late 1899, Aguinaldo tried to influence U.S. *national will* and indirectly *leadership* dimensions strategically by using tactical-level military operations to bleed U.S. forces and extend the conflict temporally—hoping to lead American citizens, and through them congressmen and the president, to tire of the war and quit the Philippines.[116] The nationalists closely monitored congressional debates on Philippines policy and collaborated with the Anti-Imperialist League, which Americans founded in Boston in early 1898 to oppose the Spanish-American War and which refocused its concerns as U.S. troops occupied the Philippines.[117] The nationalists accelerated attacks on U.S. troops in September 1900 and initiated some pitched battles in an effort to generate U.S. casualties Aguinaldo thought would help Bryan unseat McKinley in the presidential election of 1900.[118] While it was poorly executed and failed, Aguinaldo's concept of using tactical military operations to help achieve a strategic political objective was sound and anticipated more effective such strategies used repeatedly against the United States in later wars.

Aguinaldo and his supporters saw the U.S. Democratic Party and the Anti-Imperialist League as their champions and used the prospect of a Bryan victory in the U.S. elections of November 1900 to spur greater Filipino nationalist efforts in 1900.[119] While working with American anti-imperialists who were mainly Democrats, Aguinaldo's agents approached Bryan's campaign in mid-October 1900 to promise a cease-fire if Bryan defeated McKinley; Bryan's campaign officials declined to respond because they feared they would be accused of treason if they dealt with Filipino insurgents—a charge the Republicans leveled at Democrats anyway with some accuracy given the considerable collaboration between Aguinaldo's people and league members who also were Democrats and the league's overt, avowed efforts to demoralize U.S. troops in the Philippines.[120] The Anti-Imperialist League conducted its peak level of activities during the election campaign of 1900 in alliance with Aguinaldo and largely in support of Bryan, was deflated by Bryan's defeat, and was moribund by 1904.[121]

Aguinaldo's agents worked with sympathetic American journalists and the Anti-Imperialist League to publicize the relatively few atrocities U.S. troops committed and to fabricate others designed to degrade American popular and congressional support for the war.[122] Based in British-controlled Hong Kong—equipped with telegraph links to the technologically modern world and safely beyond American ability to interdict it—the nationalist propaganda apparatus told Americans that Aguinaldo and his movement were fighting imperialists like McKinley and racist soldiers who committed atrocities against Filipino civilians—not Americans as a whole.[123] Civilian casualties were an appreciable concern to Americans during the Philippine War given that Spain's use of "concentration camps" in Cuba as a counterinsurgency technique, which led to civilian deaths, was a popular justification for war against Spain; the contemporary deaths of Boer civilians in British camps in South Africa also generated considerable negative publicity.[124]

Aguinaldo's people successfully produced and disseminated some propaganda. They repeated anti-American propaganda the Anti-Imperialist League wrote for use on domestic American audiences and U.S. troops in the Philippines, which prominently featured themes of American troops' barbarism and racism toward Filipinos.[125] By focusing on real and imagined American atrocities, the nationalists also generally succeeded in damping coverage of the widespread terrorism and crimes their own soldiers committed against Filipinos and the atrocities they perpetrated against U.S. prisoners and dead American soldiers.

But other nationalist propaganda was heavy handed and ineffective. For example, Aguinaldo's propagandists told Filipinos that the Americans aimed to enslave all Filipinos and to destroy the Roman Catholic Church in favor of Protestantism—charges the Spaniards made in 1898 that were so obviously false that they discredited the insurgents.[126] The propaganda campaign did not prominently address U.S. military casualties—a major theme of later U.S. adversaries—and the topic generated little sensitivity among the American public. As during World War II but unlike in Vietnam and later wars, U.S. military commanders in the Philippines War had censorship powers of defense in this arena, which they used with apparently modest success.[127]

By placing so much emphasis on the importance of a Bryan victory over McKinley in 1900, Aguinaldo set up his supporters for major disappointment when Bryan lost; they never again were generally confident of success and guerrilla desertions to the Americans rose considerably in early 1901, soon after Bryan's defeat.[128] Aguinaldo and his propagandists thereby unwittingly created a new political vulnerability for their own cause in the *national will* and *leadership* dimensions, which U.S. troops and civilian administrators well exploited in late 1900 and early 1901 through increased military pressure and the amnesty program.

Press reports of U.S. atrocities on Samar in late 1901 prompted anti-imperialist Republican Senator George Hoar to request Senate hearings on U.S. military behavior in the Philippines, which were held intermittently from late January to June 1902.[129] Chaired by imperialist Senator Henry Cabot Lodge, the investigating committee found individual wrongdoing but no administration culpability for American atrocities on Samar or elsewhere.

Despite some domestic popular and congressional unhappiness with the conduct of the war and some unhappiness with American colonial policies in general, Congress and the American people as a whole remained fairly solidly supportive of the war until the end. The war was effectively over after little more than three years of serious fighting at a cost of 4,234 U.S. combat fatalities, another 2,818 troops wounded, and about $400 million, many times the islands' purchase price.[130] By various estimates, the U.S. military killed 16,000 to 20,000 Filipino combatants while some 100,000 to 200,000 Filipino civilians died in 1899–1902 due to war-related outbreaks of disease, collateral damage, nationalist retribution against collaborators, and atrocities committed by both sides.[131] These costs did not generate popular American and congressional opposition to the war significant enough to affect the course of the war, its conduct, or its outcome.

Americans' unhappiness with U.S. troops' behavior peaked in April and May 1902, when the Samar courts-martial occurred and when the war was nearly won. By the summer of 1902, the minor domestic crisis had passed. Democrats chose not to make the Philippines an issue in the congressional elections of 1902.[132] And, although at one time he worried about the implications of popular discontent with the war for his political future, President Theodore Roosevelt, an ardent imperialist who succeeded McKinley upon his death in September 2001, won the presidential election of 1904 easily. Filipinos' efforts to attack the United States in the national will and leadership dimensions failed.

In sum, Filipino nationalists including the Army of Liberation and guerrilla units were weak political and military actors with critical vulnerabilities in several dimensions—a reflection of the unusually extensive weaknesses of the nationalist movement—most of which the Americans found and generally well exploited. And, the nationalists failed to identify or exploit American vulnerabilities in a narrower range of dimensions. This dreadful combination from the nationalists' perspective produced an unusually rapid (by the standards of modern insurgency wars) and total defeat for Aguinaldo and his fighters.[133]

U.S. LEARNING AND ADAPTATION

In the Philippines, U.S. forces without the benefit of national intelligence agencies, sophisticated technical collection systems, or a large military intelligence corps employed what we now call tactical human intelligence (HUMINT) to gather information about the communities where they were posted. They appear to have generally accurately analyzed the information, thereby also learning. They learned local languages and stayed put for long periods of time, facilitating both learning and a desire to learn. The learning techniques they used seem to have been similar to those U.S. troops used in the Civil War and Indian wars.

Troops passed accurate observations and lessons to commanders who used them to generate archipelago-wide and local policies that effectively attacked well-understood insurgent vulnerabilities—through means like recruiting local policemen and scouts, prisoner policy, securing towns, more aggressive patrolling, and so on. Commanders effectively adapted to address local conditions in ways generally consistent with existing doctrine, especially G.O. 100, and institutional lessons of the Civil War and Indian wars. The operational adjustments of late 1900 did not require many organizational or doctrinal changes. The army maintained its conventional structure but conducted small-unit, counterguerrilla operations in ways similar to those of the Indian campaigns. Revised operational practices posed no threat to established military cultural, doctrinal, or institutional interests—speeding adaptation and making it more effective.

The military successfully ameliorated its vulnerabilities to Aguinaldo's guerrilla operations through MacArthur's adaptations of December 1900, which also enabled better attacks on the guerrillas. Politically, censorship was minimally effective against insurgent and Anti-Imperialist League propaganda, but the latter posed little threat. At home, the McKinley and Roosevelt administrations faced relatively weak political opposition to the war and successfully limited the effects of scandals like Samar by blaming military officers, not administration policies, for the atrocities. In sum, the Americans learned well and effectively adapted, leading to excellent vulnerability identification, exploitation, and amelioration.

SUMMARY

The war pitted a materially challenged but operationally effective U.S. expeditionary force against such a weak opponent that we can identify potentially fatal vulnerabilities in the nationalist movement in most dimensions—a rare case. The United States had the beginnings of political-military vulnerabilities in the national will and leadership dimensions (as evidenced by the Senate hearings in 1902 on U.S. atrocities), which Filipino nationalists accurately identified and which many U.S. enemies since have tried to exploit. But in this war, Filipino nationalists were unable to exploit them, and they had little effect on the strategic outcome. Against a competent opponent, American military-related institutional weaknesses might have cost the materially weak U.S. military its victory or forced a much larger American material commitment. We can only speculate about whether, if Filipino nationalists been more competent, the war might have dragged on and whether domestic U.S. politics might have helped determine the strategic outcome.

The case is inconsistent with Arreguín-Toft's theoretical construct in both form and substance in the unconventional second of two strategic interactions—during which the United States sealed its victory. And, while the strategic result is consistent with basic materialist theory, the mechanisms of U.S. success as illustrated by the details of the local campaigns indicate that ideational and institutional factors and *skill* were more important than the moderate amount of military materiel used, moderately superior U.S. military technologies, and the large latent material *capabilities* Washington chose not to use.

The case is consistent with some points of Shimshoni, Biddle, and Merom. American troops demonstrated the importance of good leadership and imagination, consistent with Shimshoni. No major use of combined arms teams occurred except in the early conventional battles in which artillery and naval gunfire were important, consistent with Biddle's assertion that the modern system emerged during World War I. The U.S. domestic influences that would cripple American war efforts starting in Vietnam appeared in the Anti-Imperialist League but were too weak to alter the strategic outcome—consistent with Merom's argument that such pressures became strategically important after 1945. The processes Mack discusses in the context of America's limited war in Vietnam for the most part did not occur in the Philippines in 1899–1902—an inconsistency with his theory unless one argues that the nationalist movement was so weak that there really was not much interest asymmetry between the Americans and most Filipinos.

CHAPTER 3

The Allied Bomber Offensive against Germany (1943–1945)

The United States and the United Kingdom conducted a strategic bombing campaign against Germany during 1943–1945, which contributed to the Allies' victory over Germany in World War II. This case study focuses on one aspect of a major war because airpower theorists of the U.S. Army Air Forces (USAAF) and the Royal Air Force (RAF) argued extensively before and during the war that strategic airpower alone could win the war by attacking political and economic vulnerabilities—specifically identified as such—they thought they understood and could exploit. This chapter covers American and British aspects of the bombing campaign because they were fairly well integrated, illustrate different perceptions of the vulnerabilities of the same adversary that in other wars appear as different U.S. military service perspectives, and show how different capacities to exploit them yield different tactics and operational results.

Strategic bombing originated in World War I, when German dirigibles and, starting in 1917, Gotha and Giant heavy bombers killed 1,414 British civilians in about 100 raids on cities, and aircraft of the newly created RAF killed 641 German civilians in 22 cities.[1] The French and German air forces also exchanged raids on cities. While the number of fatalities was small by the standards of the Great War, air raids were new, terrified many civilians, and fostered deep fears among political and military leaders that apparently unstoppable attacks from above would lead restive factory workers to revolt, thereby slashing defense industrial production, crippling armies dependent on large and steady supplies of munitions, and shattering governments.[2] The RAF was developing a new long-range heavy bomber when the war ended.

Beginning soon after World War I and energetically applying its apparent lessons, General Hugh Trenchard, who was the RAF's chief of air staff during 1919–1929 and was influential on military matters in retirement during World War II, developed doctrine that held that strategic bombers could win wars quickly by delivering "knock-out blows" that so discouraged enemy civilians, especially factory workers who produced munitions, that their governments would capitulate; this work was consistent with, but independent of, Italian General Guilio Douhet's influential work on strategic airpower first published in Italy in 1921.[3] The RAF called such attacks on civilians' attitudes "moral," or morale in American English, bombing.[4] The claims of airpower partisans fostered in Britain in the interwar years a widespread mixture of fascination with, and fear of, airpower; many works of popular fiction in the period featured apocalyptic aerial bombardments.[5]

The theory and rhetoric suited many ideals and interests in the interwar years. Many Britons saw use of airpower as a way to avoid repetition of the carnage of battles like the Somme in 1916 and Passchendaele in 1917. Some, like Basil Liddell Hart, based on lessons of western front battles, saw defensive operations in land warfare as systemically more advantageous than offensive operations, meaning Britain could maintain a small army because France could alone defend itself against an economically and demographically faster growing, irredentist Germany. This argument rationalized aversion to a "continental commitment," or alliance with France, that would again lead to dispatch of a British expeditionary force to France in the event of war; airpower allegedly gave Britain military capabilities without need for a continental commitment.[6] Treasury officials saw airpower as a relatively inexpensive form of effective military power compared with manpower-intensive land forces, an important policy consideration given the severe fiscal constraints Britain faced in the 1920s and 1930s.

The RAF used its war-fighting theories to argue in concert with Treasury that it could alone defend Britain. The RAF's Bomber Command claimed it could effectively and inexpensively replace the other services, including the army, the Royal Navy and its nascent naval aviation arm, and even the RAF's own fighter aircraft.[7] The argument had immediate and obviously negative implications for interservice rivalries.

The RAF's bomber theoreticians and commanders thought bombers, especially when dropping the chemical munitions the RAF expected to use on any adversary, were strategic offensive weapons of enormous power. They argued that bomber forces should be dedicated to strategic operations and should never be diverted to tactical attack or defensive roles—which would be both unnecessary and wasteful. Trenchard, like many later airpower partisans, claimed that control of strategic airpower should be entrusted to professional airmen because only airmen were capable of understanding the "fundamental principles" of air warfare.[8] Airmen should strenuously resist pernicious efforts to misuse airpower—like army generals' desires to use aircraft in interdiction and close air support roles. Such arguments were logically coherent, if factually inaccurate and operationally questionable, and provided a formula by which the young RAF hoped to solidify its organizational independence and achieve bureaucratic ascendance over its army and Royal Navy rivals. In the language of offense-defense theory, the RAF believed there was a systemic, technology-based offensive advantage to strategic bombing operations.

RAF bombing advocates' focus on strategic offensive operations against enemy morale meant they intended to kill and to terrorize noncombatant civilians in large numbers.[9] As late as 1942, before the Allied bomber offensive began in earnest, RAF contingency plans called for raids on German cities that planners expected to kill 900,000 civilians.[10] At the senior political level, the cabinet on May 7, 1942, authorized the RAF to destroy 58 German cities, thereby damaging German morale by "de-housing" 22 million people, or about one-third of the population of Germany.[11]

In contrast, the U.S. Army Air Corps, which was renamed the USAAF in June 1941, in the interwar years recognized widespread American popular opposition to attacking civilians and published doctrine that variously argued that bombing civilians was morally undesirable, militarily ineffective, and counterproductive.[12] While a minority of officers, notably Brigadier General William (Billy) Mitchell in his later writings, especially, agreed with Douhet and Trenchard that civilian morale should be the primary target of strategic bombing, aviators at the Air Corps Tactical School (ACTS), which developed army air doctrine in the 1920s and 1930s, by the mid-1930s called for priority to attacks on carefully chosen economic targets. The ACTS's variously named "industrial web" or "industrial fabric" or "key-node" theory was based on the idea that military organizations dependent upon large defense industrial bases would collapse if their economic foundations could be made to fail, thereby shattering nations' will to fight.[13] The ACTS recognized that such bombing would kill civilians but civilian deaths were not the purpose of the bombing.[14]

If effective, American air theorists argued, strategic bombers would win wars by themselves without targeting either enemy troops in the field or civilians—and save lives by ending wars more quickly.[15] The army's ground forces and the navy would be unnecessary, justifying establishment of an independent air force.[16] With war approaching, the Air War Planning Document (AWPD-1) of August 1941 confirmed USAAF doctrinal preference for targeting industrial bottlenecks in four areas—electricity, oil-based fuel, transportation, and aircraft manufacturing.[17] But AWPD-1 also recognized that bombing could affect enemy morale and might be strategically effective in some circumstances, a caveat that later would help justify American bombing of enemy cities.[18]

Thus, using similar logic, RAF and USAAF theoreticians argued that bombing focused on enemy vulnerabilities would win any war and generate respect, money, and bureaucratic independence for their institutions. The USAAF and the RAF differed on only one, albeit major, detail—the identity of the critical enemy vulnerability.

Having identified enemy vulnerabilities, each organization recognized a need to acquire means of exploiting them. For each service, this meant acquisition of fleets of bombers that could carry large amounts of munitions long distances.[19] But the two organizations again took different paths.

Because air defenses had been weak in World War I, the RAF and many British civilian officials thought strategic bombers inherently were invulnerable to enemy action. Because heavy bombers seemed superior to all current and future air defense technologies, RAF bombers needed few active defenses or other means of vulnerability amelioration. This view was summarized in the much-quoted language of Prime Minister Stanley Baldwin that a RAF propagandist might have been proud to write, "[T]he bomber will always get through."[20]

The RAF in the 1930s procured a series of technologically weak bomber aircraft, most of which were abandoned early in World War II in favor of Halifax and Lancaster heavy (and Mosquito light) bombers.[21] Emphasis on the war-winning value of strategic bombers, backed by senior politicians like Chancellor of the Exchequer and later Prime Minister Neville Chamberlain, who accepted the RAF's financial efficiency argument, led the RAF to devote few resources to other kinds of aircraft. But in the late 1930s as the German air force, or *Luftwaffe*, rapidly expanded, Britons recognized that Germany might try to defeat them with the same sort of "knock-out blow" they previously presumed they would inflict on others.[22] As a result, the RAF began to develop short-range Spitfire and Hurricane fighter aircraft for air defense purposes and ground-based air defense and related command-and-control technologies, which were to be very useful in the Battle of Britain in 1940. But Air Chief Marshall Sir Charles Portal, who headed Bomber Command early in the war and then was chief of air staff, told Prime Minister Winston Churchill that long-range fighters were not feasible, ending RAF research on them and precluding the RAF from developing British-made long-range escort fighters during the war.[23]

The Americans initially shared the RAF's faith in the invulnerability of bombers. Air Corps maneuvers in 1929 convinced senior officers that fighters posed no threat to bombers and prominent ACTS instructor and later Major General Laurence Kuter dismissed ground-based antiaircraft artillery (AAA) as irrelevant for bombers.[24]

The Air Corps in 1934 asked aircraft makers for bids on a bomber that could carry 2,000 pounds of bombs and fly at least 1,020 miles at 200 miles per hour.[25] As a result, in 1937 the Air Corps received the first of the Boeing B-17s that were to be the workhorses of the U.S. bombing campaign against Germany. By the standards of the day, the B-17 could carry much ordinance long distances and was fast. Dubbed the "Flying Fortress," the B-17 was heavily armed, originally with five and later six antiaircraft machine-gun firing positions, and armored. It outranged all contemporary American fighter aircraft. Consistent with doctrine calling for strikes against economic targets that often were small, the Air Corps adopted the Norden bombsight, developed originally for the U.S. Navy, which allowed optically guided "precision" strikes on small, or "point," targets under ideal visual conditions—including daylight and clear weather.[26] By the late 1930s somewhat worried about air defense fighters, the Air Corps began to develop fighter escorts, but with little enthusiasm, effort, or success.[27]

Therefore, both the U.S. Army Air Corps and the RAF built bombers that airmen expected to be largely invulnerable to enemy air defenses and to be able to effectively attack the critical enemy vulnerabilities they identified. Immediately before the war, they worked ineffectively to improve the defensibility or invulnerability of their bombers and worked little to produce long-range fighter escorts that could make, in essence, aerial combined arms teams that could both conduct effective strategic offensive operations and actively defend bomber formations. That is, before the war they saw few vulnerabilities in their own air forces that they perceived a pressing need to ameliorate.

OVERVIEW OF THE BOMBING CAMPAIGN

The realities of air warfare differed sharply from airpower advocates' prewar expectations. Bomber Command soon proved itself incapable of performing as it promised. With Britain alone after the collapse of France in June 1940 and under attack, only bombers could directly strike Germany, and Bomber Command was eager to demonstrate its institutional worth. But the RAF initially had a hard time hitting even the vicinity of the centers of German cities, where it presumed politically unstable industrial workers lived. Neutral Denmark and Switzerland received unintended British bombs in 1940 as aircraft veered far off course. German air defenses often killed more RAF crewmen than RAF bombing missions killed Germans on the ground. Churchill complained in September 1941 that only one-quarter of British bombers hit their targets and suggested that, far from being a war winner, Bomber Command was only an "annoyance" to Germany; Bomber Command itself in 1941 concluded that less than one-third of its crews delivered bombs within five miles of their targets.[28]

Why was Bomber Command so incompetent? Commanders seem to have assumed Bomber Command could bomb accurately and little worked to develop combat capabilities.[29] Bomber Command little trained its crews and ran unrealistic exercises over Britain that did not well address the difficulties of combat conditions at night and in the often-inclement weather of northern Europe; early raids produced bombing accuracies only about one-third as good as those achieved in exercises over Britain.[30] RAF navigation aides were poor and improved marginally during the war with development of technologies like airborne radars and navigation beacons, which Germany jammed and also used to track and target bomber formations.[31] The RAF initially had small and ineffective bombs and did not even issue clothing adequate to protect airmen from hypothermia while flying at 30,000 feet in winter in poorly heated aircraft.[32]

RAF faith in the invulnerability of its bombers also soon evaporated. British bombers were slow, lacked maneuverability, and were highly vulnerable to *Luftwaffe* fighters and large caliber AAA during daylight raids.[33] Losses often were heavy, leading Bomber Command to retire obsolete aircraft types, to scale back attacks on Germany, and then to shift to night operations in 1941 to keep from being decimated. But in giving up most daylight raids—the RAF still conducted some daytime raids on key German point targets—for the modest increase in protection afforded by flying at night, the RAF abandoned any pretense of disrupting German war production by hitting workers' residences and sapping their work ethics. Instead, from February 12, 1942, it focused on bombing civilians in general in cities as area targets, consistent with the more broadly political variant of morale bombing theory.[34] This change little challenged RAF doctrine; if anything, it further entrenched the doctrinal primacy of morale-oriented bombing because Bomber Command's night operations were incapable of consistently hitting anything smaller than cities.[35]

The United States entered the war in Europe well after it began. While American forces fought German troops in 1942 in North Africa, the USAAF was not ready to conduct significant bombing operations against Germany proper until 1943. But how would the British and Americans coordinate their strategic bombing efforts? Reflecting different bombing philosophies and capabilities, the Allies at the Casablanca summit of January 1943 agreed to a coordinated bombing offensive but disagreed about targets and tactics. Policy disagreements and bureaucratic infighting would continue to plague the bombing campaign through the end of the war.[36] The RAF, in the view of Air Chief Marshal Sir Arthur Harris, chief of Bomber Command from February 1942 through the end of the war, was allocated night bombing of area targets—a euphemism for city bombing—and the United States agreed to conduct daylight "precision" bombing of economic point targets as its part of the "Combined Bomber Offensive," which was to be supervised by the Combined Chiefs of Staff, Air Marshal Sir Charles Portal, and later supreme Allied commander General Dwight Eisenhower; the Casablanca agreement also established a list of ranked target priorities—17 categories of economic facilities.[37] American airmen argued for economic targets primarily because they thought such targets were more lucrative than morale bombing, not for ethical reasons.[38]

Neither general approach was conducted consistently over the course of the war, however. Very high loss rates and changing target priorities periodically led to changes in American and British tactics and targets. For example, Allied bombers occasionally also attacked German naval bases, V1 rocket launch sites, ground force targets, and the logistical infrastructures of Germany and German-occupied countries.

Despite such shifting targeting policies, as a general rule in 1943–1945 RAF Bomber Command's preference was to hit cities as morale targets, American bomber commanders preferred to attack economic targets, and both air forces grudgingly attacked tactical military targets when they could not evade orders to do so. American airmen knew their bombs killed many German civilians but they rationalized, to themselves and to the American public, that they aimed for legitimate military targets, that civilian deaths were unavoidable, and that civilian deaths therefore were morally acceptable.[39] American and British leaders were much more reluctant to bomb cities in occupied third countries, where victims mainly were presumably innocent non-Germans, but did so anyway if they perceived military necessity.[40]

The USAAF's 8th Air Force, formed in June 1942 and based in England, began major bombing attacks on Germany in January 1943.[41] Consistent with their doctrine and the Casablanca agreement, American bomber units concentrated on targets chosen to cause general economic collapse, like ball-bearing factories, and on targets chosen to disrupt war-making capabilities, like synthetic fuel plants and factories that made key items like fighter aircraft.[42] From July 1943 to April 1944, the Allies attacked German aircraft plants under a program code-named Pointblank.[43] Other prominent operations targeted oil facilities and transportation infrastructures.[44] Heavy bombers of the 15th Air Force, based in Italy, attacked targets in southern Germany, Austria, and elsewhere that were harder for England-based aircraft to strike.

The USAAF discovered quickly that the B-17 Flying Fortress was good at flying but was not much of a fortress—and less well-armed bombers like the B-24 Liberator were even more vulnerable to German fighter aircraft. German fighters and AAA shot down many American bombers.[45] To improve bomber survivability, the Americans assigned fighters to strategic air forces to escort the bombers. But with northwestern Europe in German hands in 1943, England-based fighters did not have the range to accompany bomber formations all the way to targets deep in Germany—like Berlin. *Luftwaffe* fighters therefore attacked American bomber formations extensively after Allied fighter escorts returned to their bases. German fighters dogged bomber formations on their way to targets, landed to rearm and refuel, and then attacked them again on their way home.

Inadequately defended without fighter escorts, bomber groups attacking targets the *Luftwaffe* chose to defend extensively often suffered heavy losses. For example, during a modestly effective raid on ball-bearing plants at Schweinfurt and an aircraft plant at Regensburg on August 17, 1943, the 8th Air Force lost 147 bombers—some 40 percent of the force dispatched—which were shot down, crash landed, or written off as too badly damaged to warrant repair.[46] Many other raids also suffered heavy losses. A USAAF study in early 1944 found that of 1,000 airmen assigned to the 8th Air Force, only 216 could expect to complete a 25-mission tour without being killed, wounded, or shot down and captured.[47]

By year-end 1943, USAAF and RAF bomber forces were operationally ineffective, were badly bloodied, and after another costly raid on Schweinfurt in October had largely withdrawn from deep strikes on Germany.[48] The Germans were winning their campaign against British and American bombers.[49]

The heavy losses finally forced the USAAF to reevaluate its assumptions about the defensive capabilities of bombers and encouraged alternative thinking.[50] Incomprehensibly, given heavy *Luftwaffe* losses over Britain in 1940 and Bomber Command's unpleasant experiences over Germany that prompted the shift to night bombing in 1941, as late as October 1942 Brigadier General Ira Eaker, commander of VIII Bomber Command and then of 8th Air Force, thought USAAF bombers could operate alone over Germany with minimal losses.[51]

A major, belatedly obvious problem was a lack of long-range fighter protection. German fighters' 20-millimeter and 30-millimeter cannon and 21-centimeter aerial rockets were more powerful than, and outranged, the .30 and .50 caliber machine guns the B-17s and B-24s carried.[52] German fighters could badly damage bomber formations at only moderate risk so long as Allied fighters did not engage them.[53] Therefore, the USAAF perceived a technical challenge: how to acquire long-range fighters with speed and altitude characteristics suitable to accompany the bombers and the fighting capacity to defeat German interceptors in aerial combat while keeping German planes away from the bombers.

An effective solution was found in use of two relatively minor technological innovations that modified existing aircraft. First, a British test pilot working with British operated, American-built P-51 Mustang fighters had the idea of replacing the P-51's Allison engine with the more powerful British-made Rolls Royce Merlin engine.[54] Finding the Merlin produced a much better fighter, the Americans reequipped their P-51 airframes with Merlin engines.

Second, the USAAF experimented with a variety of "drop tanks" to extend fighters' ranges.[55] Drop tanks are externally mounted fuel tanks that can be jettisoned on pilot command—normally either when they are empty or when enemy aircraft appear and pilots want better aircraft flight performance characteristics to engage them. Drop tanks had long been used to ferry aircraft long distances. By late 1943, the USAAF in Europe belatedly adopted a new internal tank and a 75-gallon drop tank for the P-51B model that gave the Mustangs a combat radius of some 850 miles—far enough to accompany bombers from southern England to Berlin and back—and capable P-47 Thunderbolt fighters with drop tanks had a combat radius of 475 miles.[56]

The low-tech solutions to the fighter quality problem made a major operational difference starting in January 1944, when the long-range P-51 entered squadron service with the 8th Air Force; indeed, they altered the course of the bomber campaign.[57] By mid-1944, the 8th and 15th Air Forces had nearly 2,000 fighters for bomber escort duties.[58] With continuous fighter escort coverage, bomber loss rates dropped sharply.[59] In addition, in January 1944 Lieutenant General Jimmy Doolittle replaced Eaker as commander of 8th Air Force and immediately launched an aggressive campaign of raids on Germany, including six consecutive days of heavy attacks on aircraft plants in late February that delivered more bombs than 8th Air Force dropped on Germany in all of 1943.[60]

The more aggressive attacks enabled escort fighters to destroy many more German fighters than gunners on bombers previously had been able to down. During the extensive aerial combat of early 1944, the *Luftwaffe* lost as many as half of its fighters each month.[61] In the first three months of 1944, Germany lost some 4,000 fighter aircraft.[62] The Pointblank campaign against aircraft factories only restrained the growth of German aircraft production—overall production peaked in July and monthly fighter production peaked at 2,900 aircraft in September 1944—but it achieved a major, initially unintended success by killing experienced German fighter pilots faster than they could be replaced; some 1,684 were killed in January–April 1944, degrading the quality of the German pilot corps and thereby the effectiveness of the *Luftwaffe* as a whole.[63] Later, the USAAF chose targets it hoped would draw German fighters into combat.[64] In April 1944, General Adolf Galland, chief of the *Luftwaffe*'s fighter arm, warned his superiors that his command was near collapse.[65] After April 1944, the *Luftwaffe* became more conservative in its air operations, and the Allies thereafter lost more aircraft to AAA than to enemy fighters.[66] By June 1944, the Allies had swept most German fighters from the skies of northern France, simultaneously preventing their use as ground attack aircraft against Allied forces landing in Normandy.[67]

As Allied fighters achieved air superiority over Germany, bomber strength continued to grow. From a combined force of some 1,000 bombers in March 1943, the Allied heavy bomber force focused on Germany rose to about 3,000 aircraft in February 1944 and about 5,250 planes in July 1944—a level that remained fairly stable for the rest of the war.[68] The strength was concentrated in England; in April 1944, 8th Air Force had 2,788 heavy bombers, and 15th Air Force had 1,279 bombers.[69]

But even after the Allies solved their tactical vulnerability problem and substantially seized control of the skies, bomber forces dropping massive quantities of bombs on Germany did not damage German civilians' morale in ways that induced them to demand peace or disrupt defense industrial production enough to prevent the remnants of Germany's armies from fighting tenaciously while suffering huge losses in late 1944 and 1945. Despite lingering internal USAAF debate about the morality of city bombing, with the support of President Franklin Roosevelt, Assistant Secretary of War for Air Robert Lovett, and General Eisenhower, the USAAF in September 1944 agreed to join the RAF in striking German cities for purposes of damaging Germans' morale without pretense of striking industrial targets. These raids included attacks on cities in enemy Bulgaria and Romania.[70] With the war's end in sight, American airmen grasped at German morale as a way to win the war for airpower—thereby also advancing the cause of an independent air force and impressing the Soviets.[71] General Doolittle objected on ethical grounds to orders to bomb civilian targets in Berlin but was overruled; his bombers struck Berlin on February 3, 1945, killing thousands of civilians—and attacked other cities later.[72]

Only destruction of German armies by Allied ground forces and physical occupation of Germany ended the war. The ground forces' success simultaneously handed the bomber commands of the USAAF and the RAF both theoretical and bureaucratic political defeats.[73]

THE BOMBER CAMPAIGN AND MATERIALIST THEORY

The materially dominant United States won the war—consistent with basic materialist theory. But how did the Americans win? By most analyses, the Allies won World War II by applying material superiority to bludgeon Germany into defeat by destroying much of the country.

But aspects of the campaign that led to the attritional victory depended on tactical force symmetries as well as a dominant mass of military materiel. German fighter aircraft held the materially dominant Allied bomber forces at bay so long as bomber raids lacked fighter escorts—that is, the opposing forces were asymmetric in structure and bombers flying alone were vulnerable to German fighter attack, giving Germany the asymmetric advantage. When the Allies restored fighter symmetry over Germany in early 1944 by fielding long-range fighters, the *Luftwaffe* quickly lost the air war under the weight of Allied numbers. It was the symmetry of fighter force structures and attritional warfare that killed irreplaceable, experienced German pilots, not mass of aircraft assets or aircraft production capabilities, which ultimately won the air war for the Allies and enabled more secure bomber formations to pummel Germany. General Galland warned of the possible collapse of his fighter command several months before German fighter aircraft production peaked.

The bombing campaign illustrates that superior technology alone does not win wars and that technology is a not systemic variable available to all belligerents.[74] Germany's jet aircraft, aerial rocket, and some other technologies were superior to those of the Allies, who used some German technologies in the decade after 1945 in their own air forces, but the technologies did not generate force structures or tactics that could generate German victory. The technological innovations that helped win the campaign—integration of the P-51 airframe with the Merlin engine and the use of drop tanks—were not major technological breakthroughs. Allied airmen understood each technology well and the USAAF fielded drop tanks with combat units in the Pacific well before the end of 1943. But recognition of the combat value of drop tanks slowly reached Europe even within the relatively homogenous USAAF.

The case contradicts airpower advocates' prewar claims for the offensive capabilities of strategic bombers and some IR theorists' assertions that World War II was a period of offense dominance—a view usually justified by the development of the tank and what non-German analysts called "blitzkrieg strategy," a technology and a doctrine irrelevant to the development and use of strategic airpower.[75] Bombers helped the Allies win World War II, but success in the air war was slow, the result of costly attritional fighting, and partial. The strategic offensive power of bombers did not fully develop until technologies that provided tactical defensive capabilities—the long-range escort fighters—were fielded in large numbers. There were no conceptual breakthroughs.

The Allied bombing campaign offers little support for materialist theory's normal assumption that military resources are used rationally—or skillfully. Bombing directed against civilian morale was wholly ineffective at achieving Allied political objectives because, as discussed below, there were no political vulnerabilities exploitable by bombing—as rational and well-informed intelligence analysts and leaders not afflicted by biases driven by parochial interests should have known. Against more reasonable economic targets, most bombs did not come close to hitting their intended targets. To some degree this reflected weak bombing technologies like poor navigational aids, but Allied air forces' poor intelligence collection and analyses and institutional weaknesses significantly affected whether material assets were used skillfully. For example, Allied strategic air forces did not know where good targets were despite accurate publicly available information published before the war began; displayed poor skills in bombing due to poor training; did not understand the nature of the damage they inflicted or German industrial recuperative capabilities; and declined to work cooperatively in an integrated command structure with each other or Allied ground and naval services.[76]

The bomber war illustrates that the skills of even advanced militaries, especially when modified by differing force structures, are not equal. Skill is a contingent variable, not a constant or a global normative influence. The Allies lost the morale war because their institutional skill in identifying political vulnerabilities was very poor (the RAF was worse than the USAAF), and their deficiencies in identifying economic vulnerabilities, lack of patience, and parochialism led to shifting, incomplete attacks on good targets (the USAAF was worse). German camouflage and dispersion activities designed to degrade the Allies' understanding of Germany's defense industrial economics were at least partially successful. Eventually, the squabbling Allies arrived at workable compromises on targets that helped end the war—but the effort took longer and was less operationally effective and efficient than airpower partisans expected.

Organizations' critical skills were both individual and institutional. The *Luftwaffe* lost the bomber campaign in early 1944 and later the air war as a whole because too many skilled pilots died, damaging the skill of the organization defined as the sum of its parts, not its institutional skills. As variously illustrated by the advocacy of USAAF General Carl Spaatz, British civilian advisor Solly Zuckerman, and Air Chief Marshall Harris for oil, coal, and morale targeting, respectively, individuals also can improve or damage collective institutional skills (see below). Spaatz and Zuckerman worked effectively to institutionalize their insights, but Harris's obstinacy in attacking cities illustrates that even good intelligence information will not improve the collective skill of an organization if its leaders' beliefs or parochial institutional interests preclude dispassionate analysis. Rather than display the learning propensities of German military innovators who broke the stalemate of trench warfare and restored mobile warfare to the western front in 1917, for example, Harris stuck with his convictions to the bitter end.[77]

Airmen's determination to claim victory for airpower as a means of achieving organizational goals compromised effective leadership of the bombing campaign. Institutional drives and related mental blinders contributed to nation-specific patterns of force structure, operational practices, and "skill" that had major operational implications.[78] The RAF and USAAF bomber commands, despite being fairly closely cooperating allies, maintained to the end of the war ethical and military operational perspectives appreciably different from each other, the tactically oriented elements of their own military services, and their fellow national military services.

THE BOMBER CAMPAIGN AND STRATEGIC INTERACTION THEORY

Because Arreguín-Toft does not address any aspect of World War II, I assess in this section the bombing campaign using his analytical approach only.[79] After early 1944, the United States (and Britain) used virtually the same force structure of bombers and fighter escorts to attack three general targets—German civilians, the German economy, and enemy military forces. In these three interactions, the materially dominant United States used barbarism, conventional attack, and conventional attack options, respectively. In each interaction, materially weaker Germany responded with a conventional defense. The interactions occurred for most practical purposes simultaneously, not sequentially or in different geographical areas.

The Allies lost their bid to win the war by using barbarism—or morale bombing. Germany did not capitulate in response to large numbers of civilian casualties. This opposite-approach interaction was an American (and especially British) defeat—consistent with Arreguín-Toft's expectations.

The Allies' economic campaign also failed to produce victory on its own, although it slowly and inefficiently degraded Germany's military capabilities late in the war. Because American airmen's original expectations of an economic "knock-out blow" were dashed, I rate this same-approach, conventional-conventional interaction as a U.S. defeat—contrary to Arreguín-Toft's expectations.

The Allied air forces convincingly won their campaign against German military forces—victory in a same-approach, conventional-conventional interaction consistent with Arreguín-Toft's expectations. The Allies began to win in early 1944 with the introduction of long-range escorts, which soon achieved air superiority. Therefore, in part unintentionally during the bombing campaign and in part due to Eisenhower's orders to attack German naval and ground forces, the USAAF and RAF defeated the *Luftwaffe* in the air and helped Allied ground forces win the war. The fact that the effectiveness of the Allied strategic air campaign depended largely on the presence of tactical aircraft, not just heavy bombers, is too finely grained a point for Arreguín-Toft's analytical construct.

Arreguín-Toft's analytical approach therefore successfully predicts two of three strategic interactions. Germany could not possibly have fought an air campaign literally miles above the surface of the earth with GWS, meaning Arreguín-Toft's model in this case effectively is reduced to three strategic options and the weak actor has no choice at all—not a theoretically helpful proposition. As in the Philippine War case, strategic interaction theory is inconsistent with the structure and part of the outcome of the Allies' strategic bombing campaign. It does not explain why the bomber war occurred as it did.

DIMENSIONAL ANALYSIS

The Allies, on the offensive, attacked perceived German vulnerabilities in four dimensions and achieved mixed levels of limited success. Germany, on the defensive, had many fewer, and no good, options.

The Allies

The same Allied strategic air forces tried virtually simultaneously during 1943–1945 to attack Germany in the *national will, resource mobilization, resource conversion,* and *operational execution* dimensions. These were, respectively: (1) morale bombing of cities conducted by the RAF and to a lesser extent the USAAF; (2) targeting critical general economic installations, mainly by the USAAF; (3) attacking key defense industries like aircraft factories, mainly by the USAAF; and (4) strikes by both air forces on military targets in conjunction with Allied tactical air forces.[80] Initially, Allied bombers had little success with any of the four, but by 1944 the Allied force structure that would win the air war over Germany was in place:

British Lancaster and Halifax and American B-17 and B-24 heavy bombers accompanied by long-range fighters.

This force structure had markedly different successes performing operationally similar missions against different perceived vulnerabilities of the same enemy. Allied efforts to exploit German vulnerabilities in the *national will* dimension simply failed. The bombing indeed affected civilian morale, but it did not degrade workers' work ethics or produce political instability in Germany. Most Germans worked and fought to the end without rising against the Nazi regime. Having experienced the social upheavals of 1918 that contributed to Germany's defeat in World War I, Germany's leaders worked hard to avoid a repetition in World War II.[81] The *Sicherheitsdienst*, or "Security Service," a unit of the *Schutzstaffel* (SS), or "Protection Squadron," of the Nazi party tasked with monitoring Germans' attitudes, found no mood changes of political or economic significance after air raids.[82] Adolph Hitler and other senior Nazi officials read *Sicherheitsdienst* reports closely. Even after the devastating RAF raid on Hamburg on July 24, 1943, that produced a firestorm and killed some 40,000 civilians, depressed workers kept working hard.[83] The Allies, especially RAF Bomber Command, made a fundamental analytical error: they identified a German political vulnerability that did not exist.

Morale bombing advocates misunderstood basic human psychological responses to air attacks. In Germany and Britain during World War II, and in many other wars before and since, civilians consistently respond to air strikes with anger against their attackers, not rebellion against their own leaders.[84]

Morale bombing advocates ignored evidence readily available to them. RAF folklore notwithstanding, although air raids on the United Kingdom during World War I led Britons to be angry at their government, they demanded better air defenses, better air raid warnings, and retaliation against Germany—not capitulation.[85] The RAF should have known that during the Battle of Britain in 1940, when Germany extensively bombed British cities and killed thousands of civilians, hospital admissions for neuroses declined, the suicide rate fell, drunkenness apparently declined, and civilian morale rose.[86] Moreover, even after its defeat in the Battle of Britain, Germany periodically launched bomber raids on British cities and in 1944 and 1945 fired thousands of V1 and V2 rockets at British cities—all without significant effect on Britons' determination to win the war. Prime Minister Churchill repeatedly cited German air raids to rally Britons to work harder to win. In addition, Alexander De Seversky, a well-known Russian émigré aircraft designer in the United States and friend of Billy Mitchell, in 1942 published a book that noted that civilian resolve in the face of bombing was much greater than airpower proponents expected before the war.[87] Hence, on a variety of grounds, reasonable analysts might have anticipated before the Allied bombing offensive started in 1943 that air attacks would fail to significantly influence German civilians. But British officials, including Sir Charles Portal, convinced themselves without evidence that German civilians were more vulnerable to morale bombing than were Britons.[88]

Morale bombing failed to cow Germans for addition reasons. Germany was not a democracy in which a free press could accurately present the costs of bombing and popular discontent could generate policy changes. The *Sicherheitsdienst*'s surveillance led to growing numbers of arrests and about 100 death sentences per week by 1943 for defeatism and sabotage, nipping actual and latent defeatism in the bud before it became a political threat to the Nazi regime.[89] The German government offered carrots and sticks to workers—special inducements to valued workers and harsh disciplinary measures against laggards; some economic managers were court-martialed for poor performance.[90] German industry relied extensively on foreign and forced laborers whose political attitudes and morale were largely irrelevant.[91] In addition, Nazi propagandists exploited Allied demands for unconditional surrender to conjure up dire implications of surrender in order to encourage Germans to fight on; German civilians especially feared the consequences of Soviet occupation of Germany—a fear Red Army troops later demonstrated by their extensive looting and abuse of civilians to have been justified.[92] Thus, while German civilians were physically highly vulnerable to Allied bombs, the danger did not constitute a political vulnerability for the Nazi regime.[93]

Why did the Allied air forces so badly misread German vulnerabilities in the national will dimension? Captured by the mystique of the bomber and its value for advancing parochial organizational interests, the bomber commands as institutions refused to reconsider whether their prewar doctrinal concepts, technologies, force structures, and tactics could attain the strategic military results they hoped to achieve.[94] Most notably but not uniquely, Bomber Command's Harris throughout the war and in postwar defenses of his command consistently argued that more morale bombing alone would have won the war; he stubbornly refused to heed intelligence reports inconsistent with his preconceptions.[95] Despite evidence that morale bombing was not working, U.S. political and military leaders, including Roosevelt and Eisenhower, in 1944 endorsed morale bombing in hopes of ending the war quickly; American airmen still looking for a way to win the war for airpower embraced city bombing as a last chance to markedly enhance prospects for an independent air force.[96]

Attacks in the *resource mobilization* and *resource conversion* dimensions were slowly effective, but inefficient, in damaging the German war economy. They also failed to alone determine the war's outcome. The Allies did not well identify key economic vulnerabilities or exploit the ones they accurately identified. Allied intelligence officers little understood the economic linkages of complex, interdependent economies in general, and they knew less about how Germany had engineered its economy and modified it during the war to reduce vulnerabilities to attack and otherwise to ensure production of war materiel.[97]

American expectations that industrial bottlenecks like ball-bearing factories could be attacked occasionally with appreciable military effect were incorrect. Germany had numerous sources of supply of such products and inventories sufficient to weather the temporary production interruptions Allied bombers caused without broader consequences. Even when they accurately identified vulnerabilities, the Allies had poor bomb damage assessments, meaning they little understood the nature and significance of the damage they inflicted; were impatient, shifting targets before the sometimes serious damage they caused showed tangible economic results due to German product reserves, workarounds, and repair efforts; and suffered chronic bureaucratic infighting at intelligence and command levels that led to rapid shifts in targets in a largely vain search for what Harris derisively called "panacea" targets before concentrated attacks could shatter a critical sector of the German economy beyond the Germans' ability to make timely repairs.[98]

Still, the bombers achieved some successes. Of particular note, air attacks on German oil targets, advocated by General Spaatz, chief of all U.S. air operations in Europe, and some British leaders like Chief of Air Staff Portal, along with Soviet conquest of Romanian oil fields, had appreciable effect in late 1944.[99] Spaatz saw oil as the opening for airpower to win the war.[100] After the bombing of fuel-related targets began, German production of natural and synthetic oil fell by two-thirds between January and December 1944, and fuel inventories were miniscule by the end of 1944, limiting German air and ground combat operations.[101] The 15th Air Force effectively ended production at Germany's only natural oil field, the Ploesti field in Romania, after 24 major raids over several months ending in August—just before the Red Army captured the field; the raids, some of which involved more than 1,000 heavy bombers, in total cost 305 bombers shot down and more than 3,000 American lives.[102]

And, the Americans, whose military intelligence was not as good as that of the British, also achieved notable success after Air Chief Marshal Sir Arthur Tedder, Eisenhower's deputy for air operations, ordered them to attack rail and water transportation of coal beginning in September 1944. Tedder, on the advice of his civilian advisor Solly Zuckerman, realized that interdiction of coal-related transportation facilities would shut down munitions factories for lack of fuel.[103] Coal, not oil, was Germany's key industrial fuel. By striking railroad bridges and rail marshalling yards, bombers by November 1944 effectively shut down the *Reichsbahn,* Germany's rail system, and severed rail links between the Ruhr coal fields, Germany's biggest, and the rest of the country; by early 1945, Germany's economy had virtually collapsed for lack of coal.[104] A key node in the resource mobilization dimension finally was found! But it was too late to change the course of war.

Even had the Allies found critical German vulnerabilities sooner, their bombers had limited capabilities to exploit them. The RAF's night bombing remained very inaccurate throughout the war and even USAAF "precision" bombing was inaccurate. The U.S. Strategic Bombing Survey (USSBS) concluded after the war that as late as the winter of 1944–1945, after bombing proficiency had improved considerably but inclement winter weather hampered navigation, some 42 percent of American bombs fell further than 5 miles from their targets, and the rest were on average 2.48 miles off target; only 2.2 percent of American bombs dropped on high-priority oil-related targets hit "damageable facilities," and more than 87 percent were scattered over the countryside.[105] The USAAF's official history of the war concluded that in 1945 the 8th Air Force's bombing missions had an average circular error probable (CEP) of two miles and that "many of its attacks depended for effectiveness on drenching the area with bombs."[106]

Despite its many internal political and industrial problems, Germany's vulnerability amelioration efforts exacerbated the Allies' poor bombing capabilities and helped protect its defense industrial production. In addition to active defenses, the Germans employed passive defenses like camouflage and used smoke generators to obscure facilities from bombardiers' view. Dummy factories drew as many as one-third of the bombs dropped on some missions.[107] Economic managers dispersed factories into small structures in towns and around the countryside and built numerous underground factories.[108] But while these measures protected factories, they also prevented large-scale production required to the generate production efficiencies the Reich badly needed to redress its material weaknesses vis-à-vis the Allies.[109]

The macroeconomic results of Allied airmen's poor vulnerability identification and exploitation capabilities were predictably unimpressive. General economic bombing inconvenienced German economic managers, caused considerable damage, and diverted resources to air defenses and bomb-damage repairs, but German industrial output rose through 1944; production of key war materials peaked in July 1944 and declined only as Allied land armies captured German-held territory and critical resources—like Soviet capture of Ukrainian iron ore in February 1944 and Romania's oil in August 1944.[110] Adam Tooze figures bombing retarded the *growth* of Germany's production of war materiel after mid-1943.[111]

Analyzing more broadly, John Kenneth Galbraith, head of the USSBS unit that studied the effects of USAAF bombing on Germany's economy, concluded that the bombing may actually have helped the German war effort by permitting German armaments minister Albert Speer to streamline the economy using stringent measures that may not otherwise have been politically possible and that the cost to the Allies of bombing was greater than the value of the damage inflicted on Germany.[112] Decades later, Galbraith called the Allied bombing effort a "disastrous failure."[113]

From the German perspective, Speer thought Allied strategic bombing had no major economic effect before 1944; he seems to have become notably concerned only in May 1944, when bombers attacked synthetic fuel plants at Leuna and Merseberg.[114] Speer said after the war that the Allies did not attack Germany's electric power grid to the extent he feared, while the German Quartermaster General Department was pleased that the Allies did not much disrupt explosives production.[115] Field Marshal Erhard Milch, chief of *Luftwaffe* procurement for much of the war, thought the Allies could have shortened the war by striking oil facilities earlier and Speer thought the Allies could have ended the war in 1943 by concentrating on arms production instead the "vast but pointless area bombing."[116] General Galland saw no appreciable impact of the bombing on aircraft production until 1944; he thought attacks of fuel production facilities were "the fatal blow for the Luftwaffe" and marveled that the Allies did not attack them sooner.[117] And, Hitler recognized in 1942 the critical importof rail transportation of coal—a vulnerability the Allies did not attack until the war's outcome was already determined by ground fighting.[118]

Robert Pape similarly argues that strategic bombing of economic targets was ineffective for three reasons: (1) German workers, even if demoralized, continued to work hard; (2) German economic managers generated efficiencies that kept the war machine supplied; and (3) Allied targeting of cities missed many industries that were located outside city centers.[119] Pape further argues that strategic bombing is only effective in long wars of attrition in which productive capacity, not existing war stocks or external supply sources, supply combat units.[120] In World War II, German armies were seriously affected by supply shortages only in late 1944, several months after industrial production peaked.

Airpower partisans unsurprisingly are much more positive about strategic bombing. They claimed then, and some still claim, much for the economic bombing. More balanced views credit strategic bombing with increasing the cost of the war to Germany and diverting resources to air defenses, civil defense, repairs, and strategically useless "retaliation" raids on British cities that could better have been devoted to fighting the Red Army, but they do not confirm the more grandiose claims of airpower partisans and leave untouched the issue of cost-effectiveness.[121] The debate goes on.

Attacks against German armed forces in the *operational execution* dimension were much more effective. By early 1944, with the introduction of long-range escort fighters, Allied airmen killed experienced German pilots faster than Germany could replace them. With growing air superiority and the invasion of France looming, in May 1944 General Eisenhower shifted bomber forces from strategic bombing of Germany to preparing for the invasion—over the strenuous objections of unhappy bomber generals—by attacking German logistical lines and disrupting troop movements in France.[122] Heavy bombers effectively attacked rail-marshalling yards and bridges, cutting supplies to forward units and complementing tactical air forces targeting German ground combat units in northwestern Europe. After the war, U.S. Army General Omar Bradley credited airpower with helping secure victory, but he emphasized the contribution of tactical airpower—particularly fighters' roles in achieving air superiority and securing the Normandy beachhead.[123] Pape similarly argues that "denial" strikes that destroyed German combat forces, not "punishment" raids against German civilians, alone were successful.[124]

Germany

On the other side, the Germans within the context of the strategic bombing campaign alone had many fewer options than the Allies. They tried to exploit perceived Allied vulnerabilities in only the *national will* and *operational execution* dimensions.

German propagandists tried to convert the massive bombing of German civilians and cultural centers into an exploitable vulnerability in the Allies' *national will* dimension without appreciable success. Josef Goebbels's propaganda ministry angered Germans by publicizing the effects of bombing throughout the war—undoubtedly contributing to the many episodes in which enraged civilians attacked downed Allied flyers—but Goebbels's propaganda theme that the Allies, and especially the British, conducted *Terrorangriff,* or terror attacks, for most of the war had little impact on neutral countries, let alone on Allied audiences.[125] Ongoing, if relatively modest, German bombing of British cities undoubtedly did not help. American leaders were privately sensitive to such charges, however, and occasionally changed target packages in Italy and the Balkans, especially, in response to, or in order to avoid, German and even Soviet propaganda attacks.[126]

A partial break in this pattern occurred after 796 RAF Lancaster bombers and a follow-on American raid largely destroyed Dresden on February 13–14, 1945. A cultural center sometimes called "Florence on the Elbe," Dresden also was an industrial center and a transportation hub. Parts of the city therefore were legitimate military targets. It also was a refuge for civilians fleeing the advancing Red Army. The RAF used incendiary bombs that created a firestorm that destroyed 13 square miles of the city's center; the American attack did little additional damage.[127] Some 25,000 to 35,000 people, mainly civilians, died.[128]

The German propaganda apparatus predictably decried the raids, reported inflated casualty figures, outraged Germans, and, for a change, also angered many neutral observers. But Allied military censors did the worst public relations damage by clearing for publication an article by an American journalist that said the Allies purposefully used "terror" against civilians. The article ran under front-page headlines in many American newspapers, upset many readers, and prompted a letter-writing campaign that put protests on the desks of Generals Eisenhower and H. H. Arnold, commander of the USAAF.[129] Most Americans, and a larger majority of Britons, saw the bombing as justified to protect Allied troops, and the United States later would more pointedly target Japanese civilians, but the protesting minority was vocal and its arguments resonated.[130]

The destruction of Dresden helped inspire the widespread popular revulsion with bombing civilians that led to postwar tightening of the rules of war and concerns about *collateral damage*—inadvertent damage to civilians' lives and property—that have become fixtures of modern war. Britons after the war reviled "Bomber" Harris and his Bomber Command for their methods of war, embittering him. Even Churchill questioned the need for the Dresden raid. Dresden has become a metaphor for senseless killing of civilians in war, but it came too late to meaningfully affect American and British popular support for war against Germany.[131] Despite German propagandists' considerable efforts and American leaders' awareness of potential vulnerability, civilian deaths were not an American political vulnerability during World War II.

Military casualties similarly did not constitute a political vulnerability for the United States. U.S. leaders and Americans in general were willing to take casualties to win but accepted foreign civilians' deaths to protect their troops if they could.[132] Many Americans, including airmen, saw the atomic bomb blasts that destroyed Hiroshima and Nagasaki in August 1945 as justifiable to avoid the invasion of the Japanese home islands planned for 1946 that U.S. officials expected to be very bloody.[133]

Germany actively engaged the Allied bomber offensive in the *operational execution* dimension. German fighters exploited bombers' tactical vulnerabilities by attacking them asymmetrically when the bombers had no fighter escorts. They had good success foiling Allied strategic bombing until the long-range P-51s produced fighter force structure symmetry and eventually defeated the *Luftwaffe* in attritional force-on-force aerial combat. Even after its fighter capabilities were broken in the first half of 1944, the *Luftwaffe*'s AAA, totaling some 30,000 heavy guns in late 1944, continued to destroy bombers until the end of the war.[134]

In sum, the bitter debates within British and American air forces about the proper use of bombers—morale attacks versus precision economic strikes versus support of ground operations—indicate clearly that the airmen recognized the importance of identifying and attacking vulnerabilities but could not agree about the nature of German vulnerabilities or find effective means to exploit them.[135] Weak Allied intelligence organizations identified German political vulnerabilities that did not exist and did not well identify actual economic ones. Allied bombers did much damage and contributed to winning the war but could not deliver a "knock-out blow" in resource dimensions due to the inaccurate bombing technologies of the day and because German economic planners considerably ameliorated the vulnerabilities of the large and complex German economy. Because neither German civilians' psychologies nor the German war economy displayed the vulnerabilities that bomber theorists believed before the war, the Allies were reduced to using symmetrical force-on-force attritional warfare in the operational execution dimension to achieve very costly victory on the battlefield.

The analytical errors contributed to the deaths of more than 600,000 German civilians, 8,953 RAF bombers downed, and 55,564 RAF bomber crewmen killed—some 51 percent of all the crewmen who flew RAF bombers during the war.[136] The 8th Air Force lost some 26,000 killed, another 18,000 wounded, and 20,000 shot down and captured; the USAAF lost 6,537 bombers in the European theater.[137] The total U.S. cost of the air war against Germany was about 40,000 lives and $43 billion—a big chunk of the war's entire financial cost to the United States.[138] These casualty rates were among the highest of all Allied armed services, confirming in another way that the victory, though real, reflected attritional warfare and confirming that, unlike in later American wars, the U.S. military and U.S. government were willing to take very heavy casualties to achieve a military objective.

This history contrasts sharply with German use of airpower and the RAF's use of its air defense assets in World War II. The 1919 Versailles Treaty abolished the nascent German air force. As German air forces were surreptitiously reconstituted in the 1920s and early 1930s, and after the *Luftwaffe* was established overtly in 1935, the German air force cooperated with other German military services and was largely dedicated to supporting ground operations.[139] The *Luftwaffe* procured fighters and light strike aircraft like the Ju-87 Stuka useful for close air support and interdiction, and it developed with the army the successful combined arms, air-land battle doctrine we now call blitzkrieg, which for a time in 1939–1941 exploited with devastating effectiveness the tactical- and operational-level vulnerabilities of European armies of the sort Biddle discusses.[140] For these reasons and because the German aircraft industry had limited capabilities, the *Luftwaffe* developed no heavy bombers like the B-17. The *Luftwaffe* used only medium bombers in its unsuccessful campaign against Britain in 1940.

The *Luftwaffe* had no Douhets, Mitchells, Trenchards, or Harrises committed to strategic bombing as an institution-building device. Despite some leadership problems and interservice rivalries, it was better integrated into national-level command-and-control networks than the Allied bomber commands, which were notoriously unwilling to cooperate with each other and with other services of their own nations, and even to take direct orders from very senior officers like Eisenhower that they did not like.[141] The *Luftwaffe* was generally effective when focusing on tactical operations. The Germans therefore had fundamentally different views than British and American airmen about the best way to use airpower. Although Germany lost the war and occasionally strayed from its military aviation forte—like in 1940 over Britain—German ideas about military aviation in retrospect seem to have been more coherent than Allied views.[142]

The Allied bomber experience also contrasts sharply with the history of the RAF's Fighter Command, the hero of the Battle of Britain. While overshadowed politically and within the RAF by Bomber Command, Sir Hugh Dowding's Fighter Command, created only in 1936, developed an effective air defense system that integrated a series of new technologies. Supported more by civilian politicians than by the RAF's bomber community, Dowding integrated fighters, radar, and radio communications into an organization that identified incoming German aircraft, then rapidly and effectively vectored fighters against them.[143] Dowding built training programs and logistical infrastructures that enabled Fighter Command to take appreciable losses while continuing to function. Dowding's success rested not on one or a few technologies or a mass of material assets, but in creating an integrated system of technologies, people, and processes well designed to meet the air threat Britain faced. Fighter Command thus represents a stark contrast to the much poorer, simultaneous performance of the more materially richly endowed Bomber Command.

U.S. LEARNING AND ADAPTATION

In the bombing campaign, the USAAF poorly identified German vulnerabilities and attacked them with limited success. Allied military vulnerabilities in this campaign were modest, and the USAAF effectively, if belatedly and inefficiently, addressed them. The USAAF better ameliorated its political vulnerabilities.

American understanding of its German enemy was weak. The ACTS in the 1930s developed a logically sound doctrine of economic bombing, but the USAAF did not work out how to conduct such campaigns, and it little understood before the war the details of the workings of Germany's peacetime economy. It also did not monitor very well the many ways Germany changed and camouflaged its war economy. The bitter debates over the efficacy of targeting aircraft production plants versus oil versus coal indicate a collective inability to identify critical economic vulnerabilities—apparently reflecting data limitations and analytical errors that partly reflected training inadequacies.

Before the war, the Americans seemed to understand the limitations and drawbacks of morale bombing, but they abandoned the reservations in 1944 in a vain hope that more city bombing might finally break German will. The hope seems to have been driven mainly by worries about prospects for the USAAF's parochial organizational ambitions, not sound data or dispassionate analysis. The USAAF examined targeting issues through the lens of prewar ACTS-written air doctrine, sometimes limiting acceptable exploitation options for institutional reasons. Allied airmen eventually found no critical vulnerability susceptible to air attack, leading to the de facto adoption of an attrition strategy in 1944 when airmen selected targets they expected German fighters to heavily defend.

The USAAF adapted modestly and slowly. The major, and ultimately effective, technological adaptations were those that produced the effective long-range fighter escorts that produced the force symmetry that enabled the attrition strategy to work. These came late but were effective. Other technological adaptations were more modest and had less operational effect. Even "precision" bombing capabilities were very limited, and the Americans relied on aircraft masses for bombing effectiveness—not efficiency—explaining the numerous raids of 1,000 or more heavy bombers. Doctrinal beliefs also drove airmen to resist adjustment of target priorities to tactical targets that helped ground troops win the war but did not prove the indispensability of airpower or justify an independent air force.

The USAAF had few vulnerabilities, and it successfully, if slowly, addressed them. Militarily, the major one was the bomber vulnerability problem solved in early 1944 by deploying long-range fighters. The USAAF throughout the war was keenly aware of the potential political vulnerability that city bombing posed, and it worked successfully to ameliorate it during the course of World War II in at least three ways: altering targeting periodically in response to actual or threatened German and Soviet propaganda; focusing on economic bombing and justifying civilian casualties as unintentional even when they were not; and blaming the RAF for city bombing.[144] These worked well enough to help win the war and to gain an independent U.S. Air Force in 1947, but they did not stop the development of long-term popular unhappiness with collateral damage that dogged airmen in later American wars, beginning in Vietnam in 1965.

The case shows how hard it is for individuals and institutions to accurately identify enemy vulnerabilities. Bright people worked hard to find exploitable enemy vulnerabilities with modest success. Data and analytical challenges, and institutional parochialism, combined to keep the morale-bombing concept attractive to some American (and especially British) bomber partisans even in the face of massive amounts of contradictory evidence. Factual, analytical, and cognitive errors, and wishful thinking encouraged in substantial part by service parochialism, were great enough to prompt Tami Biddle to pointedly distinguish bomber zealots' rhetoric from the reality of strategic bombing in World War II.

The net result of relatively poor vulnerability identification and exploitation capabilities was an attritional war of the sort basic materialist theory expects. World War II was the last American war to be so decided. While limited available information about Germany played a role, many of the causes of poor learning and adaptation reflected institutional factors—air force culture, air doctrine, and the need for operational successes to further ambitions for an independent air force.

SUMMARY

The bombing campaign illustrates well how belligerents sought to identify and exploit vulnerabilities that could both win the war and advance parochial air force interests. But no "panacea" targets were found or exploited. Allied air forces had only modest ability to exploit the vulnerabilities they accurately identified. The consequence of a lack of strategic imagination was default to attritional warfare in which the residual, and ultimately critical, vulnerability was Germany's material deficiency.[145]

Robin Neillands argues that Allied airpower partisans fundamentally misunderstood the consequences of bombing civilians and that morale bombing has never been effective—a view Michael Horowitz and Dan Reiter support statistically.[146] There was no exploitable vulnerability in the national will and leadership dimensions on either side. Neither civilian nor military casualties deterred each side from fighting to a decisive, bloody conclusion. Eighth Air Force and RAF Bomber Command suffered very high casualties, as did the *Luftwaffe*. The legacy of the massive numbers of civilian casualties was a very large sensitivity to such casualties that in subsequent American wars, beginning in Vietnam, quickly became a widely recognized and exploitable political vulnerability in America's national will and leadership dimensions. Put differently, revulsion to the 1945 city bombings generated an expansion in the scope of later warfare and increased the salience of ideational dimensions as critical operational dimensions.

Vulnerability theory accounts for the disparate outcomes of simultaneous fighting in different tactical and strategic arenas by the essentially identical opposing forces—in this case by explaining why the vulnerabilities airmen theorized were hard to find and exploit and why attritional warfare was the result. Arreguín-Toft's model, in contrast, is structurally limited in that one of its four actor options clearly does not apply to air warfare and it accurately predicts only two of the three strategic interactions I identify. Basic materialist theory successfully predicts the war's outcome but cannot explain the conduct of the bomber campaign and the primary reason for the Allies' eventual victory in the air—the attrition of Germany's pilot skills, not its material war-making capabilities.

This case supports both Shimshoni and Biddle to considerable but differing degrees.[147] The inability of airmen to agree on the most effective strategy supports Shimshoni's argument that identification of vulnerabilities and producing means of exploiting them—a leadership function—is important but difficult work. The key to successful targeting lay in intelligence collection and analysis, and the acceptance of intelligence findings by open-minded commanders, not the brute force of machines and munitions. The Allies' struggle to find successful tactics and techniques is consistent with Shimshoni's entrepreneurship argument. The Allies and Germans in dynamic interaction, and the Allies among themselves, tried new approaches to targeting and the operational conduct of raids until the Allies finally found a way to prevail. While technology did not determine the conduct of the war or its outcome—and Allied airmen's prewar claims for the systemic offensive advantages of strategic bombing were debunked—senior leaders on all sides were at pains late in the war to find ways to improve the quality of their crews' skills and operational techniques and had some modest successes.

Biddle is less directly relevant to strategic air war because he focuses so heavily on tactical ground operations, but his argument about the importance of combined arms teams is applicable. Adding long-range fighter squadrons to strategic (bomber) air forces made combined arms teams that reduced bomber vulnerability and enabled Allied formations to effectively fight German fighters, and thence to conduct more effective strategic offensive operations. The *Luftwaffe*'s deep defenses also were consistent with Biddle's argument that forces that operate in depth can defeat attacking forces that overextend themselves—just what Allied bombers did before large numbers of modified P-51s started escort duty.

The morality of civilian bombing was a major topic of internal USAAF and senior civilian policy debates before and during the war, and it modestly affected targeting decisions. Reaction to the Dresden raid enables Merom to argue reasonably that 1945 marks a turning point in Western views of warfare.[148] The bombing campaign differs from Mack's argument about the negative consequences of interest asymmetry that was designed with the U.S. experience with limited war in Vietnam in mind. In this near-total war in which both sides struggled for survival, interest asymmetries were small.

CHAPTER 4

The United States in Vietnam (1961–1975)

The United States lost a long war in Vietnam. Despite initially good relations with the Viet Minh, a national front organization founded and run by Ho Chi Minh's Marxist-Leninist *Lao Dong* (Vietnam Worker's Party), relations soured as the Truman and Eisenhower administrations supported French claims to Indochina and Cold War containment policy led Washington to oppose the Viet Minh as a communist-led organization allied with Moscow. France's defeat at Dien Bien Phu led to the Geneva conference of 1954 that established two Vietnamese states—the Democratic Republic of Vietnam (DRV or North Vietnam) and the Republic of Vietnam (RVN or South Vietnam)—divided at roughly the 17th parallel, not the 13th parallel as the northerners wanted. *Lao Dong* leaders felt betrayed by their Soviet and Chinese sponsors and determined to pursue in other ways the victory they thought they had earned.[1] The United States supported the government of South Vietnam (GVN).

After hoping vainly that the elections envisioned at Geneva might peacefully unify Vietnam on its terms, Hanoi authorized resumption of the armed struggle in January 1959.[2] These efforts accelerated after North Vietnam established in South Vietnam in December 1960 the National Liberation Front (NLF), a front organization like the Viet Minh.[3] The NLF called its military arm the People's Liberation Army (PLA), later changed to People's Liberation Armed Forces (PLAF).[4] In 1962, Hanoi formed the People's Revolutionary Party (PRP), an overtly Marxist party, to lead the NLF.[5] South Vietnam's President Ngo Dinh Diem called the NLF's political and military arms the Viet Cong—a contraction of Vietnamese communist—which Americans contracted further to VC and which some view as pejorative.[6] In this chapter, I generally use American terms for Vietnamese institutions.

The DRV's senior military officer, Minister of Defense and General Vo Nguyen Giap, employed a strategy based on Mao Zedong's "people's war" doctrine and Vietnamese experiences fighting France in 1946–1954. Vietnamese communists emphasized organizing the people and they subordinated violence to political objectives.[7] DRV leaders recognized Mao's stages of insurgency starting with political organizing and culminating with conventional military victory, but added uniquely Vietnamese characteristics to their theory and practice, including a willingness to use regular and guerrilla forces simultaneously (vice sequentially in Mao's experience) and devotion to the general uprising.[8] The North Vietnamese emphasized political proselytizing against enemies and expected a long struggle to wear out the Americans and cause them to quit, just as they had worn out the French.[9] And, they focused on the peasants, not the urban proletariat, as key political and military actors.[10] Hence, the war in South Vietnam was mainly fought in the countryside.[11]

Consistent with this doctrine, the VC especially attacked honest and effective GVN civilian officials, including schoolteachers, in villages and hamlets who, because of their competence and integrity, were the NLF's biggest political threats.[12] Figures on political murders vary by source but are in all cases significant. Over the course of the war, allied records indicate that the VC assassinated 36,725 GVN officials and other civilian opponents of the communists.[13] The VC often justified the deaths as revolutionary justice for GVN crimes.[14]

Militarily, the VC used guerrilla tactics similar to methods the Viet Minh used against France. The VC increased military pressure on the Army of the Republic of Vietnam (ARVN) as they formed larger armed units and conducted progressively bigger operations in the early 1960s. A VC battalion defeated a larger ARVN force at Ap Bac in January 1963 in the first of the big fights that would continue for years.[15] VC pressure grew in 1964, aided by Hanoi's decision to dispatch troops of the People's Army of Vietnam (PAVN), or, as Americans called it, the North Vietnamese Army (NVA), to the south beginning in late 1964.[16] The NVA fought well as conventional and as guerrilla forces, meaning the NVA and VC together presented the ARVN, U.S., and other allied forces with a competent enemy capable of fighting a mix of large and small, conventional and unconventional engagements.

U.S. aid to the GVN before 1961 largely consisted of money, arms, and training. President John Kennedy, however, expanded the role of U.S. military advisors as the VC increased pressure on the GVN. The U.S. military advisory presence rose from about 700 in January 1961, when Kennedy became president, to more than 16,000 at the time of his death on November 22, 1963.[17] Americans also began to fight in this period as, for example, advisors to ARVN combat units and pilots of two-seat attack aircraft in which a Vietnamese rode in the back seat. In 1964, the United States began to support South Vietnamese commando raids on the north.

On August 2, 1964, North Vietnamese torpedo boats attacked a U.S. Navy destroyer on an intelligence-gathering mission in the vicinity of one of the South Vietnamese raids on the north.[18] After a second naval incident on August 4, U.S. military forces for the first time overtly attacked North Vietnam—naval air strikes on patrol boat bases and fuel storage facilities at Vinh.[19] On August 7, Congress authorized President Lyndon Johnson to use force in Vietnam in the so-called Gulf of Tonkin Resolution.[20]

President Diem was a controversial leader. A Catholic refugee from the north, Diem was arrogant and unresponsive to many of his people, including Buddhists and members of the large and politically important Cao Dai, Hoa Hao, and Binh Xuyen sects, some of whose militiamen eventually became VC fighters. While capable and successful in many respects, he and his brother Ngo Dinh Nhu ineffectively ran some important institutions and programs, including the so-called agroville and Strategic Hamlets programs that were attempts to wrest control of the countryside from the VC (see below). They also alienated some senior American officials. After disgruntled generals killed the Ngo brothers on November 1, 1963, with U.S. complicity, a series of generals provided weak leadership and South Vietnam was politically unstable until General Nguyen Van Thieu took power in June 1965.

In 1964, as the GVN was largely dysfunctional, the VC and NVA made steady political and military progress. Washington regarded the deteriorating situation in Vietnam with increasing alarm. President Johnson on December 1, 1964, authorized U.S. military action against North Vietnam upon arrival of a suitable excuse.[21] That excuse came on February 6, 1965, when the VC attacked the American air base at Pleiku, destroying or damaging a score of aircraft and killing nine Americans.[22]

In "response," Johnson authorized air strikes on the DRV and deployment of ground combat troops to South Vietnam to protect U.S. airbases. U.S. troops soon began conducting offensive combat operations. As more NVA units infiltrated south, the U.S. commander in Vietnam from 1964 to 1968, army General William Westmoreland, in June 1965 asked for and got 44 maneuver battalions. U.S. troop strength rose rapidly to about 184,000 by the end of 1965 and peaked in mid-1968 at about 543,000 troops.

NVA and VC troop strength also rose in this period despite heavy combat losses. U.S. intelligence estimated that enemy troop strength rose from 160,000 in 1965 to 220,000 a year later; infiltration and local recruitment then added the equivalent of about 15 battalions per month to VC and NVA force structures.[23] NVA/VC troop strength increased again in 1967.[24] As result, allied armed forces never achieved numerical superiority greater than about 3:1, well below optimal counterinsurgency ratios of 10:1.[25] Westmoreland thought superior American technology made success with a lower ratio feasible.

With the NVA and VC more than replacing battle losses, allied military forces obviously were not winning the war. Americans became increasingly uneasy with the war as it dragged on and American casualties mounted. By year-end 1967, domestic opposition to the war was significant and growing. The Johnson administration nevertheless argued that progress was being made. But in late January 1968, a major VC offensive coincident with the Tet lunar New Year holiday seemed to contradict administration claims, prompting more popular unhappiness. Johnson's advisers soon thereafter recommended disengagement from Vietnam. On March 31, 1968, Johnson announced he would not seek reelection, severely restricted bombing of the north, and offered again to negotiate an end to the war.

Richard Nixon won the presidency in 1968 in part by promising to resolve the war. He wanted "peace with honor"—a goal easier stated than accomplished. Under Nixon's "Vietnamization" policy, the United States increased economic and military aid to South Vietnam, delivering large amounts of fairly modern American equipment to the ARVN and the Vietnamese Air Force (VNAF), hoping the RVN could survive without a U.S. ground presence.[26] Nixon began to withdraw U.S. troops unilaterally in 1969 and effectively ended a U.S. ground combat role in 1972. In January 1973, Washington and Hanoi reached an agreement that ended the U.S. military role in Vietnam. But Congress in mid-1973 banned U.S. military operations in Southeast Asia, abrogating Nixon's promise to aid the RVN if necessary and rendering the agreement with Hanoi unenforceable. Congress reduced material aid for the RVN in 1974. In early 1975, when the NVA launched another major offensive, Congress ignored President Gerald Ford's request for $300 million in emergency military aid.[27] The GVN collapsed in April 1975.[28]

The North Vietnamese victory was an unambiguous defeat for the United States. American policy from the Truman years was clear and consistent: the United States wanted an independent, non-communist Vietnam.[29] Under domestic political pressure to end the war, Nixon in 1969 modified U.S. objectives to include a U.S. withdrawal—but on terms that gave the GVN a reasonable chance of survival.[30] The fall of Saigon meant the United States failed to achieve a goal that six presidents pursued.

OVERVIEW OF THE WAR

Given the defensive nature of U.S. goals in Vietnam, President Johnson reluctantly escalated American involvement in Vietnam in 1965.[31] Assistant Secretary of Defense for International Security Affairs John McNaughten, a close advisor of Defense Secretary Robert McNamara, said "70 percent" of America's purpose in Vietnam was "to avoid a humiliating defeat."[32] Paul Kattenburg, a State Department official during the war, argues that Washington pursued a policy of "stalemate."[33] In his memoirs, McNamara confirms that the United States only sought to avoid defeat.[34]

Although committed to the survival and independence of South Vietnam, Johnson did not want to strongly pressure North Vietnam because, with memories of the Korean War fresh, he did not want to risk a wider war with the Soviet Union or China.[35] Johnson therefore ruled out conventional ground force incursions onto DRV territory even though U.S. intelligence officials believed there was no chance of direct Chinese military involvement unless the Chinese border was threatened.[36]

The result was chronic policy ambivalence that emphasized compromise and half-measures largely in response to enemy activities.[37] Johnson increased American commitments when they seemed necessary to prevent collapse of the GVN but was never willing to commit material resources or political capital to a military campaign that could have decisively defeated North Vietnam. Yet Johnson also believed Hanoi would not seriously negotiate until it was convinced it could not win militarily.[38] Johnson thereby created for himself a daunting challenge: how could Washington convince a strong-willed Hanoi that it could not prevail while severely restricting U.S. military options consistent with a transparent unwillingness to bear more than limited political, economic, and human costs?

Because the threat to the RVN involved both external (North Vietnam) and internal (VC) components, Johnson ordered attacks on both. Determined to keep the war limited, he also restricted U.S. conventional attacks on Laos and Cambodia, which were NVA staging areas, to air raids. In South Vietnam, where a U.S. presence posed less risk of a wider war, Johnson authorized a major commitment of ground troops, supporting air and naval forces, and civilian agencies that focused mainly on pacification activities—which were commonly called the "other war."

Operation Rolling Thunder—the U.S. air campaign against North Vietnam conducted from March 2, 1965, to October 31, 1968—had two focuses: one primarily military and one mainly political. Both had the strategic objective of persuading the DRV to stop trying to absorb the RVN. First, American airmen sought permission to devastate North Vietnamese military bases, defense-related industries, and transportation networks or, in military jargon, lines of communication (LOCs) supplying combat forces in the south.[39] Destruction of these targets, airmen believed, would break North Vietnam's will to fight by destroying its ability to fight.[40] They thought the DRV was the primary supplier of insurgent fighters in the south—even though U.S. intelligence estimated that in 1964 infiltration from the north contributed only 15 percent of the personnel increment to VC strength—and they believed North Vietnamese defense industries produced much of the war materiel the NVA and VC used.[41] Air doctrine held that political decisions follow altered material facts on the ground; airmen wanted to change the facts drastically with a "sudden sharp blow."[42]

In contrast, Johnson's senior civilian advisors, including national security advisor McGeorge Bundy and Walt W. Rostow, sought to threaten damage to the DRV to induce the North Vietnamese to quit.[43] Influenced by limited war theory developed in the context of the U.S.-Soviet nuclear competition in the 1950s, they argued that it was important to leave potential targets untouched, which they assumed the DRV wanted to protect even at the cost of abandoning its war objectives, as aids to bargaining.[44] Threats of strikes constituted "signals" that the United States could inflict more damage but that a favorable response from Hanoi—defined as a willingness to talk about, but not a requirement seriously to negotiate, an end to the war—would prevent additional attacks.[45] Gradual escalation—or what became known as the "slow squeeze" as opposed to the military's preferred "fast" or "full squeeze"—would, these people thought, keep America's negotiating posture credible.[46]

The two approaches were incompatible because they worked with the same list of targets. Nevertheless, in a continuing give-and-take between civilian and military leaders that consistently produced compromise, the air forces conducting Rolling Thunder were given three objectives: (1) strategic persuasion of the DRV by both imposing physical damage and leaving other targets untouched as bargaining chips; (2) bombing to raise the morale of the South Vietnamese; and (3) interdiction of North Vietnamese infiltration routes to the south—what became known as the Ho Chi Minh Trail.[47]

The goals addressed different levels of war. The first goal was strategic in nature; the Americans wanted to win by persuading North Vietnam to abandon its political/military goals and accept America's. The third was tactical. Bombing the Ho Chi Minh Trail meant destroying bridges, depots, trucks, and the like, which could only have strategic implications over a long period of time. The second sounded nice but its value was never clear. While focusing initially on the strategic goal, after mid-1965 the main effort of Rolling Thunder was interdiction.[48]

The bombing was conducted under significant political constraints. Rolling Thunder increased in intensity slowly to keep the signaling option credible.[49] Military commands drew up target lists that Johnson and McNamara personally approved.[50] They prohibited raids against many targets including Hanoi, where attacks on military targets might generate significant civilian casualties or hit Soviet diplomats; areas near the border with China; and the port of Haiphong, where Soviet ships regularly delivered economic and military aid.[51]

Rolling Thunder's initial target list consisted of 94 military, industrial, and transportation-related facilities—all standard military targets.[52] The first targets struck were ammunition storage depots.[53] When bombing most of the original targets made little military difference and Hanoi did not respond politically, the military expanded its target list.[54] In 1966 the list contained 242 targets.[55] In July 1967, it had 359 targets, including some minor facilities.[56] Civilian leaders never approved some targets located near Hanoi and Haiphong. The number and variety of targets U.S. aircraft actually struck was greater, however, because airmen could attack some targets of opportunity while on "armed reconnaissance" missions, especially if they received ground fire.[57] The limited and incrementally increasing bombing infuriated airmen, especially U.S. Air Force officers, for violating air doctrine.

As destruction of industrial, military, and transportation targets continued to fail to work as expected, airmen searched for other enemy vulnerabilities. In early 1966, Rostow, the Joint Staff, and the Pacific Command suggested attacking petroleum, oil, and lubricants (POL) storage facilities despite intelligence assessments that bombing them would make little difference. After a lengthy public discussion in the United States, which enabled the DRV to disperse its POL stores, Johnson approved attacks on them in June 1966.[58] U.S. aircraft destroyed 7 percent of North Vietnam's POL storage capacity; the rest of its estimated 25,000 tons of POL was scattered about the country in buried tanks and in 55-gallon drums that literally lined streets in Hanoi, which DRV leaders knew was a U.S. exclusion zone.[59] That is, they put assets in areas they knew American aircraft would not strike due to presidential constraints imposed largely out of concerns about civilian casualties. In 1967, Johnson approved attacks on electric generating plants designed primarily to shut down industry, including munitions factories and fertilizer plants, but also to demoralize North Vietnamese civilians; the raids knocked out about 85 percent of the DRV's small generating capacity.[60]

None of these attacks had more than minimal impact on NVA infiltration rates, military operations in the south, or DRV diplomacy. The CIA, several Defense Department analytical organizations, and independent research centers like the RAND Corporation and the Institute for Defense Analyses (IDA), repeatedly estimated that because the NVA and VC had minimal logistical requirements, and because North Vietnam's logistical system was redundant and easily repaired, a strategic air campaign would be ineffective.[61] Senior officials who influenced Vietnam policy participated in some of these analyses. For example, in the fall of 1963 a war game called SIGMA-I—involving General Maxwell Taylor (later ambassador), George Ball and Alexis Johnson of the State Department, and Director of Central Intelligence John McCone—concluded that North Vietnam could match any U.S. ground force buildup in the south despite U.S. bombing; they also forecast that a U.S. troop buildup in Vietnam would precipitate an uproar at home.[62] In a game in 1964, the Red (enemy) side sought to protect its factories by moving children and prisoners of war into them as human shields—frustrating the Blue (U.S.) players and forcing escalation of American bombing.[63] McCone in April 1965, as he was leaving office, argued against what he saw as a feeble Rolling Thunder effort that would little bother Hanoi but would increase domestic opposition to the war.[64] Airmen rejected all such analyses; the president generally backed them.

In October 1968, after repeated bombing pauses failed to convincingly "signal" Hanoi that Washington wanted negotiations, lame-duck President Johnson halted most raids on the north.[65] Rolling Thunder ended, not by defeat at the hands of North Vietnamese air defenses, but because it failed, despite one million sorties flown and about three-quarters of a million tons of bombs dropped, to persuade North Vietnam to quit the war while simultaneously imposing substantial material and political costs on the United States.[66]

Thereafter, American air forces increased their efforts to interdict logistical lines stretching from southwestern North Vietnam through Laos and Cambodia and into South Vietnam under the code name Commando Hunt. U.S. aircraft also shifted to strikes in the south, where B-52 bombers and tactical aircraft already were attacking enemy troops and bases. The heavier interdiction raids also failed to prevent NVA troops and supplies from reaching South Vietnam.[67] North Vietnam gradually built an extensive network of all-weather roads, bunkers, and a fuel pipeline, which increased the capacity and improved the resilience of the Ho Chi Minh Trail.[68] This logistical infrastructure enabled major NVA/VC offensives in 1968, 1972, and 1975. In short, both U.S. air campaigns failed.[69]

On the ground, the United States had two goals: (1) kill enemy troops in combat and (2) support South Vietnamese pacification efforts. They were related goals because a basic requirement of pacification is security. General Westmoreland focused U.S. forces on combating main force VC and NVA units because he saw big enemy units as the major security threats. Westmoreland's goal was to "break the will of the DRV/VC by denying them victory"—virtually the same as airmen's goal for Rolling Thunder.[70] He recognized the importance of pacification in principle but concentrated on destroying large enemy units because he reasoned that there could never be local security so long as they roamed the countryside.[71] To support this view, he cited battles in Binh Dinh province in November 1964 in which VC regiments mauled smaller ARVN units supporting pacification; Andrew Krepinevich argues that Westmoreland rationalized what he wanted to do anyway for reasons consistent with doctrine based on the army's successes against Germany in 1944–1945 and the way the army then organized and trained to fight Warsaw Pact armies in Europe—which Krepinevich calls the Army Concept.[72]

Westmoreland refused to commit significant numbers of U.S. troops to local security roles except around major U.S. bases. He left local security chores primarily to the ARVN and local Vietnamese security forces.[73] Westmoreland therefore rejected a direct U.S. role in population control activities, which U.S troops embraced in the Philippines and which then-prominent counterinsurgency experts like David Galula and Robert Thompson considered essential to defeating insurgencies.[74]

Instead, Westmoreland hoped to pacify the south by killing enough NVA and VC soldiers to eliminate the military threat to the GVN. He thought that enemy attrition beyond a certain point—a "crossover point" at which communist casualties exceeded infiltration and recruitment in the south—would prevent the NVA and VC, respectively, from regenerating their forces, leading gradually to smaller and less threatening enemy troop levels and thus to success defined as a secure RVN.[75] Westmoreland understood that his attrition strategy was politically unpopular but thought the constraints Johnson imposed left him no other choice.[76] The Joint Chiefs of Staff (JCS) and civilian leaders accepted the decision.[77]

In one of the first big battles involving U.S. combat units, elements of the army's 1st Cavalry Division in November 1965 fought three NVA regiments in the Ia Drang Valley, killing some 1,200 North Vietnamese at a cost of about 200 American dead.[78] Westmoreland thought the kill ratio acceptable. With his attrition strategy seemingly validated, Westmoreland devoted most of his troops' efforts to searching for and destroying enemy combat forces.

Lao Dong leaders erred in sending NVA troops into South Vietnam in 1964. Instead of continuing to gradually push the rapidly deteriorating ARVN and GVN into collapse, their aggressive attacks provoked the major U.S. military commitment in 1965 that bolstered the GVN, extended the war, and increased their costs markedly. But the U.S. presence also gave them new opportunities. Shifting focus, in 1965–1967 NVA and VC units fought U.S. units according to an attrition strategy designed to attack Americans' will to fight. Because they recognized that American sensitivity to casualties was far greater than their own, North Vietnam's leaders accepted chronic tactical defeats and heavy losses in order to inflict enough U.S. casualties to sap Americans' will to fight and lead them to quit.[79] Giap's efforts to draw American forces away from population centers, which Westmoreland accommodated, enabled VC recruiters and the Ho Chi Minh Trail to give him adequate human resources.[80] Giap and the *Lao Dong* leadership had few qualms about spending manpower, and the North Vietnamese people accepted casualties that would finally total as many as 1.4 million NVA and VC troops dead and missing.[81]

While they took heavy casualties, the VC and NVA retained operational flexibility. They could bloody American units in conventional fights and through ambushes, using mines and booby traps, and by shelling U.S. bases. Their flexibility, good tactical intelligence, and mobility enabled them to initiate some 80–90 percent of engagements as times and places of their choosing; per Maoist doctrine, they could retreat strategically to small guerrilla engagements to avoid susceptibility to major defeats.[82] They therefore controlled the tempo of fighting and limited their losses sufficiently to ensure that Westmoreland never achieved his crossover point.[83]

Giap's strategy was apparent to some American officials. In July 1965, as the U.S. ground force buildup was beginning, senior State Department officials argued that Giap would adopt this approach.[84] McNamara says he recognized the strategy in January 1966 but could not figure a way to counter it.[85] Westmoreland, however, stuck with his approach.

In 1967, undeterred by losses and ignoring persistent American efforts to entice negotiations, Giap prepared a major offensive that he hoped would (1) cause a general uprising consistent with Vietnamese political/military doctrine and (2) influence American public opinion about the war.[86] In October 1967, he launched a series of preparatory attacks, mainly by NVA units, designed to pull ARVN and U.S. troops away from urban areas.[87] In January 1968, he launched the main effort—the Tet Offensive of mostly VC attacks on most major South Vietnamese towns and cities. The VC and NVA for a time held parts of several cities, most notably the large northern city of Hue.

The offensive failed to achieve Giap's first goal but was spectacularly successful in the second despite being a devastating defeat in narrowly military terms. The VC and NVA lost around 40,000 soldiers killed and several thousand more captured in a month of heavy fighting—most of them VC soldiers and cadre personnel used in combat roles.[88] The Tet battles also claimed about 1,100 American, 2,300 ARVN, and 12,500 civilian lives, including about 2,800 people, many of them GVN officials, whom the VC/NVA executed in Hue.[89] U.S. officials emphasized the severe military defeat the VC suffered, but most Americans were shocked by the scale of the fighting, especially given recent administration claims of progress, and many paid no attention to their by-now largely discredited political and military leaders.[90]

It was the turning point of the war.[91] Americans' willingness to continue the war crumbled. The DRV's official history of the war considers Tet a major victory because thereafter "[t]he American imperialists' will to commit aggression began to waver."[92]

The VC never recovered from their losses of Tet and two smaller "mini-Tet" offensives that began in May and August 1968. Hanoi dispatched North Vietnamese to refill VC combat units and to replace southern cadre. But VC combat forces and the VC's political infrastructure remained weaker than before Tet until the end of the war, when North Vietnam demonstrated where power really lay by absorbing the NLF into DRV institutions.[93]

As Washington drew down U.S. troop strength and provided substantial quantities of modern weaponry to the ARVN and VNAF as part of Nixon's Vietnamization program, fighting continued to be inconclusive. The Nixon administration aggressively engaged diplomatically the DRV's patrons, especially in Moscow, to pressure Hanoi for negotiations and to produce greater U.S. freedom of action than Johnson feared he had. Nixon's primary negotiator, national security adviser Henry Kissinger, thought he was making progress.[94] But in 1970 Giap began to prepare another major offensive—this time by regular NVA divisions.[95] North Vietnam reasoned that the ARVN was too weak to resist without a U.S. ground presence and thought an offensive would damage Nixon's reelection prospects.[96] The offensive began on March 30, 1972, the Thursday before Easter, prompting its name—the Easter Offensive.

With U.S. ground troops largely gone and no new ground involvement politically possible at home, Nixon threw airpower at NVA units. After some initial defeats, the ARVN fought fairly well. American aircraft devastated NVA units that, concentrated in conventional fights with ARVN troops, were visible and thus vulnerable to airpower. U.S. intelligence estimated that American aircraft alone inflicted 120,000 casualties between Easter and August 1972.[97] It was a substantial North Vietnamese military defeat, even though NVA troops retained some of the territory they took early in the offensive. But the offensive contributed to a significant diplomatic win when Nixon in May 1972 reduced U.S. negotiating demands to merely a cease-fire and a return of American prisoners in exchange for a U.S. withdrawal from Vietnam. The DRV was on the verge of political victory despite its military setback.

While demolishing NVA formations in the south, U.S. airmen also renewed attacks on the north in an operation code named Linebacker. They used newly developed precision munitions, including laser- and television-guided bombs, which improved bombing accuracy and reduced risks of collateral damage.[98] As a result, American aircraft attacked some targets the North Vietnamese had placed in exclusion zones to exploit American targeting policy.[99]

The attacks hurt North Vietnam, and Hanoi returned to the negotiating table, leading to a tentative agreement in October 1972 that called for U.S. withdrawal in exchange only for North Vietnamese restraint on infiltrating more troops into the south. The agreement permitted Hanoi to maintain its troop presence in South Vietnam—a concession that angered President Thieu. Nixon threatened to abandon South Vietnam completely if Thieu did not agree but softened the blow by promising aid if Hanoi broke its promises and attacked again.[100] Thieu was unhappy but had little choice. But the North Vietnamese again hesitated; in frustration, Nixon launched new air attacks on the north in December 1972 under the code name Linebacker II.[101] Using B-52s on Hanoi for the first time, the bombing raids did much damage. The DRV agreed to basically the same terms it accepted in October, and Linebacker II ended after 11 days—producing an apparent coercive victory for airpower.

But Linebacker II was not a U.S. military victory. Hanoi got what it wanted—a U.S. withdrawal with minimal limitations on its own military capabilities, including U.S. acceptance of some 200,000 NVA troops in South Vietnam.[102] Nixon knew that leaders of the Congress elected in November 1972 planned to cut funding for the war soon after taking office in January 1973, tying his hands. Nixon took his last best chance to salvage "peace with honor" ahead of a congressionally mandated, unilateral

American withdrawal that would seal defeat. As astute students of American politics, North Vietnam's leaders knew that all they had to do then was to wait.[103]

Linebacker II achieved no victory, but it aided airmen's enduring folklore that heavy bombing could quickly have produced victory if only, as commander-in-chief Pacific in 1964–1968 Admiral U.S.G. Sharp put it, meddlesome and incompetent civilians had not forced airmen to fight "with one hand tied behind our back."[104] Robert Pape also mistakenly calls Linebacker II a success for coercive airpower.[105]

Meanwhile, the United States aided the GVN's efforts to consolidate its political control of the countryside by backing a series of pacification programs that many critics charged were poorly conceived and executed. With Westmoreland little interested in committing U.S. troops to it, the U.S. pacification effort was limited and fragmented among civilian agencies, including the CIA, the Agency for International Development (AID), and the U.S. Information Service (USIS).

In April 1967, President Johnson appointed his White House advisor for pacification issues, Robert Komer, to head the U.S. pacification effort in Vietnam, which was reorganized as Civil Operations and Revolutionary Development Assistance (CORDS)—a joint civilian-military effort housed organizationally within the Military Assistance Command, Vietnam (MACV), Westmoreland's command.[106] The performance of the pacification effort quickly improved markedly for several reasons: (1) MACV control unified the American effort; (2) as Westmoreland's deputy, the abrasive but dynamic "Blowtorch" Komer, and later William Colby, were effective leaders; (3) MACV provided material resources the civilian agencies did not have;[107] (4) General Creighton Abrams, who replaced Westmoreland as commanding general in mid-1968, was more favorably inclined toward counterinsurgency operations than was Westmoreland;[108] (5) President Thieu supported an enhanced GVN pacification effort; and (6) not least, the slaughter of VC cadres in their 1968 offensives substantially weakened VC opposition to the more competent allied pacification effort.

By 1970, according to CORDS metrics and the assessment of most reputable observers, the countryside was largely secure and most peasants were satisfied with, if not enthusiastic about, GVN administration.[109] MACV's Hamlet Evaluation System rated more than 90 percent of the hamlets in the country as secure and fewer than 2 percent as being under VC control.[110] During Tet and the 1972 and 1975 NVA offensives, most South Vietnamese fled, or tried to flee, to GVN-controlled areas. During the Tet fighting, many townspeople helped allied troops by identifying VC/NVA positions. Many South Vietnamese thus repeatedly backed the GVN against the NVA and VC. But by this time, the American public and many congressmen, especially Democrats, were tired of the war and did not care about progress in the "other war."

Success in the counterinsurgency part of the war made little difference in the end. Giap used regular NVA units to conquer South Vietnam in the final, largely conventional campaign of 1975.[111] Hanoi recognized that American will had evaporated after Congress banned U.S. military operations in Southeast Asia on June 30, 1973; Congress passed the War Powers Act on November 7, 1973; and their archenemy Nixon was felled by the Watergate scandal in mid-1974.[112] In October 1973, the *Lao Dong* central committee decided to launch another offensive; formal planning began in early 1974.[113] The DRV could plan another conventional attack without having to worry about American airpower. Congress gave Hanoi the green light for its final offensive.

THE VIETNAM WAR AND MATERIALIST THEORY

The materially dominant U.S. military lost this war decisively, in part because the United States used its material wealth and technologies in strikingly cavalier, inefficient, and operationally ineffective ways. The U.S. force structure used in Vietnam was poorly specialized for the war at hand. Conventionally minded U.S. airmen and soldiers used forces and doctrine they developed for other purposes, most notably a prospective war against the Warsaw Pact in central Europe, in ways highly inappropriate for counterinsurgency operations.[114] The U.S. Army insisted on using large conventional units and firepower when finesse would have been more effective. American firepower often was wasted or actively counterproductive. For example, heavy use of artillery and air strikes alerted the VC/NVA to impending U.S. actions, enabling them to evade U.S. troops. *Reconnaissance by fire* and *harassment and interdiction* artillery fires were wasteful, destroyed civilian lives and property, alienated local people, and provided material for anti-American propaganda. Army Chief of Staff General Harold Johnson said:

> We were indiscriminate in our application of firepower, in the true sense of discriminating, because too much of it went out on a relatively random basis.[115]

American air forces were as egregiously ill suited for this war. The F-105 fighter-bombers the U.S. Air Force used extensively against North Vietnam were designed to drop nuclear weapons on the USSR and were ill-suited for close air support, interdiction, or fighter roles.[116] When loaded with conventional bombs, the F-105 maneuvered clumsily and needed extensive amounts of clear air to enable pilots to bomb accurately—a demanding requirement in Southeast Asia, especially in the rainy season. The air force resisted acquisition of slow, armored aircraft with long-loiter times that are especially useful in counterguerrilla operations because they were no good against the Soviet Union; the obsolete A-1 Skyraider propeller-driven aircraft the United States gave to the VNAF was among the most effective allied aircraft. Impressed with its own equipment, the air force initially flew unarmed B-57 medium bombers over North Vietnam because mission planners arrogantly and foolishly thought the mere presence of such bombers would scare North Vietnam into submission.[117]

In addition to failing to achieve its strategic goals, airpower was very cost-ineffective.[118] By October 1968, when Rolling Thunder had inflicted an estimated $600 million in damage on the north, Alain Enthoven, a former deputy controller of the air force and assistant secretary of defense for systems analysis, figured the United States spent $6 billion to replace the 990 aircraft lost to North Vietnamese air defenses—not to mention some 800 dead or captured airmen, the costs of ordnance expended, and normal operating costs.[119] The CIA estimated that it cost U.S. air forces $6.60 to inflict $1.00 worth of damage on the DRV in 1965 and $9.60 in 1966.[120]

The inappropriate force structure and changing military cultural standards led human resources to be misused. Of 543,000 U.S. troops in Vietnam in 1968, only about 80,000 were combat troops.[121] The rest were support troops satisfying the huge logistical and maintenance requirements of the equipment-heavy U.S. effort as well as providing infrastructure that gave substantial material comfort to support troops, at least.[122]

Air force personnel, especially, were organizationally culturally disposed to rely on engineering solutions to military problems.[123] Because they thought their technology superior, many airmen were convinced they inevitably would win the war.[124] When conventionally oriented technology was inappropriate for the counterinsurgency mission, airmen used it anyway. The air force came up with some good ideas—like AC-47 and AC-130 gunships and the A-37 light attack bomber—but American air forces relied mainly on technologies procured to meet conventional needs.[125] Defoliants—an air force favorite—generated few tactical benefits but alienated the Vietnamese and caused long-term health and environmental damage.[126]

Nevertheless, airmen claimed in this war, as in ones before and after, that the offensive advantages of strategic airpower could have been realized if only it had been unleashed.[127] While airpower partisans claim that civilians cost them the victory that Linebacker II showed was possible, the nature of the DRV and VC targets suggests instead that there was little possibility of decisive victory.[128] Tactical airpower was effective only when the NVA chose to concentrate, meaning there was no consistent offensive advantage to airpower. U.S. faith in technology also was operationally very debilitating because it discouraged learning and adapting to meet military and political challenges.

The Americans made many other errors that can broadly be placed under the rubric of skill. Most critically, there was a continuing inability to understand the nature of the war. Hampered initially by a dearth of Southeast Asian expertise in the U.S. government, senior military and civilian decision makers relied on irrelevant Cold War–derived theories and lessons from wars not similar to the one in Vietnam, but which were consistent with institutional interests.[129] The U.S. government poorly focused its considerable intellectual resources on understanding the situation in Vietnam.[130] The military bureaucratically punished officers who criticized conventional policy and who opted to fight the "other war" by serving in advisory roles.[131] Moreover, McNamara notes that the United States failed to link its military actions with its diplomatic initiatives—effectively delinking politics and military actions.[132] The U.S. government as a whole thus was institutionally incapable of learning from adversity or adapting effective new approaches.[133] As Lieutenant General Wallace Nutting said of the American conduct of small war, "[A]s a nation we don't understand it. As a government, we are not prepared to deal with it."[134] In short, the United States was institutionally incompetent in Vietnam.[135]

These learning and adaptation failures in turn precluded effective or timely alterations in force structure and doctrine that might have improved over time both the big-unit and pacification aspects of the war. Westmoreland went to attrition as a default strategy—understandably to some degree given his presidential constraints—but his doctrinal blinders and a lack of imagination led to no significant attempts to find new strategies or tactics despite the fairly obvious deficiencies of material-heavy operations from late 1965 to mid-1968, when he left Vietnam. Staff studies within the military like PROVN, and by the CIA, RAND, and IDA that generated both good assessments and sound recommendations, were quashed or rejected. When, at a press conference, Westmoreland was asked how to win the war, he replied "firepower."[136] By the time the wiser Abrams assumed command, U.S. goals were evolving toward Vietnamization and withdrawal, not victory.

Reflecting institutional perception errors, the training of individual troops throughout the war also was poor. Like many conventional soldiers, airmen thought that if they were well prepared for total war that limited wars would be easy to fight. As a result, bombing accuracy early in the war was poor because the air force had trained its pilots mainly to deliver nuclear weapons, not to attack ground targets.[137] With some practice, the average circular error probable of about 750 feet at the beginning of Rolling Thunder declined to about 365 feet late in the war.[138] But the air force rejected the idea that their people needed special training on counterinsurgency techniques, undoubtedly fostering vulnerability to NVA/VC efforts to generate politically useful collateral damage incidents, and counterinsurgency was little mentioned in air doctrine.[139] General Curtis LeMay, who was air force chief of staff in 1961–1965, told Congress in 1961 that air doctrine written in 1935 remained applicable.[140] The air force decided in midwar that it would not assign pilots to an involuntary second tour in Vietnam for morale reasons; it did not want to lose pilots to the airline industry.[141] As a result, pilots' bombing skills throughout the war were poor.

Retired army General Maxwell Taylor, an adviser to President Kennedy and ambassador to Saigon, worried that army troops were not adequately trained in counterinsurgency techniques.[142] Even advisers were poorly equipped with Vietnamese language and cultural awareness training and rotated too soon to become familiar with local conditions and their supported units—a subject of chronic ARVN complaints. Troops went to war as individuals on short tours, ensuring a chronic lack of unit cohesion and contributing to institutional learning problems. Late in the war, retired army officer and CORDS civilian official John Paul Vann quipped ruefully, "The United States has not been in Vietnam for nine years, but for one year nine times."[143] Military intelligence assets focused on tactical intelligence on big enemy units, ignoring the cultural and local characteristics essential to effective counterinsurgency campaigns. For domestic American political reasons, the army refused to teach recruits in the United States about actual fighting in Vietnam, leading the 1st Special Forces Group to establish a remedial jungle warfare training program in northern Okinawa for soldiers fresh out of army schools in the United States on their way to Vietnam; graduates reportedly suffered appreciably lower casualty rates in Vietnam than other soldiers.[144]

There was a fundamental failure of leadership. Senior civilians and the generals never developed a coherent strategy for the war—a source of aggravation for both conventionally minded troops unhappy with the limited U.S. effort and those critical of conventional methods who favored emphasis on counterinsurgency operations. Among those who thought conventional victory was squandered were army Colonel Harry Summers, air force General William Momyer, and Admiral Sharp.[145] Andrew Krepinevich well represents the counterinsurgency school.[146] Both camps agree the generals failed to conduct coherent operations consistent with presidential constraints—a topic of no small criticism by, and unhappiness among, soldiers after the war.[147]

Military culture was a daunting obstacle to effective use of military resources. Brian Jenkins, who served in Vietnam as a junior army officer, quotes an unnamed "senior American officer in Saigon" as saying:

> I'll be damned if I permit the United States Army, its institutions, its doctrine, and its traditions, to be destroyed just to win this lousy war.[148]

Thus, while senior civilians starting with presidents made many geopolitical strategic and military policy mistakes, military leaders who per the division of labor of the U.S. civil-military tradition run the military under general civilian guidance, bear appreciable responsibility for dysfunctional decisions about preparing and conducting the war. Edward Luttwak summarizes the failures of military commanders that were independent of alleged civilian meddling:

It was not the civilians who insisted that the war be shared among all the bureaucratic segments of the armed services. It was not the civilians who allowed each force to perform in its own favored way of war making—relevant or not, or even counterproductive. It was not the civilians who willed the hundreds of daily sorties of the fighter-bombers and the almost *4 million* helicopter gunship sorties of 1966–1971, whose bombs, rockets, and cannon shells would have destroyed all the armies in history had even a small fraction been aimed at worthy targets . . . It was not the civilians whose poverty of operational thinking and atrophied tactics were revealed by such futile use of so much firepower. It was not the civilians who condemned the enlisted men to fight and die among strangers by making every unit a mere transit pool for individual soldiers, each on his own twelve-month Vietnam tour. It was not the civilians who laid down six-month duty tours for commanders, thus ensuring . . . the constant renewal of inexperience . . . . It was not the civilians who impeded the improvement of Vietnamese forces by denying promotion to officers who chose to serve as advisers instead of "punching their tickets" in the customary sequence of command slots needed for career advancement. Finally, it was not the civilians who decided that every service unit and base, every headquarters and depot, be built on a lavish scale and administered by crowds of desk-bound officers, so very few of whom were concerned with any aspect of warlike endeavor, whether tactical or leadership, operational planning or strategy.[149]

The U.S. military displayed a consistent unwillingness or inability to recognize the realities of the war and a chronic tendency to subordinate the collective pursuit of national goals to parochial personal and institutional interests. Military personnel often were at odds with civilian intelligence officers and the Office of the Secretary of Defense over threat assessments, policy, and civilian oversight.[150] Turf fights among the military services were chronic and debilitating. The U.S. military commander in Vietnam, who always was an army general, never directed naval air forces that attacked VC in the south, aircraft striking most of the north, or B-52s. The services and the U.S. Pacific Command in Hawaii insisted on controlling pieces of the war. The navy and air force squabbled over which parts of North Vietnam each could bomb—eventually dividing the country into seven geographical areas called "route packages" the services effectively owned for parochial bureaucratic purposes, not for reasons of operational effectiveness or efficiency.[151] MACV drew Special Forces teams away from the effective Civilian Irregular Defense Group (CIDG) program that helped secure the central highlands in part because Special Forces troops were working for the bureaucratically rival CIA—damaging GVN control of the central highlands.[152]

Even the painfully obvious defeat was not scrutinized for lessons. Instead, the military largely blamed civilians for their defeat.[153] George Herring marvels at the paucity of reflective military thinking about what went wrong.[154] Krepinevich and Luttwak go so far as to say the U.S. military learned little or nothing of value from Vietnam.[155] But military personnel did not so much try to learn and fail as they purposefully forgot what they, and their institutions, learned. Because the war was a bad experience, they resolved to not fight such wars again and refused to prepare for them.[156] The 1976 edition of the army's basic doctrine on operations, Field Manual 100–5, contains no mention of counterinsurgency.[157] It is little wonder that the United States made similar mistakes in Iraq in 2003.

THE VIETNAM WAR AND STRATEGIC INTERACTION THEORY

Arreguín-Toft addresses the American war in Vietnam as one of his five cases.[158] He identifies four strategic interactions—each consistent with his theoretical expectations:

1. *Rolling Thunder (1965–1968).* The United States in its Rolling Thunder air campaign employed barbarism against a conventional defense—and lost the opposite-approach interaction.
2. *The main force war (1965–1969).* U.S. conventional ground forces conventionally fought regular units of the NVA in "a series of pitched battles" in 1965–1969, leading to U.S. victory in a same-approach interaction.
3. *The guerrilla war in the South I (1965–1973).* The conventional U.S. military fought VC guerrillas using GWS and lost the opposite-approach interaction.
4. *The guerrilla war in the South II (1965–1973).* The United States successfully attacked the VC with its Strategic Hamlets and Phoenix programs, employing barbarism against GWS and winning the same-approach interaction.

Interaction One: Rolling Thunder (1965–1968)

Arreguín-Toft says that the "main goal" of the American bombing of the north was to destroy the willingness of North Vietnam to support the guerrilla campaign in the south.[159] Because, he argues, the Americans accepted collateral damage after bombing inadequacies became clear, Rolling Thunder constituted barbarism. The barbaric attacks against a conventional North Vietnamese defense failed, leading to a DRV victory in an opposite-approach interaction.

The history of Rolling Thunder is appreciably different. U.S. airmen initially aimed in March 1965 to destroy standard military, industrial, and logistical targets while "slow squeeze" advocates wanted to threaten the same North Vietnamese assets. Both groups wanted to degrade *Lao Dong* leaders' will to fight by damaging (or threatening damage) to the DRV's ability to fight—not to harm civilians. Put differently, airmen and policymakers meant to conduct "denial," not "punishment," bombing.[160]

U.S. ROE in 1965 prohibited air strikes near the Chinese border and within 30 and 10 nautical miles, respectively, of the centers of Hanoi and Haiphong—the DRV's two largest cities.[161] The restrictions ensured the safety of many civilians at the price of rendering many defense-related industries and military facilities invulnerable. When, after some strikes on other ostensibly strategic military and economic targets Hanoi showed no inclination to quit, in mid-1965 Rolling Thunder's emphasis shifted to tactical-level interdiction targets—logistical facilities, transportation infrastructure like bridges, and cargo trucks—mainly in rural areas far from population centers. Primary targets were the thinly populated Mu Gia Pass region of southwestern North Vietnam, which was the head of the Ho Chi Minh Trail, and sparsely populated southern Laos.[162]

The attacks on electric power plants in 1967, which mainly aimed to shut down industrial production, were the only part of Rolling Thunder remotely consistent with Arreguín-Toft's thesis and Pape's assertion that Rolling Thunder aimed to affect civilian morale.[163] But the DRV's generating capacity was small and most North Vietnamese used little or no electricity.[164]

The narrow list of targets senior civilians approved and restrictive ROE reflected a very strong U.S. effort to limit civilian casualties. They well understood that killing civilians would not generate useful pressure on DRV leaders. They knew that civilian casualties instead would generate politically negative repercussions for the United States at home, in Vietnam, and around the world.[165]

Both belligerents knew that it was politically helpful for the DRV to be perceived as a victim of enemy barbarism. North Vietnam exploited collateral damage incidents through a sophisticated worldwide propaganda effort. It exaggerated American mistakes and fabricated additional collateral damage incidents to get more political mileage from barbarism-related ideational attacks on U.S. policy. It also produced civilian casualties by placing normally legitimate military targets near noncombatants. North Vietnam thus used a variant of barbarism against its own people for purposes of garnering political advantages vis-à-vis the United States.

The North Vietnamese knew very well that the United States was not trying to kill Vietnamese civilians. That is why they located sensitive installations like surface-to-air missile (SAM) launch sites and supplies like POL overtly in residential neighborhoods; they took advantage of America's highly restrictive ROE and targeting policy.[166] Knowing U.S. ROE prohibited attacks on dikes unless weapons on them were actually firing at U.S. aircraft, the North Vietnamese placed guns on them; because the U.S. Air Force grilled its pilots when they attacked weapons on dikes that were firing on them, American airmen eventually avoided all attacks on dikes in order to minimize U.S.-imposed bureaucratic hassles—the greater threat to pilots' well-being.[167]

American military leaders and statesmen of the time, including Secretary of State Dean Rusk, regretted that restrictive ROE forced changes in U.S. operations that led to U.S. casualties. They nevertheless were willing to accept some American deaths to protect Vietnamese civilians—and to blunt DRV propaganda attacks.[168]

U.S. intelligence estimated that Rolling Thunder killed about 52,000 North Vietnamese civilians.[169] While obviously unfortunate for the victims, Rolling Thunder evidently imposed only moderate inconvenience on most North Vietnamese civilians; living standards apparently declined little from the relatively low levels the DRV's own poor economic policies imposed on the country, and many people lived safely in Hanoi throughout the war.[170] Rolling Thunder also gave North Vietnam's leaders means to rally their people of the sort Winston Churchill used to rally Britons during the Battle of Britain.[171] And it gave Hanoi's propagandists useful material. RAND analyst Oleg Hoeffding wrote in 1966, "In terms of its morale effects, the U.S. campaign may have presented the [North Vietnamese] regime with a near ideal mix of intended restraint and accidental gore."[172]

Arreguín-Toft claims Rolling Thunder exemplified barbarism because the campaign continued after Secretary McNamara concluded that it produced no military value, thereby allegedly purposefully inflicting civilian casualties.[173] This logic is very poor. While some American officials shared McNamara's pessimism about the effectiveness of Rolling Thunder, many very senior ones—including President Johnson, the JCS, theater commanders Admiral Sharp and General Westmoreland, and most of Johnson's senior civilian advisors—did not. When Johnson realized the extent of McNamara's defeatism in late 1967, he transferred McNamara to the World Bank.[174] Before and after effectively firing McNamara, Johnson retained tight personal control over war policy and tactical-level bombing decisions. Said the earthy Johnson of the U.S. military, "They can't bomb an outhouse without my approval."[175]

Senior officers like Admiral Sharp and General Momyer, commander of the 7th Air Force during much of Rolling Thunder, judged Rolling Thunder a failure, not a defeat, because they believed political constraints on targeting imposed from Washington severely degraded the inherent offensive capabilities of airpower against military and economic targets, not noncombatant civilians.[176] They viewed the narrow target lists, slow target approval process, and what they believed to have been excessively restrictive ROE as debilitating constraints. In this view, the North Vietnamese did not so much win the Rolling Thunder campaign as the United States chose to lose it—hence the title of Sharp's book *Strategy for Defeat*.

Given the flaws of Arreguín-Toft argument, what is a better explanation? Rolling Thunder amounted not to barbarism but to a direct, conventional-conventional interaction that the United States lost because it did not achieve its goal of altering Hanoi's strategic policies. American airmen certainly saw it that way. The materially dominant power lost a same-approach interaction—contrary to Arreguín-Toft's expectations.

Interaction Two: The Main Force War (1965–1969)

Arreguín-Toft argues that the United States won the same-approach, conventional-conventional war with main force VC regiments and NVA regulars because it won battles and successfully ran large-scale sweep operations. Americans indeed won many battles.[177] But tactical successes did not win the war for the United States. Instead, the NVA and VC pursued their successful attrition strategy in the *main force war*. Their strategic success came without appreciable battlefield success defined as either taking ground or inflicting more enemy casualties than they suffered. But it required combat effectiveness sufficient to make U.S. units pay appreciable human prices for their tactical victories.

For example, in May 1969 the 101st Airborne Division found an NVA regiment on Ap Bia Mountain, near the Cambodian border. At a cost of 50 American dead, the 101st said it killed 597 NVA soldiers.[178] At the time, U.S. commanders expressed pleasure at the enemy body count and kill ratio. But American troops soon abandoned the mountain, generating a storm of protest in the United States about the battle's allegedly costly uselessness.[179] The name of the battle U.S. commanders thought went well has entered American political lore as emblematic of the war's waste—"Hamburger Hill." When President Nixon visited Vietnam in July 1969, he told General Abrams to avoid such battles in the future.[180]

The casualties of this and many other fights were not sufficient to defeat the NVA, but they helped to destroy Americans' will to fight the war, giving North Vietnam the strategic victory. Given the sharply asymmetrical national political environments and the DRV's substantial force generation capabilities, the kill ratio favored the NVA's variant of attrition strategy.[181] The point is made succinctly in a conversation U.S. Army Colonel Harry Summers had with an NVA officer in 1975, soon after the war was over:

> "You know you never defeated us on the battlefield," said the American colonel [Summers]. The North Vietnamese colonel pondered this remark a moment. "That may be so," he replied, "but it is also irrelevant."[182]

The Vietnamese officer was right. Contrary to Arreguín-Toft, the United States lost the main force war.

Interaction Three: The Guerrilla War in the South I (1965–1973)

Arreguín-Toft argues that guerrillas using GWS defeated the conventional U.S. military in an opposite-approach interaction.[183] From 1965 the NVA and the VC conducted military operations in the south; both often used small-unit, guerrilla tactics. While they clearly did not defeat the Americans, one might argue that the VC won in 1965–1967 according to the common view that guerrillas who are not defeated win. But Giap expended much of the VC's fighting strength in the Tet and smaller offensives of 1968. Thereafter, the VC's capacity to fight declined sharply, and Hanoi had to rely on NVA divisions. As a result, in 1969–1973 U.S. and ARVN forces faced a much reduced guerrilla threat they handled fairly well—leading in part to the oft-made assertion that the allies won the "other war" only to throw away the victory in 1975. Arreguín-Toft's argument that the United States lost the guerrilla interaction in 1965–1973 therefore is incorrect, especially about the latter half of the period.

Interaction Four: The Guerrilla War in the South II (1965–1973)

Arreguín-Toft claims the United States won a same-approach interaction by using barbarism against GWS through use of two programs—the Strategic Hamlets and Phoenix programs.[184] He rates Strategic Hamlets a failure, however, meaning he rests his argument on the alleged effectiveness of Phoenix. In essence, Arreguín-Toft characterizes as barbarism two aspects of a large and complex set of activities collectively known as pacification—multiple efforts to win the "hearts and minds" of the people, keep them safe from VC predations, and bind them psychologically to the GVN. His "barbarism" incorrectly characterizes the pacification program, regardless of how much violence occurred as allied forces fought armed enemies to gain control of the countryside.

Strategic Hamlets

The Strategic Hamlets program was designed to foster pacification by producing local security by concentrating peasants' homes in hamlets with modest physical fortifications and modestly trained, armed, and organized local self-defense forces designed to counter low-level security threats like VC assassination squads and local guerrilla units.[185] The ARVN and regional security forces were to take on the main force VC and NVA units that could easily overwhelm a hamlet's small security force.

There is some disagreement in the literature about the main authors of the Strategic Hamlets program, but the bulk of the evidence indicates that the Ngo brothers drove the program. President Diem decreed the effort on February 3, 1962, and Douglas Pike calls Ngo Dinh Nhu its "mastermind"—a view other knowledgeable contemporary observers support.[186] Even though Americans, including a presidential mission headed by Eugene Staley and the CIA, proposed similar programs before 1962, the U.S. Embassy in Saigon and the U.S. military mission were surprised when Diem announced the program.[187] The U.S. military, while formally backing Strategic Hamlets, encouraged the ARVN to conduct operations independent of it. Civilian officials in Washington were more sympathetic, however, and by September 1962, the United States funded much of the program.[188]

The six-person British advisory mission led by Sir Robert Thompson, a former senior defense official in colonial Malaya who participated in Britain's successful counterinsurgency campaign there in the 1950s, also played a role—albeit a smaller one than either the GVN or the Americans in that it was wholly advisory.[189] Thompson dealt directly with Diem and Nhu, but neither the Ngo brothers nor the Americans seem to have paid much attention to him.[190] The program looked superficially like a British counterinsurgency tool used successfully in Malaya to separate ethnic Malay and Chinese villagers from mainly Chinese insurgents, leading some observers to think, incorrectly, that the British provided the initiative for Strategic Hamlets.[191]

The GVN announced six formal goals for the program that focused on security.[192] But Diem and Nhu also seem to have designed the program to meet personal political needs and philosophical outlooks.[193] They created some new institutions, including the Republican Youth, which resembled function-based NLF social organizations (for farmers, women, youth, and others) but were weaker and less focused. The organizations helped secure hamlets and supported the Ngos politically.

Diem applied the plan to the entire country. In August 1962, the GVN announced plans for construction of 11,316 fortified hamlets and claimed at year-end 1962 that 3,225 of them had been completed.[194] Nhu claimed success in early 1963 and shrill VC propaganda attacks on "concentration camps" at the time lent support to his claims.[195] The GVN reported that 5,900 strategic hamlets were built by April 1963; 7,200 were built by July 1963; and, 8,737,000 people lived in strategic hamlets in late 1963.[196]

Some American and British officials thought the effort was going fairly well in 1963.[197] While the VC destroyed some of the more isolated and poorly built and manned strategic hamlets, there was tangible evidence of progress.[198] Communist documents from the period indicate that the communists also saw a well-run Strategic Hamlets program as a major threat to their political base and thought their counteroffensive was not going well.[199] As the program expanded in 1963, the communists continued to fight it with limited success and continued to worry about it.

The VC could not defeat it, but the Ngo brothers' deaths on November 1, 1963, ended Strategic Hamlets virtually instantly by eliminating its patrons.[200] The generals whose troops killed Diem and Nhu dismissed many province and district chiefs, who ran the program, because of concerns about their loyalty to the new government. ARVN units that provided security for strategic hamlets redeployed to the cities to protect the insecure new rulers. Social organizations like the Republican Youth that aided the effort but also backed the Ngos politically were dissolved.

The communists could not believe their luck and wondered how the Americans could have been so stupid as to overthrow Diem; they internally called Diem's death a "gift."[201] They accelerated their activities in the hamlets in late 1963 in response and quickly achieved considerable success. By March 1964, the VC believed they had destroyed 80 percent of the strategic hamlets—helping generate the VC gains that triggered Johnson's decision in late 1964 to expand American involvement in the war.[202]

American observers later were harder on the Strategic Hamlets program than were the communists.[203] Despite some successes, the program had many inadequacies and some outright failings.[204] By early 1963, even some GVN officials favorably disposed to the effort saw major, albeit potentially fixable, problems. For example, even though the program was designed to avoid a key failure of Diem's so-called agroville program of 1959–1961—the fact that many Vietnamese peasants did not want to move from their homes to more defensible locations—many peasants in the Mekong delta area, especially, had to do just that.[205] The program called for more and better-armed security personnel, with Civil Guard units available to reinforce village self-defense forces in the event of VC attacks, but Nhu expanded the program faster than the GVN could adequately train and arm local security personnel and the VC developed tactics, including ambushes of reinforcing Civil Guard units, that enabled them to overrun some hamlets. The GVN did not provide adequate numbers of competent GVN officials to serve villagers. Money and building materials shortages sometimes led to minimal fortifications; some hamlets had only a row of bamboo stakes as a security barrier, making them hard to defend.[206] Nhu also claimed way too much for the program and statistics on completed hamlets were inflated.[207]

Knowledgeable observers place blame for inadequacies on the American side as well. MACV damaged the program by gutting the CIDG, which the CIA and Army Special Forces operated in the central highlands to protect hamlets, by converting CIDG units into mobile strike forces in 1963 and using them in offensive operations well away from their home villages.[208] There was little coordination among U.S. agencies; the military sought to engage VC military units while civilian agencies worked primarily pacification-focused agendas. The military by this time already used firepower excessively and President Kennedy approved use of napalm and defoliants—adding to grievances about the Americans even as the GVN, CIA, AID, and USIS worked to help the peasants and produce loyalty to the GVN.

Thus, Arreguín-Toft is right to say Strategic Hamlets was not very successful. But it was not an American program. Nor was it close to qualifying as barbarism by most definitions, including Arreguín-Toft's. And, it occurred outside the time period he claims to assess.

Phoenix

The Phoenix program was part of the joint U.S./GVN pacification effort run on the American side by CORDS. In 1968, Komer and the Vietnamese established the first relatively coherent joint pacification effort, which was a comprehensive security, social, economic, and political effort to wrest the countryside from the VC and to consolidate GVN control and legitimacy.[209] Phoenix has received much inaccurate publicity as an assassination or torture or mass-murder program. Because Phoenix fit into a set of existing U.S. and GVN programs and organizations, some background information is needed to place Phoenix in context—and thence to assess its ability to do what Arreguín-Toft says it did.

Phoenix was an intelligence program that focused on improving the effectiveness of security forces' attacks on the VC's political leadership.[210] It began as an idea of CIA officer Nelson Brickham, who envisioned an organization that would improve U.S. and GVN efforts to root out the so-called Viet Cong Infrastructure (VCI)—meaning PRP members, other NLF cadres, and North Vietnamese staffing the communist shadow government in the south—by improving intelligence sharing.[211] The many agencies that worked against the VCI were intensely competitive with each other.[212] For bureaucratic competitive reasons and to protect sensitive information from VC spies, who were thought to be numerous, agencies were reluctant to share their information with other agencies, often preventing successful exploitation of time-sensitive information. Robert Thompson, who was close to both the GVN and the U.S. government, says that in 1966 he counted 17 separate U.S. and GVN intelligence organizations—a number that alone explains Brickham's motivation.[213]

Recall that in the Vietnamese communists' strategic scheme the political struggle was more important than the armed struggle. Senior VC cadres were key actors because they were political leaders.[214] Therefore, it was logical to target VC cadre and some GVN and American organizations did so from the beginning of the war. Many cadres periodically fought as soldiers and VC military security details often accompanied personally unarmed senior VC officials—making them effectively armed.[215] Because many VC cadres were armed and "illegal" VC often fought to the death rather than be captured, allied security agencies targeting them often used the tactics and weaponry of commando forces.[216] Traditional police methods usually were sufficient to arrest "legal" VC. The VC similarly targeted and killed some CIA officers—civilians who also were armed and intimately involved in fighting the war, broadly defined.

The CIA evaluated Brickham's idea, found it attractive, and pitched it to CORDS chief Komer, who in mid-1967 won approval for it from General Westmoreland and Ambassador Ellsworth Bunker. Komer then created the Intelligence Coordination and Exploitation (ICEX) program, which the CIA ran within CORDS.[217] President Thieu approved the idea in December 1967, naming the corresponding Vietnamese program *Phung Hoang*—the name of a mythical Vietnamese bird.[218] Komer then renamed ICEX "Phoenix," an Americanization of *Phung Hoang*.[219]

ICEX/Phoenix began to create in late 1967 intelligence coordination centers at district and provincial levels—known respectively as District

Intelligence and Operations Coordination Centers and Province Intelligence and Operations Coordination Centers.[220] Their purpose was to do just what their names indicated—coordinate intelligence activities in support of field operations against the VCI.[221] The centers were Vietnamese-staffed organizations which Americans advised. The Tet Offensive disrupted the pacification effort for a time but by mid-1968 the situation had stabilized. According to William Colby, a former CIA station chief in Saigon who was first a deputy to Komer and then head of CORDS while on leave from CIA, Thieu reconfirmed support for it in mid-1968, although he seems to have lost interest in it thereafter.[222] Thus, the program became fully functional only in late 1968. Colby especially wanted to root out the VCI because the NLF in April 1968 formed a new front organization—the "Alliance of National, Democratic and Peace Forces"—which Colby accurately saw as a Hanoi-directed effort to boost control of the countryside before a possible political settlement of the war.[223]

The formal purpose of Phoenix, according to Colby's instruction to American personnel working the program, was "one of advice, support and assistance to the GVN Phuong [sic] Hoang program, aimed at reducing the influence and effectiveness of the Viet Cong Infrastructure in South Vietnam."[224] Colby wanted to generate intelligence information adequate to take senior VC cadres alive so they could be interrogated in order to gather additional information that could be used to round up still more cadres. In his written directives to U.S. personnel, Colby prohibited Americans from assassinations and required them to report such activities by Vietnamese officials. Brickham similarly ordered his American subordinates not to engage in torture and says he knows of no cases in which Americans did so.[225] Motives for keeping Americans' hands clean were undoubtedly heavily political but they also were functional—corpses do not provide information to interrogators.

Phoenix was closely linked to allied intelligence agencies and field operational units over which it had little influence. Phoenix/*Phung Hoang* provided information to Vietnamese police and paramilitary units that pursued the VCI, and to American military and ARVN intelligence units. Among the most effective Vietnamese-manned security forces were the Provincial Reconnaissance Units (PRU) and the Special Branch police, an arm of the National Police.[226] The CIA created and ran the PRU for several years before 1968 and also advised the special police. While operations varied, the PRU specialized in using small teams to conduct night raids on VC cadre in rural areas. Because they typically targeted senior cadre in contested or VC-controlled areas who were either armed or who had armed bodyguards, there often were firefights. In these raids, the PRU both killed and captured VC before evacuating operational areas. Failure to quickly evacuate would have been fatal to small teams operating in enemy territory.

Such efforts contributed appreciably to the "neutralization" of VCI before Phoenix was created. Neutralization was accomplished in one of three ways: (1) killing VC personnel; (2) capturing or arresting them; or (3) inducing them to defect, or rally, to the government's side. GVN organizations specialized in different types of neutralization. For example, the PRU and special police mainly killed and captured hard-target "illegal" VC in raids.[227] The National Police often peacefully arrested legally resident South Vietnamese or tried to induce them to work as GVN agents. And, the *Chieu Hoi* (Open Arms) program focused on encouraging VC to defect and then on resettling them.

During the 1968 to mid-1972 life of Phoenix, CORDS counted 26,369 VC cadres as killed, 33,358 as captured, and 22,013 rallied—for a total of 81,740.[228] Robert Thompson, after examining neutralization data for 1970 and 1971, concluded that conventional allied military units conducting combat operations accounted for 87 percent of VCI kills in those years; police and other security services contributed the rest.[229] Colby told Congress the figure for 1971 was 87.6 percent.[230] Assuming a constant 87 percent rate for military kills for the entire life of Phoenix, the remaining 13 percent of kills amounted to 3,428 people—many of whom clearly died in military-like firefights with organizations like the PRU.

Estimates of the neutralizations accomplished by each allied organization are subject to data quality problems that tend to both raise and lower aggregate estimates.[231] For example, some agencies, apparently including the National Police, tended to inflate neutralization figures in order to look good. The PRU, however, evidently reported inaccurately small figures—an apparent CIA effort to make its effective program appear innocuous.[232] Some people killed in fights against targeted VC may have been innocents, inflating estimates of neutralizations, while VC cadre killed in combat while fighting as soldiers went unreported because allied combat units could not identify VC corpses. Many apparent defectors in fact remained VC agents or redefected. And, because GVN police and courts frequently quickly released arrested VCI, many were not neutralized for long.

Because the purpose of Phoenix was to make operational field forces more effective by providing them with better information, the best way to judge the effectiveness of Phoenix is to evaluate the extent to which it collected more and better intelligence, improved the timeliness and quality of analyses, expediting the dissemination of intelligence, and improved the counter-VCI field work of other agencies—that is, Phoenix's marginal contribution to the effectiveness of the overall anti-VCI effort. In general, former U.S. government employees' retrospective assessments of Phoenix take roughly this analytical approach.

Phoenix generated and disseminated some useful information but the program had so many problems that most knowledgeable American officials were critical of its performance.[233] Participants identify problems of competence and execution—not abuses like assassination and torture.[234] For example, after he left government, Komer said Phoenix was "poorly managed, and largely ineffective."[235] The GVN's many security agencies remained bureaucratic rivals and Phoenix evidently had little effect on information sharing.[236] At Komer's unwise request, the GVN gave security agencies quotas for arrests, leading them to pick up low-level sympathizers relatively easy to capture, not the harder-to-find senior cadres the Americans wanted to neutralize.[237] The quotas drove security officers to claim incorrectly that VC killed in combat or surrendered for any reason were due to *Phung Hoang* activities.[238] And, after CIA in mid-1969 pulled most of its people out of Phoenix to focus on unilateral strategic intelligence collection versus what CIA considered to be the tactical focus of Phoenix, their replacements were mostly junior army officers who often were more interested in tactical military intelligence than intelligence on the VCI—a major change in perspective.[239]

GVN policies and practices influenced the effectiveness of Phoenix. For example, due to weak senior-level GVN command emphasis after Thieu's initial commitment, there seems not to have been much of an improvement in the quality of information the allied intelligence agencies produced.[240] General Nguyen Ngoc Loan, chief of the National Police, opposed Phoenix because he distrusted the CIA and was a political opponent of President Thieu, who he saw gaining by a successful Phoenix effort.[241] The GVN personnel management system treated military personnel better than policemen, leading to lower-quality personnel in the nonmilitary security services and lower pay scales, which encouraged corruption. Some policemen unsurprisingly were corrupt—leading to arrests of innocents in pursuit of bribes and VC cadres' chronic ability to quickly bribe their way to freedom.[242] Some prisons nevertheless became overcrowded as a result of Phoenix-related arrests, contributing to premature release of VC cadres.[243] And, the Vietnamese legal system, subject of other American reform efforts, by late in the war produced judges who routinely dismissed cases against actual VCI as too weak to take to trial.[244]

Other GVN programs were started or improved in ways that increased the effectiveness of the allies' attacks on the VCI independent of Phoenix/*Phung Hoang* activities. For example, after Tet the GVN passed new laws that better mobilized the country for war. Coincident with Nixon's Vietnamization program, the GVN sharply increased the size of its territorial military forces—the Regional Forces (RF) and Popular Forces (PF)—and at U.S. expense for the first time armed them with modern M-16 rifles instead of the World War II–vintage M-1s they had for years. The GVN also created the village-level People's Self-Defense Force (PSDF). As district- and village-level organizations, the PF and PSDF increased pressure on the VC cadre, leading to some arrests and evidently forcing others to curtail their activities—a form of functional neutralization that personnel-focused statistics cannot capture.

The bottom line is that while Phoenix had some successes, American participating officials were not happy with its overall results. Colby says that public claims of numbers of VC cadre neutralized are both exaggerated and misleading due to the low level of individuals picked up.[245] The GVN also seems not to have been very happy with *Phung Hoang*; in 1971–1972, it gradually merged the program into the National Police.[246]

Communist sources thought more highly of Phoenix—apparently because they saw the effects of all allied operations against them and could not discern the marginal effectiveness of Phoenix alone.[247] Apparently on the basis of information from communist sources, Gabriel Kolko says Phoenix was more effective than American officials thought and Stanley Karnow's former VC sources reported after the war that Phoenix hurt the VCI.[248] Communist documents produced during the war similarly credit counter-VCI programs with hurting them.[249] Former NVA Colonel Bui Tin thinks Phoenix was effective—but as part of the larger pacification effort.[250]

These understandable perceptions notwithstanding, the effectiveness of Phoenix assessed as its marginal contribution to related U.S. and GVN security programs seems to have been modest. Its effects on the pacification effort as a whole were even smaller. Major contributors instead included Thieu's land reform program of 1970, economic prosperity, and enhanced village security provided by larger and better equipped local military forces. Together these efforts are widely and reasonably credited with helping to win the "other war."

So why are there such spectacular claims about the allegedly nefarious effectiveness of Phoenix? CORDS's sloppy language and statistics undoubtedly contributed. Substantively, the concept behind Phoenix was a potentially serious threat to the communists. Vietnamese communists and their American fellow travelers therefore attacked the program propagandistically, generating inflated and flatly incorrect claims typical of Hanoi's propaganda messages.[251] *New York Times* reporter Tad Szulc reported that some 20,000 suspected VC were "murdered" under the auspices of Phoenix and compared the program to the My Lai massacre in which American soldiers killed several hundred villagers.[252] Many such claims still circulate, including Arreguín-Toft's erroneous assertion that Phoenix was a "sustained effort to kill noncombatants."[253]

Such accusations led in 1970 to U.S. Senate hearings on Phoenix, which found no evidence of U.S. complicity in assassinations or abuse of prisoners.[254] The Senate later confirmed Colby as director of central intelligence, effectively absolving him of wrongdoing. Mark Moyar more recently investigated allegations by former U.S. military personnel of improprieties in Phoenix and found all of them bogus.[255]

Instead, it was the VC, on the defensive, who continued their campaign of violence against GVN civilian officials and other civilians, as Arreguín-Toft (p. 148) recognizes. In 1969, even after the heavy casualties of 1968, the VC were strong enough to assassinate another 1,200 GVN civilian officials and their civilian supporters, including 23 village chiefs and 126 hamlet chiefs.[256] Former Provisional Revolutionary Government Minister of Justice Truong Nhu Tang says Phoenix prompted the VC to respond with a heightened campaign of "terror."[257] Given U.S. estimates that the VC killed more than 36,000 civilians in 1957–1972, the VC evidently assassinated far more genuinely noncombatant civilians than the number of VC cadre all allied efforts killed in any way during the 1968–1972 life of Phoenix.[258]

As in the Philippines in 1899–1902, the VC used terror or "barbarism" extensively as a matter of policy. U.S. military and civilian agencies did not. A formal VC security service conducted many of the assassinations. While successful in the early 1960s, by 1970 the tactic generated a backlash among Vietnamese peasants who did not appreciate increased VC taxes, drafting young men for VC combat units, and greater VC use of summary executions for progressively less good reason.[259]

Summary

While rightly arguing (implicitly) that the "other war" was an important interaction and that counterinsurgency programs were effective late in the war, Arreguín-Toft's argument contains several important errors. The Strategic Hamlets program was a Vietnamese program, not American, and was designed to foster rural security for civilians largely through defensive measures, not to attack civilians. Phoenix was a small part of a fairly sophisticated economic, political, social, and military pacification effort. Phoenix attacked only "civilians" who were intimately involved in the war as participants and did not produce anything close to the violence propagandistically attributed to it.[260]

The Bottom Line

Arreguín-Toft's descriptions of aspects of the war frequently are incorrect, his analysis is questionable, and my evaluation of his strategic interactions yields results that are largely incompatible with the structure of his theoretical framework. My analysis indicates that Arreguín-Toft is wrong about the barbaric aspect of Rolling Thunder; the communists, not the Americans, won the big-unit war; the VC may have won the 1965–1967 phase of the guerrilla war, but they were a spent force in 1969–1973; and neither Strategic Hamlets nor Phoenix came close to Arreguín-Toft's definition of barbarism. Moreover, many of the activities Arreguín-Toft calls strategic were in fact tactical in nature, and he does not well link them to achievement of strategic objectives.

Arreguín-Toft says the United States won two and lost two interactions. He explains the strategic outcome of the mixed results of subordinate interactions by saying merely that the DRV won by "delaying its defeat long enough . . . to force the United States to abandon the GVN" (p. 157). This reasonable observation little integrates American and Vietnamese communist interactions dynamically or weighs their conflicting influences. Elsewhere, he (p. 232) rates the war as a whole an opposite-approach interaction in which the strong actor "loses/tie"—hardly an accurate characterization of a clear U.S. defeat. In short, Arreguín-Toft does not explain the war's conduct or its outcome.

DIMENSIONAL ANALYSIS

The United States and North Vietnam identified different vulnerabilities in their enemies and had different levels of success in exploiting them. The DRV did far better.

The United States

The United States focused mainly on attacking North Vietnam in the *leadership, resource,* and *operational execution* dimensions. But it did not understand North Vietnamese vulnerabilities and did not design or execute its attacks well, meaning its overall performance was poor.[261] Moreover, the war lasted long enough that DRV exploitation of U.S. political vulnerabilities affected U.S. policy—about bombing the north, for example—accounting for part of the U.S. diplomatic and military ineffectiveness.[262] The United States ultimately failed to find or exploit any critical North Vietnamese vulnerability as the DRV effectively attacked U.S. vulnerabilities.

American officials largely simply assumed North Vietnamese vulnerabilities. The U.S. government as a whole made little effort to understand Vietnam or to test its assumptions about DRV/NVA/VC vulnerabilities.[263] The theoretical and doctrinal constructs that guided U.S. operations little recognized the realities of Vietnamese politics, economics, and social and military affairs. Even when government agencies and advisors like RAND produced reasonably good assessments, normal mechanisms of government led to poor military and presidential decisions. Leslie Gelb and Richard Betts may be right that the U.S. governance system worked as designed, but it did not lead to wise policies or effective policy implementation.[264] Instead, airpower partisans, soldiers committed to the Army Concept, and White House officials devoted to inapplicable theories brought unhelpful preconceptions and a desire to do as little as possible to bear on the problem of how to convince a strongly motivated and competent DRV to stop trying to unify Vietnam on its terms. The resulting patchwork of limited operations largely reflected reaction and compromise, not a well-conceived plan. In short, the United States had no national strategy worthy of the name.[265] Robert McNamara makes a different but equally damning assessment of the cause of American incompetence;

he says the United States had a coherent strategy but all of its underlying assumptions were wrong.[266]

The United States first tried to attack the will of North Vietnam's *leadership* by destroying or threatening its military, economic, and logistical capabilities though military action. It did not try to influence North Vietnamese *national will* by killing or otherwise influencing civilians, with the minor exception of the electric power plant attacks of 1967. Airmen tried to persuade Johnson and McNamara to let them rapidly destroy military, industrial, and transportation targets; that is, by affecting domestic North Vietnamese *resource mobilization* and *resource conversion* for the purpose of influencing *force generation* and thence *leadership*. Johnson rejected their arguments. His civilian advisors thought mere threats would produce the same result.

In any case, the DRV was not vulnerable to such attacks. North Vietnam had no significant defense industry and imported modest amount of mainly weapons, food, and POL from its Soviet and Chinese patrons, who Johnson chose not to attack. Contrary to American expectations, the north showed little desire to protect its meager industry. The DRV was not a closed economic system, so attacking the DRV's small defense industrial production capabilities could never achieve what the Allies accomplished after a massive effort against Germany in 1944 and 1945 and what the Americans did much more efficiently in the Philippines in 1899–1902. When American aircraft destroyed military-related equipment—like trucks plying the Ho Chi Minh Trail—China and the USSR replaced it through land routes from China and the ports of Haiphong and Kampong Sam, Cambodia, which were off limits to American airmen for most of the war. China and the USSR served as giant warehouses in sanctuaries Hanoi could tap as needed.[267] Therefore, as Earl Tilford observes, the bombing was not so much ineffective as it was irrelevant.[268]

The logistical interdiction objective that was the biggest part of Rolling Thunder, all of Commando Hunt, and much of the tactical bombing in the south similarly never stood a chance of substantially affecting the communists' war effort and thence, using U.S. logic, their will to fight. The CIA told MACV in 1966 that most of the supplies the VC used were obtained in the south from sympathizers, the black market, and captured ARVN stocks.[269] Other U.S. analyses, including military studies, noted that the light NVA forces required few supplies and that the north's redundant logistical network could supply NVA/VC needs even under intense bombing.[270] Airmen said the solution was more bombing.

In the south, Westmoreland's attrition strategy was designed to wear down NVA and VC units to the point that their ability to fight was destroyed. Here, the logic was to defeat the enemy in the *operational execution* dimension in order to produce losses greater than the DRV's *force generation* capabilities, reducing NVA/VC force levels and thereby influencing *leadership*'s commitment to the war. But even while killing tens of thousands of enemy combatants every year, U.S. and ARVN forces never reached the crossover point and never seriously diminished NVA/VC capacity to fight.

American leaders consistently failed to understand the strength of North Vietnam's commitment to national unification—and correspondingly the price it was willing to pay to win.[271] President Johnson seems to have simply assumed that *Lao Dong* leaders would respond to American actions in ways amenable to U.S. interests.[272] American officials thought there was a price in casualties—what they called a "breaking point"—at which the north would quit, which they assumed they could impose with the limited means they were willing to use. They thereby made the fundamental error of attributing to their opponent their own values and resultant notions of bearable political/military costs, which intelligence analysts call "mirror-imaging."[273]

Instead, North Vietnam's leaders willingly spent very large numbers of lives to achieve their political goals. Against the big human price the DRV paid, White House threats to bomb a few small factories, interspersed with repeated bombing halts, chronic pleas for negotiations, and offers of bribes in the form of aid packages, must have appeared to North Vietnam to be almost comical demonstrations of the feckless weakness of American leaders.

While U.S. attacks on the DRV in the *resource mobilization* and *resource conversion* dimensions that focused on defense industry and transportation were limited and ineffective, the DRV had two key types of domestic resources that, if damaged, would have affected the NVA's war fighting potential: people and food. U.S. targeting of people, as the USAAF and RAF did in 1943–1945, would have destroyed human resources supporting the war effort and presumably led North Vietnam's leaders to increase efforts to defend and recuperate in the north. Dikes in the Red River valley provided irrigation water and flood control and were potentially lucrative targets; their destruction would have flooded farmland and damaged food crop production, hindered transportation, and killed people and livestock.[274] But American targeting policies throughout the Vietnam War protected both from attack.[275]

In the south, the United States belatedly attacked the VC in the *national will* and *resource mobilization* dimensions, but only after crumbling U.S. national will rendered that success strategically irrelevant. Beginning in 1968, the United States and the GVN conducted counterinsurgency operations in South Vietnam that had considerable success by 1972. These efforts cut peasants' support for the VC, which in turn affected VC ability to mobilize resources by recruiting soldiers and collecting taxes. CORDS and South Vietnamese efforts, along with the decimation of VC combat forces in 1968, amounted to a considerable defeat for the DRV. But rather than rebuild the VC, Hanoi used PAVN divisions in largely conventional offensives in 1972 and 1975, winning the second decisively.

And, U.S. military conduct of the war in the *operational execution* dimension was mixed. Despite the Americans' consistent ability to hurt the NVA/VC if they could find them and bring firepower to bear, they could not often do so, and American performance in counterguerrilla warfare and counterinsurgency operations more broadly was often poor. American airmen and soldiers did not have the skills or desire to fight an unconventional war and had little inclination to acquire them. Military leaders resisted President Kennedy's efforts to get them to focus on counterinsurgency in 1961–1963. General George Decker, who was army chief of staff in 1960–1962, responded to a presidential lecture on the importance of counterinsurgency capabilities by saying, "any good soldier can handle guerrillas"—effectively dismissing Kennedy's concerns.[276] A devastating indictment of the U.S. way of war is the fact that despite vast numbers of helicopters and trucks and sophisticated electronic intelligence gathering capabilities, largely foot-mobile NVA/VC forces consistently held the operational initiative.

The U.S. military insisted that the crux of the problem in Vietnam was military, not political, and their doctrinal, training, and organizational responses to troubles in combating capable NVA/VC opponents were minimal. This inflexibility contrasts markedly with the reassessment General MacArthur made in the Philippines in December 1900, which led revisions in the political and military campaign that, as discussed in chapter 2, were quickly successful in 1901.

In sum, the United States did not identify or exploit any critical enemy vulnerability. U.S. military personnel both failed to appreciate the capabilities of their enemies and grossly overestimated their own.[277] And, after the war, many senior American civilian leaders said their biggest mistake was to underestimate the will of their adversary.[278] The result was an incoherent strategy, inept operational execution, and defeat.[279]

North Vietnam

The DRV's primary target was American will.[280] The North Vietnamese saw the United States, powerful but committed around the world, as both willing to sacrifice its South Vietnamese ally and unwilling to make a major commitment to win.[281] They correctly believed the United States pursued a "limited war" strategy that meant the DRV would not have to face the full brunt of U.S. military power.[282] They accurately recognized and effectively exploited massive American political vulnerabilities, especially Americans' sensitivity to U.S. military casualties and civilian casualties inflicted by U.S. military forces.[283]

Far from North America and possessing little international economic leverage, North Vietnam could not directly influence U.S. *resource mobilization* or *resource conversion*. Nor could it directly influence physical aspects of U.S. *force generation*. It therefore opted to use military force in the *operational execution* dimension to complement its political efforts to influence Americans' *national will* and thence a *leadership* decision to quit the war.

North Vietnam employed a sophisticated strategy to exploit U.S. vulnerabilities. Politically, the North Vietnamese attacked the United States ideationally as an aggressor and brute for inflicting civilian casualties. Hanoi propagandistically exploited actual civilian casualties caused by air attacks against North Vietnam and bombing and artillery strikes in the south in order to win sympathy within the United States and internationally. And it emphasized the U.S. cost in human lives.

North Vietnam aimed to achieve its political, military, and diplomatic goals by targeting the American people.[284] It especially aimed to give political ammunition to the U.S. domestic antiwar movement.[285] Said former NVA Colonel Bui Tin, who was well connected to DRV leaders as a writer and editor for the PAVN's daily newspaper *Quan Doi Nhan Dan*, of the American antiwar movement:

> It was essential to our strategy. Every day, our leadership would listen to world news over the radio at 9 a.m. to follow the growth of the American antiwar movement. Visits to Hanoi by people like Jane Fonda and Ramsey Clark and ministers gave us confidence that we should hold on in the face of battlefield reverses. We were elated when Jane Fonda, wearing a red Vietnamese dress, said at a press conference that she was ashamed of American actions in the war and that she would struggle along with us.[286]

After the war, General Giap told U.S. Navy Admiral Elmo Zumwalt that the war was won in the United States and that the Jane Fondas of the world were very useful to him.[287]

Doctrinally focused on looking for internal contradictions within the colonialist power, the DRV sought to aid the "progressive movement" in the United States and win the hearts and minds of Americans by "motivating and exploiting" American public opinion.[288] For example, the South Vietnam People's Committee for Solidarity with the American People corresponded with sympathetic Americans.[289] On five occasions during 1967–1972, the DRV released groups of three each freshly captured and little abused American prisoners to peace activists to bolster their cause and to publicize the DRV's ostensibly conciliatory attitude.[290] The DRV placed an intelligence agent as a correspondent for Western news agencies in South Vietnam, including the U.S. news magazine *Time*, where he fed many prominent American reporters information and perspectives while also conducting espionage activities.[291] In its first public recognition of American war protesters (and indirectly the effectiveness of its work) Radio Hanoi in October 1970 broadcast a message from Prime Minister Pham Van Dong praising Americans' protests against the war as a "noble reflection" of the American public's desire to save its sons "from a useless death in Vietnam."[292] The message reflected Hanoi's key propaganda theme of enhancing Americans' sensitivity to U.S. military casualties.

Hanoi actively managed strategically important information. Internally, it limited foreigners' access to the DRV and areas in the south it controlled to people who supported its goals.[293] It invited sympathetic journalists to observe the war. For example, at the suggestion of Australian communist journalist Wilfred Burchett, Hanoi in 1966 invited to North Vietnam Harrison Salisbury of the *New York Times,* which took an editorial position strongly against Johnson's Vietnam policy; Salisbury obligingly wrote dispatches that incorrectly suggested that American bombing of the north was indiscriminate.[294]

Externally, the DRV released torrents of incorrect information hoping some would stick—a goal often attained among Americans sympathetic to its cause and others who were naïve or gullible. For example, the DRV claimed far more political prisoners in the south than existed—a claim widely accepted in the United States.[295] It claimed that it treated American POWs well and that many recanted their war crimes, when in fact the North Vietnamese brutalized many prisoners and evidently killed some.[296] DRV propaganda organs exaggerated and fabricated collateral damage incidents to amplify their propaganda value.[297] Colocated DRV and NLF information offices in numerous cities around the world spread the word.[298]

The disinformation effort featured elaborately staged political dramas. For example, the DRV in June 1969 created the Provisional Revolutionary Government in the south in an effort to convey the erroneous impression of an autonomous revolutionary government there.[299] For the same reason, Hanoi hid its sponsorship and control of the NLF. The DRV slandered its political opponents.[300] Hanoi in 1968 adopted a policy of *danh va dam*, or "talking and fighting," to convey an impression that it wanted to negotiate an end to the war without ever intending to negotiate seriously.[301] It did not meaningfully respond to any U.S. peace initiative but consistently increased transportation of soldiers and materiel south during bombing halts.[302] In 1972, the DRV engaged Democratic presidential candidate George McGovern with what former NLF official Truong Nhu Tang calls "chimerical" peace proposals.[303]

Militarily, North Vietnam used its forces to influence American will. Strategically, the Tet Offensive successfully influenced U.S. public opinion and the DRV designed its 1972 offensive in part to discredit Richard Nixon during the presidential campaign season. While unsuccessful in producing Nixon's electoral defeat, Nixon in May 1972 during heavy fighting reduced his diplomatic demands on Hanoi.[304] And, its integrated political/military attrition strategy successfully produced both continuing American casualties and mounting domestic unhappiness about them.[305]

The DRV, NVA, and VC developed many tactical means to turn the firepower American units used in vast amounts against American interests—mainly by generating collateral damage their propaganda organs exploited. In the south, the VC repeatedly occupied villages in the vicinity of American and ARVN troops, whose standard practice was to call for air strikes when in contact with enemy soldiers; after learning U.S. reaction times, they melted away unobserved just before aircraft arrived, leaving the Americans to bomb a village devoid of VC—thereby wasting U.S. ordnance, making villagers unhappy with the United States and the GVN, and generating propaganda opportunities.[306] Similarly, they sniped at U.S. troops from villages, hoping to draw return fire at villagers.[307] And, about 5 percent of American bombs dropped on South Vietnam were duds, giving the VC material to build booby traps that killed about 1,000 Americans in 1966 alone.[308]

Vietnamese communists exploited aspects of U.S. military force structures and doctrine as part of their vulnerability amelioration efforts. They developed "hugging" tactics whereby VC/NVA troops got close enough to U.S. forces to keep the Americans from using air strikes or artillery out of fear of fratricide.[309] The NVA and VC developed ways of avoiding American technological strengths, including the helicopter warfare the Americans introduced as early as 1962.[310] These techniques were most effective at countering U.S. conventional units; when American forces were used in counterinsurgency roles—as opposed to the big war sweep operations—they were both more effective at killing enemy soldiers and had lower casualty rates than conventional units.[311] Their attrition strategy succeeded in part by helping persuade Westmoreland that his own strategy was working.

American leaders understood that national will was their critical vulnerability but did not arrest its decay or effectively counter North Vietnamese efforts to exploit American vulnerabilities they accurately recognized in general terms. From the Kennedy years, public and congressional support for the war was weak, leading presidents to emphasize the modest goals of their operations in Vietnam, which contributed later to credibility problems.[312] Senator Mike Mansfield told President Johnson in mid-1965 of deepening popular discontent with the war, and Secretary of State Dean Rusk told Johnson in December 1965 that his first concern was American public opinion.[313] Johnson told his staff in late 1965:

> [T]he weakest chink in our armor is American opinion . . . . Our people won't stand firm in the face of heavy losses, and they can bring down the government.[314]

And, in a message to the Joint Chiefs in January 1966, Admiral Sharp wrote that the DRV had "staged a remarkable worldwide political and propaganda campaign. They anticipate that the VC, fully supported by North Vietnam, can inflict sufficient casualties to generate internal U.S. pressures to end the war short of the objectives we seek."[315]

Johnson repeatedly displayed discomfort with the war in his diplomacy, bombing halts, and public comments. Washington approached Hanoi as early as June 1964 about a settlement but was rebuffed; the DRV's attitude hardened further after the Gulf of Tonkin Resolution passed soon thereafter.[316] Johnson expressed reluctance in the 1964 presidential campaign to send American troops to Southeast Asia and expressed regret, not determination, when he announced dispatch of the first large contingent of ground troops to South Vietnam in mid-1965.[317] On April 7, 1965, he expressed willingness to talk with Hanoi "without posing any preconditions" and offered a bribe in the form of an aid package for all of Vietnam if peace were arranged.[318]

Johnson said publicly that he would not invade the DRV, throwing away a potentially valuable negotiating point, refused to call up reserves, and for several years even declined to raise taxes to pay for the war.[319] Johnson himself counted 72 American peace initiatives and 16 bombing pauses during his administration alone—not evidence of a determined belligerent.[320] McGeorge Bundy explained much when he wrote to the president in February 1965, before the major U.S. buildup began, that his administration was hobbled by a "widespread belief that we do not have the will and force and patience and determination to take the necessary action and stay the course."[321] McNamara says he recognized Hanoi's "strategy of attriting our national will" from mid-1966.[322]

Johnson noticed and reacted to what he called the "increasingly heavy propaganda barrage from Hanoi."[323] He saw a pattern to the messages. First, Johnson believed, communist countries picked up Hanoi's messages, which then spread to the West European press and then to the U.S. domestic press.[324] He responded to the propaganda. For example, he admits making the April 7, 1965, speech noted above in response to DRV propaganda.[325] But he could not find a way to neutralize it.[326]

Sensitivity to domestic public opinion also prompted military operational decisions. Admiral Sharp says the bombing pauses were designed to assuage domestic war critics as well as to appeal to Hanoi.[327] In 1968, soon after Johnson restarted bombing North Vietnam after one of the pauses but limited attacks to the small area south of the 20th parallel, U.S. aircraft struck a target just south of the 20th parallel, leading to protests from American antiwar activists that led Johnson to further restrict bombing to the much smaller part of the DRV below the 19th parallel. Sharp says communist propaganda and domestic complaints about civilian casualties prompted alterations in U.S. targeting decisions in 1966 and 1967.[328] The already incoherent American war strategy was further shredded, distressing military leaders.

Vietnamese communists' political/military attacks on America's national will, and thereby its leadership, were successful. The north's courting of American antiwar activists helped fuel antiwar sentiment. Sensitivity to domestic opposition to the war influenced Johnson's policies throughout his presidency and drove him from office. In the congressional elections of 1972, opponents of the war, mainly Democrats, won by demanding unilateral withdrawal over the objections of President Nixon. Even before Nixon was forced from office, Congress abandoned South Vietnam by banning U.S. military support and cutting deliveries of war materiel, substantially enabling the NVA's successful 1975 offensive. Former Vietnamese communists Bui Tin and Truong Nhu Tang reasonably maintain that the DRV effectively manipulated American public opinion and that its success is evidenced by the fact that Congress became the source of U.S. military impotence.[329]

North Vietnam's ability to identify and exploit U.S. vulnerabilities was enabled substantially by unusual internal strengths in the *leadership* and *force generation* dimensions. Ho Chi Minh declared that the DRV would unify Vietnam for reasons of nationalism, not communism, and the *Lao Dong* leadership and the North Vietnamese people embraced his vision. The DRV therefore committed to a total war effort. The *Lao Dong* central committee reaffirmed its commitment to victory at whatever cost in May 1972 even as the NVA was reeling from allied counterattacks during the Easter Offensive.[330]

The party entrusted day-to-day military operations to General Giap, an unusually able and experienced soldier who led communist Vietnamese military forces as senior general and defense minister for some 30 years—an unprecedented period of senior leadership—but it also made major decisions like commitments to strategic offensives. The stable and relatively united leadership thereby took collective responsibility for strategic errors and did not look for scapegoats. Instead, Giap and the party learned from their mistakes and did not repeat them.

With these assets, the DRV built a large and capable army. PAVN strength rose from 300,000 troops in 1963, to 700,000 in 1966, and one million in 1973 even as it lost many troops in battle.[331] The DRV appears to have improved the quality of its armed forces at the same time. Effective monitoring of operations, learning from negative experiences, and effective adaptations led to many midcourse corrections—from flawed strategic decisions like the Easter Offensive to refinement of troop training regimens. The result was a high-quality force that American combat troops widely respected as a competent and tenacious enemy.[332]

U.S. LEARNING AND ADAPTATION

In Vietnam, the United States demonstrated a comprehensive lack of effective learning and adaptation and consequently poorly identified, exploited, and ameliorated vulnerabilities. Kennedy and Johnson administration officials originally were little concerned about Southeast Asia and relatively uninformed about Vietnam. But despite much attention by the massive national intelligence complex built after World War II—including the CIA, the Defense Intelligence Agency, and the National Security

Agency—senior leaders remained ill informed about Vietnam throughout the war. While the Munich analogy, containment policy, and reliance on limited war theory played roles in policy making, they had little relevance for U.S. military conduct of the war. Senior military officers, especially air force officers, instead rejected accurate intelligence information and analysis about the nature of the DRV's economy and military forces and of their vulnerability to bombing. General Westmoreland similarly seems to have disregarded information inconsistent with his attrition strategy.

Accurate understanding of some aspects of the situation by some individuals suggests that the military's collective inability to learn about Vietnam reflected an unwillingness to learn, not cognitive failure. The air force needed a DRV vulnerable to airpower to enable it to play a major role in winning the war. The other services displayed similar if less pronounced refusals to learn for reasons of organizational culture, doctrine, and institutional interest. President Johnson generally backed the generals against the CIA, which he regarded as destructively disloyal for its independent judgments.[333] And, at the tactical level, rotation policies designed to keep hardships minimal damaged U.S. military institutional knowledge and learning by shipping troops home at about the time they were beginning to develop some local expertise.

North Vietnamese security and deception operations inhibited military and civilian learning, including learning by average Americans, about the nature of the war.[334] Hanoi deceptively convinced many Americans that the NFL was an independent entity, cultivated the erroneous impression that it was receptive to peace proposals, and persuaded many that the U.S. military tried to kill civilians—all as parts of the strategy that altered American will and ultimately changed U.S. policy.

Poor learning and dysfunctional institutional traits inhibited adaptation to exploit DRV and VC vulnerabilities that did exist.[335] Successful operations in the "other war" after 1968 suggest that the outcome might have been different if Westmoreland had not chosen an attrition strategy or if Abrams had become commander sooner. Policy constraints also played a major role. Johnson's limits on the war effort sharply constrained senior military commanders' options. These narrowed further as the services refused to refocus force structures and doctrine from those applicable to their preferred European/Warsaw Pact operational scenario. Services' fights with each other and civilian agencies over turf hampered initial development of coherent strategy and bureaucratic squabbling largely precluded timely and effective midwar adaptations.

The United States similarly failed to ameliorate vulnerabilities many Americans recognized—especially in the domestic political arena. Johnson never made a persuasive case for the war, and his deference to the military on bombing and firepower-heavy operations that alienated the South Vietnamese and American people and gave massive quantities of useful propaganda material to the DRV meant he failed to ameliorate political vulnerabilities he generally understood. In this respect, Johnson's fault was not the micromanagement the military accuses him of, but rather inadequate supervision of military operational practices that abetted the strategy the DRV used to defeat the United States.

In sum, the United States failed to adequately identify, exploit, and ameliorate vulnerabilities for many reasons that mainly reflected institutional factors. While symptoms of ignorance appear in the record regularly, their underlying causes were institutional, not information inadequacies, resource shortages, cognitive failures, or inherently incapable organizations.

SUMMARY

DRV exploitation of U.S. sensitivity to civilian and U.S. military casualties, combined with U.S. incompetence, produced a complete strategic victory over a far more materially wealthy and, by conventional military standards, more powerful United States. The DRV attacked U.S. national will with a sophisticated, integrated political/military campaign that drove Johnson from office and conquered Congress, which then defeated Presidents Nixon and Ford and the military establishment. North Vietnam's strategy as a whole was well conceived and executed. It learned from some serious errors and adapted well; the United States did not. American material preponderance and technological prowess were largely useless because they were ill-suited to the operating environment; used inexpertly for local conditions; hobbled by hubris, dysfunctional personnel policies and a lack of coherent strategy; and restricted by ethical concerns and a lack of will.

U.S. domestic sensitivities to U.S. military and foreign civilian casualties were so obvious that they altered the way the United States, and its friends and adversaries, conducted political/military operations thereafter. They set the stage for a new development in U.S. military history—internal military sensitivity to U.S. military casualties so severe that it affects operations. These factors appear repeatedly in the next three chapters.

The Vietnam case generally confirms Shimshoni, Biddle, Merom, and Mack. The case dramatically confirms Shimshoni's emphasis on the importance of leadership. American leaders were inept at strategy formulation and tactical execution. Vietnamese communist leaders, while making some mistakes, learned from them and eventually came up with, and implemented, a winning strategy. The United States demonstrated considerable military capabilities in the conventional wars in 1941–1945 and 1950–1953, but it failed badly in this unconventional war. Biddle's modern system did not apply in Vietnam because the North Vietnamese and VC refused to fight a conventional war, sharply degrading the value of hardware and skills designed for modern conventional warfare.

Merom is supported. It was ideas and the pain of casualties that drove the United States to quit. The war was after 1945—when he argues such sensitivity became operationally relevant. Mack's argument focuses on the asymmetries and political vulnerabilities in this war, and my assessment is similar to his. But U.S. strategic political and military incompetence generated many of the events that drive his analysis. One can only wonder what would have happened if American leaders had been only nearly as competent as Ho, Giap, and their comrades and designed a war plan to defeat the north, not just react to it.

CHAPTER 5

NATO's War against Yugoslavia (1999)

President Bill Clinton took the United States and NATO to war against Yugoslavia in March 1999 because he opposed Yugoslav military and Serbian internal security forces' killings of ethnic Albanian civilians in the Serbian territory of Kosovo.[1] But he was willing to do so only in ways that minimized risks to American troops, reflecting concern for what the U.S. military called "force protection." Both civilian casualty aversion and force protection had by this time achieved the status of norms.[2] The result of these motives and constraints was a highly unusual NATO force structure, severely restrictive ROE, and operational practices inconsistent with U.S. military doctrine. These created tactical military and strategic political vulnerabilities that NATO's Yugoslav enemy and its putative friend, the Kosovo Liberation Army (KLA), exploited effectively in different ways. NATO prevailed, but so many problems arose that Ivo Daalder and Michael O'Hanlon rightly characterize the result as "winning ugly."[3]

It was an odd war. Clinton launched it for ethical reasons over the objections of senior military leaders and many members of Congress. Ideational, not traditional material, concerns dominated the debate over war policy. Clinton's clumsy coercive diplomacy aimed at Yugoslav President Slobodan Milošević failed, forcing escalation into a real war Clinton initially did not want and that the U.S. military had not prepared to fight.[4] While European states contributed significantly, key leaders, operational doctrines, planning, and most of the forces employed in what NATO called Operation Allied Force were American.[5] I concentrate in this chapter on American conduct of the war.

For the United States, civilian casualty aversion became a major domestic political and military operational issue during the Vietnam War when, as discussed in chapter 4, North Vietnam skillfully used collateral damage incidents against the United States, driving U.S. commanders to limit civilian deaths via restrictive targeting and ROE that dramatically affected air operations, especially.[6] It had been a chronic, major U.S. military operational issue since, as Clinton knew.

It also was driven by major international concerns. Several NATO member states deployed troops to the former Yugoslav republics of Bosnia and Herzegovina, and Croatia, in 1992–1995 as part of the UN Protection Force (UNPROFOR) that strove to protect civilians from bloody ethnic warfare. The Yugoslav wars of dissolution strengthened international norms against crimes against civilians by, among other things, prompting establishment of the International Criminal Tribunal for the Former Yugoslavia (ICTY). Popular European approval of UNPROFOR operations and reactions in 1998 and 1999 to modest numbers of civilian deaths in Yugoslavia indicate that the norm was stronger in Europe than in the United States.

Force protection as a normative imperative in the United States dates from the 1970s when U.S. military leaders, in reaction to defeat in Vietnam, took steps to protect themselves from what they perceived to be unfair treatment by politicians.[7] Force protection is an imperative to protect troops from harm as a stand-alone objective, even at the expense of mission accomplishment.[8] It is not synonymous with commanders' efforts to minimize casualties while ensuring mission accomplishment—sound resource management that good leaders long have used.[9] It also is not the long-standing U.S. practice of using firepower as a partial substitute for manpower; commanders throughout American military history before the Gulf War of 1991 consistently accepted casualties if necessary to accomplish their missions.

Force protection also had some normative strength in European polities and militaries in 1999. For example, Rwandan extremists in April 1994 murdered 10 unarmed Belgian soldiers at the start of their genocide, calculating correctly, in part from their assessment of the then-recent U.S. withdrawal from Somalia, that Belgium similarly would withdraw its troops from the UN mission in Rwanda if some of its soldiers died, thereby simplifying the task of perpetrating genocide.[10] Belgium's withdrawal indeed wrecked the UN mission, leaving little international military presence in Rwanda as several hundred thousand people were murdered. In May 1995, British Army General Rupert Smith, commanding UNPROFOR in Bosnia and in a confrontation with the Bosnian Serb Army, received orders from UN headquarters to back off along with these words: "The execution of the mandate is secondary to the security of UN personnel," which is a classic statement of the force protection imperative.[11] Still, European electorates were willing to take casualties on the UNPROFOR mission, which suffered 213 fatalities—mainly French and British soldiers.

By 1999, then, it was clear that both casualty-related norms affected Western military operations, often simultaneously. As demonstrated in Bosnia, civilian casualty aversion acted in two ways that were related but analytically distinct. First, it drove states to act to save lives in the absence of material motives—exhibiting what later became known as the "responsibility-to-protect" (R2P or RtoP) norm. Second, once operations were underway, it restricted the actions of troops to ensure that civilians were protected as much as possible, sometimes even to the extent of abandoning the long-standing recognition that military necessity may make civilian deaths inevitable. Influence mechanisms were through domestic political and international legal channels such as ICTY.

Force protection concerns influenced state policies through domestic politics and the bureaucratic politics of military institutions. The U.S. military had become a major domestic proponent of force protection and promoted it through the long-standing American tradition whereby civilian leaders defer to generals on the details of military operations, in appeals to sympathetic congressmen and the public, and through its ability to strongly influence NATO operations.

And, not least, belligerents and third parties had learned to use casualty aversion norms instrumentally to entice adherent states to act, or not, by strategically manipulating perceived state interests, broadly defined. Vietnamese communists demonstrated effective norm-based techniques against France and the United States. In 1999, materially weak Yugoslavia and the much weaker KLA exploited norms in different ways in another kind of war.

PRELUDE TO OPERATION ALLIED FORCE

Clinton displayed clear preferences to avoid both U.S. military and civilian casualties in earlier presidential decisions. Early in 1993, he launched an ambitious multilateral nation-building effort in Somalia that used military forces in humanitarian roles despite growing congressional reservations as the mission degenerated into a fight with local warlord Mohammad Farah Aidid. A raid in Mogadishu that captured aides of Aidid but led to the deaths of 18 U.S. soldiers in fighting with Aidid's militia on October 3–4, 1993, prompted angry congressional reaction, and Clinton on October 7 announced withdrawal of U.S. forces by March 31, 1994—the earliest date military logisticians told him a withdrawal could be accomplished.[12] Clinton says in his memoirs that the battle "haunted" him and that he withdrew U.S. troops because he did not believe the "benefits of capturing or killing Aidid justified further loss of life on either side."[13]

Under domestic political pressure in 1993 to counter surging immigration from Haiti and to restore a deposed president to power, Clinton encouraged Haiti's military government to step down. The United States eventually engineered peaceful regime change and deployed a force that ultimately accomplished little of substance. Of importance for this story is the arrival of the USS *Harlan County* at Port-au-Prince a week after the Mogadishu debacle. The ship carried American and Canadian troops who planned to train Haitian soldiers and policemen under UN mandate. Lightly armed Haitian toughs blocked the ship's access to a pier, yelling "Somalia! Somalia!" The captain of the armed, tank-landing ship turned around without debarking his passengers and cargo, scuttling the mission.[14] The episode was another in a series of cases in which even threats of violence against U.S troops changed U.S. foreign and military policies.

Clinton inherited a U.S. presence in, or rather over, Iraq as a legacy of the Gulf War, but his administration quickly established its own policies. To pressure Iraqi President Saddam Hussein, American and British aircraft patrolled Iraq's "no-fly zones" and periodically bombed defense-related installations and equipment. Iraq sought to protect sensitive equipment like aircraft by placing them near facilities protected under most circumstances by international law and by local norms—like mosques. The United States responded by, among other things, developing precision-guided small bombs and concrete bombs that destroyed targets by impact alone, sparing nearby facilities and civilians from the blast and shrapnel of exploding bombs.[15] That is, the U.S. military changed its force structure, in this case through technological innovation, in response to enemy attempts to exploit U.S. commitment to international norms, in this case the Geneva Conventions. More generally, a dynamic interaction featured evolving Iraqi tactics designed to exploit U.S. norm-based political vulnerabilities and to achieve some protection for military assets, which in turn produced changing U.S. targets, ROE, and force structures, and then more Iraqi innovations.

The United States regularly in the 1990s bombed Iraq but it refrained from ground operations, thereby protecting U.S. troops. In December 1998, in a confrontation with Iraq over weapons of mass destruction (WMD) inspections coincident with Clinton's impeachment crisis, Operation Desert Fox sought to "degrade" Iraqi military capabilities with air and missile strikes over a period of less than four days.[16] Clinton also used cruise missiles to attack an alleged chemical weapons plant in Khartoum, Sudan, and al-Qaeda training sites in Afghanistan in August 1998. Al-Qaeda criticized the United States for its alleged cowardice in making unwarranted attacks that killed civilians without endangering Americans—a charge it would make again in 1999.

Candidate Clinton in 1992 chided President George H. W. Bush for inaction in then-disintegrating Yugoslavia, but as president, Clinton criticized the performance of NATO allies who provided most of UNPROFOR's troops while refusing to commit U.S. ground troops to peacemaking activities out of concern for U.S. casualties. Clinton in 1993 infuriated the allies by pushing for a "lift and strike" policy—*lifting* the arms embargo against Bosnian Muslims and hitting the Bosnian Serb army with air *strikes*—ignoring allies' concerns that the policy would make their troops, often dispersed in small groups, vulnerable to retaliation. While conducting some air strikes in 1994–1995, Clinton refused to commit ground troops until a peace agreement was reached; American diplomat Richard Holbrooke's account of the diplomatic process that ended the war in Bosnia, published in 1998, confirmed that Clinton's heart was not in a ground war in the Balkans.[17]

Nevertheless, Clinton and some of his advisors, especially Secretary of State Madeleine Albright, shared a keen dislike for violent "ethnic cleansing" that Serbs perpetrated in Bosnia and Croatia in 1992–1995 and again in Kosovo in 1998 and 1999.[18] This amounted to display of a variant of the civilian casualty avoidance norm given that Serbs systematically used murder and rape, as well as property crimes, to encourage Kosovar Albanians to leave their homes. Clinton and Albright also personally disliked Milošević, who they held responsible for many Serbian misdeeds during the earlier Balkan wars of the 1990s.

FORCE PROTECTION AND U.S. MILITARY CULTURE: A LEGACY OF VIETNAM

In 1999, U.S. military culture had been evolving for two decades in ways that produced a strong internal emphasis on force protection. Most senior officers of the 1990s had fought in Vietnam and were strongly influenced by the experience.[19] While some understood that the military made crucial mistakes, a dominant lesson of Vietnam was that civilian leaders, inappropriately micromanaging from Washington, abused the military by bleeding it badly in an ill-conceived, losing, unpopular cause—thereby immorally and needlessly killing American troops.[20] General Creighton Abrams, who became army chief of staff after he returned from Vietnam in 1972, reorganized the army to place most support functions in reserve components (the National Guard and Army Reserve), ensuring that the army could not undertake a major operation without activating reservists, which Abrams thought would protect active component personnel.[21] In 1984, Vietnam veteran Major General Colin Powell was military assistant to Defense Secretary Caspar Weinberger who, in the wake of the bombing of the barracks in Beirut that killed 241 U.S. troops on an ill-defined mission that reminded many of Vietnam, outlined restrictive circumstances that should exist before U.S. troops are committed to combat.[22] At the time, Powell worried that enemies looking for vulnerabilities in U.S. policy, or what he called "loopholes," might exploit Weinberger's criteria, but he eventually concluded Weinberger was right.[23]

Powell, who became chairman of the JCS in 1989, added a time element to Weinberger's perspectives that some saw as a pre-deployment requirement for an "exit strategy."[24] U.S. military personnel embraced his views about the appropriate use of force, which entered U.S. military culture as the Powell Doctrine or, sometimes, the Powell Corollary to the Weinberger Doctrine.[25] In practice, Weinberger/Powell criteria limited use of the U.S. military mainly to conventional wars the United States could win quickly with minimal U.S. casualties—a superficially reasonable principle that justified few operations and strongly encouraged force protection in Department of Defense policy and military culture.[26]

In 1990, Powell opposed war to liberate Kuwait because he feared U.S. casualties that planners estimated could run into the thousands; theater commander General Norman Schwarzkopf also worried about friendly casualties.[27] President Bush and Defense Secretary Richard Cheney overruled Powell, but they honored his request for a large troop deployment and use of overwhelming force designed to protect American lives.

Powell's philosophy influenced Gulf War operations.[28] Air component commander Lieutenant General Charles Horner ordered his airmen to fly at medium altitude, beyond the range of Iraqi antiaircraft guns, to protect aircrews despite the loss of bombing accuracy.[29] Losses of as few as two aircraft, as two A-10s lost on February 15, 1991, led commanders to reduce attacks and to order pilots to fly higher; the *Gulf War Air Power Survey*, the official after-action report on the Gulf air war, noted "a sensitivity to loss having few precedents in American military history."[30] Under U.S. civil/military tradition, uniformed commanders run day-to-day operations under general guidance from civilians, meaning the military crafted operations designed to minimize friendly casualties. That is, force protection concerns, which were increasingly internalized in U.S. military culture as a norm, produced in 1991 standard operating procedures designed to protect military lives independent of civilian guidance.

Thus, the core U.S. military motive for force protection was moral, although military personnel also used it instrumentally to protect institutional and personal interests. Force protection as stand-alone mission predated the expansion of U.S. peacekeeping operations in the 1990s, which usually involved little danger to U.S. troops, and the demise of the USSR.

The United States also showed aversion to civilian casualties far greater than it displayed in Vietnam. When, on February 13, 1991, U.S. aircraft struck the Al Firdos command bunker in Baghdad, killing several hundred civilians and prompting unfavorable press coverage, the White House expressed concern. Powell told U.S. Central Command (CENTCOM) that thereafter he personally would approve strikes on Baghdad and few would be allowed; CENTCOM planners complained that the move would aid Iraqi propagandists, but Powell believed that more such incidents would shatter the coalition.[31] Attacks were halted for four days, and thereafter Baghdad was struck only five times.[32] A key point again was made: civilian casualties generate changes in U.S. military behavior. When Schwarzkopf declared all military objectives achieved, Bush ended the war to prevent criticism that the United States needlessly killed even Iraqi soldiers after victory was won even though enough Republican Guard units escaped to help keep Saddam Hussein in power—leading many to see the victory as incomplete.[33] Saddam's political survival set the stage for the low-level air war of the 1990s and renewal of major fighting in 2003.

The June 25, 1996, bombing of the Khobar Towers barracks in Dhahran, Saudi Arabia, which killed 19 U.S. airmen, further strengthened the force protection norm. Secretary of Defense William Cohen asked retired army General Wayne Downing to investigate the incident. Downing blamed the senior U.S. Air Force officer in Saudi Arabia, Brigadier General Terryl Schwalier, even though he bore no direct responsibility for a truck bomb blast of then-unprecedented size. Cohen had approved a promotion for Schwalier but rescinded it after receiving Downing's report, prompting Schwalier to retire. In a rare public protest by a senior officer, the air force's chief of staff, General Ronald Fogelman, took early retirement in part over what he considered unfair treatment of Schwalier.[34] The Cohen and Fogelman decisions became widely known among military personnel; the lesson for many officers was that U.S. casualties under most circumstances would end the career of a commander whose personnel died.[35]

U.S. MILITARY DOCTRINE: INSTITUTIONALIZING CULTURAL CHANGE

The reaction of U.S. military personnel to the Schwalier decision was consistent with ongoing policy and doctrinal changes. U.S. Air Force Colonel John Warden, an architect of the 1991 air campaign plan, in 1992 argued that the Gulf War air campaign offered a model for future operations because no American government, except in dire circumstances, would accept more than a "handful" of casualties.[36] The 1993 version of the U.S. Army's premier doctrinal document on operations, Field Manual 100–5, states, "The Army must be capable of achieving decisive victory," with "decisive victory" meaning an ability to "win quickly with minimum casualties."[37] The secretary of defense's 1995 *Annual Report* states, "When the call comes, the nation wants its armed forces to be ready to respond, to succeed quickly, and to suffer the fewest possible casualties."[38] In 1996, Defense Secretary William Perry issued Directive 2012, which ordered the joint chiefs to combat terrorism, which he defined as attacks on U.S. troops. The army thereafter required force protection plans for overseas deployments. The U.S. European Command, which would provide forces to fight Operation Allied Force, said its primary objective was "To Protect and Take Care of the Force."[39] Brigadier General Richard Coleman, chief of air force security forces, in 1997 said, "Force protection has become the Air Force's highest priority . . . conducting that mission now is as important as projecting our combat power."[40] While senior officers like chairman of the JCS General John Shalikashvili warned that casualties still would occur, the Defense Department and uniformed leaders were trying to change a military culture that long had sought to limit casualties—but only in a context in which mission accomplishment was the undisputed first priority.[41]

Some military officers noted the change in emphasis with concern. They warned that robust force protection policies, rather than saving lives, encouraged attacks on U.S. interests, compromised mission accomplishment, and damaged military ethics. A U.S. Navy captain warned that potential adversaries would not miss doctrinal statements that America cannot stomach more than minimal casualties.[42] An air force major argued that Warden's views about airpower as a casualty minimizer had been widely cited at the Air Command and Staff College and could not have been missed by the foreign students there.[43] Air force lawyer Charles Dunlap said "Uniformed professionals need to ask themselves whether the military's altruistic ethos, axiomatic to its organizational culture, is being replaced by an occupationalism that places—perhaps unconsciously—undue weight on self-preservation over mission accomplishment."[44] Retired Colonel Don Snider, a professor at the U.S. Military Academy, summarized the concerns by observing that placing the safety of military personnel above that of civilians they vowed to protect had serious, negative implications for military ethics. Said Snider, the professional ethic "is built on the principles of self-sacrifice and mission accomplishment. Troops are supposed to be willing to die so that civilians do not have to." At the troop level, U.S. soldiers in Bosnia—burdened with body armor and holed up in fortified camps to enhance force protection—referred to themselves as "Ninja Turtles" who were "prisoners of peace."[45]

U.S. allies had similar concerns, generating tensions in Bosnia and at diplomatic levels.[46] In Bosnia in 1996, a Polish officer told a U.S. Army officer "you wear all that gear [body armor] because you are afraid, but don't worry, we'll protect you."[47] In 1999, a senior British officer, talking about the U.S. Army's Camp Bondsteel in Kosovo but reflecting views common in Bosnia before Operation Allied Force about similar U.S. force protection policies, argued that Bondsteel, a fortified mini-U.S. city with a Burger King restaurant, was a "Disneyland" that kept U.S. troops away from local people and detracted from the peacekeeping mission.[48]

FORCE STRUCTURE: THE NARROW AIR AND MISSILE FORCE

Precision-guided munitions (PGMs) encouraged civilians and the military to expect better operational compliance with both casualty avoidance norms and significantly influenced the U.S. force structure used against Yugoslavia. Technological advances since the 1970s enabled much more accurate and reliable U.S. air and missile strikes than were possible in Vietnam, contributing much to the victory and low coalition military and Iraqi civilian casualties in the Gulf War. By 1999 new technologies, especially

Global Positioning System (GPS) navigation satellites, further improved the accuracy of munitions, again reducing risks of collateral damage and enhancing force protection by enabling aircraft to release munitions far from enemy air defenses.[49] Unmanned cruise missiles eliminated risk to pilots.

The role of PGMs in 1991 placed airpower advocates in strong bureaucratic positions in internal Pentagon battles over budgets and operational planning, and air and missile forces dominated U.S. combat operations during Clinton's presidency. As the negative operational consequences of civilian casualties apparent in 1965–1973 and 1991 continued to grow, lawyers gained greater authority over targeting and ROE decisions, prompting military wags to quip that warfare had devolved into *lawfare*. Intelligence became more important because accurate data on the locations of targets was critical to programming precision munitions that usually hit spots targeteers wanted to hit. In early 1999, no U.S. ground troops had purposefully been committed to combat since 1993 and airpower partisans ridiculed ground forces, especially armored units, as Cold War relics.

YUGOSLAVIA'S INSTRUMENTAL USE OF NORMS

Yugoslavia's leaders understood U.S. war-fighting practices and recognized Clinton's aversion to American casualties. Some cited Clinton's reaction to the 1993 Mogadishu battle as key to understanding U.S. foreign policy.[50] Yugoslav leaders could hardly have missed comments by actual and potential U.S. enemies to the same effect. The commander of Iran's Revolutionary Guards in 1995 claimed he could take 20,000 U.S. prisoners if America invaded Iran but said Washington would be at the negotiating table long before that point.[51] In 1996, Chinese leaders who received a warning from Assistant Secretary of Defense Chas Freeman that the United States would respond militarily to a Chinese attack on Taiwan retorted, "No you won't. We've watched you in Somalia, Haiti and Bosnia and you don't have the will."[52] Osama bin Laden, Syrian President Hafez al-Assad, and Saddam Hussein made similar comments.[53] Milošević attributed similar attitudes to Europeans, telling German Foreign Minister Josckha Fischer, "I can stand death—lots of it—but you can't."[54]

Yugoslavia recognized U.S. reluctance to use ground troops in the Balkans. After NATO aircraft bombed Bosnian Serb positions around Goražde in April 1994, Mihajlo Marković, a leader of Serbia's ruling Socialist Party, told a television audience not to worry about an American ground involvement, that Clinton had domestic problems and "he can't afford to have even a few soldiers killed in Bosnia."[55] Given American preferences for airpower, Yugoslavia asked Iraq for lessons of the 1991 Gulf War and the then-ongoing, low-level U.S./UK air campaign against it. Such collaboration was not new. Yugoslav firms helped build Iraq's air defense system before 1991.[56] Milošević also received warnings personally from Ambassador Holbrooke, NATO's military commander General Wesley Clark (U.S. Army), and Lieutenant General Michael Short (U.S. Air Force) about the consequences of noncompliance with NATO demands.[57]

Yugoslavia therefore had plenty of information upon which to estimate how NATO would attack. Milošević's initial judgment evidently was that NATO was bluffing or at most would launch a few nuisance air or missile raids, declare success, and quit shooting—the pattern of cruise missile diplomacy Clinton displayed repeatedly and roughly what NATO then in fact planned, as a Yugoslav agent at NATO headquarters apparently reported to Belgrade.[58] The Yugoslav military undoubtedly made plans based on U.S. air doctrine and targeting practices, the operationally relevant, air-dominant U.S. force structure in the region, and obvious American preferences for air rather than ground operations.[59] An accurate Yugoslav threat assessment required understanding *both* the material assets NATO likely would use and normative limits on NATO forces' operational practices—which each reflected both casualty aversion norms. Neither alone provided adequate information for Yugoslav planners. Both enabled development of plans for asymmetric operations that could both exploit NATO vulnerabilities and ameliorate Yugoslav vulnerabilities to NATO's technological superiority and material preponderance.

OVERVIEW OF THE WAR

The Kosovo crisis developed early in 1998 after the outbreak of an insurgency led by ethnic Albanians of the KLA, who sought independence for Kosovo.[60] After the KLA killed some policemen, Serbian security forces badly bloodied the KLA and committed atrocities against Kosovar Albanian civilians.[61] Thousands fled their homes. Given previous Serbian mistreatment of Croats and Bosnian Muslims, and Western frustration with Milošević, NATO members aggressively attacked Yugoslavia diplomatically. On March 31, UN Security Council Resolution (UNSCR) 1160 condemned Serbs' attacks on civilians. On September 23, with winter approaching, UNSCR 1199 condemned the impending humanitarian disaster awaiting displaced, homeless Kosovars and called on both sides to stop fighting and start talking. The next day, NATO ambassadors authorized air strikes.[62]

The crisis came to a head on January 15, 1999, when Serbian security forces and the KLA fought a minor battle at the village of Račak. After the fighting ended, security personnel apparently killed some noncombatant Kosovar Albanians.[63] William Walker, a retired American ambassador and head of the Kosovo Verification Mission of the Organization for Security and Cooperation in Europe, immediately blamed the Serbs publicly without waiting for a detailed investigation. Clinton and Albright had enough; they began to take NATO to war.[64] Since early 1998, about 2,000 people had been killed—many of them combatants in what had become a real war.[65]

While tragic for the unfortunate, the toll was small in the context of modern civil wars, including the Bosnian war that killed roughly 100,000 people.[66]

With Russia and China opposed to UN Security Council enforcement action, NATO summoned the Yugoslav government and Kosovar Albanians to two rounds of talks in Paris and at Rambouillet chateau near Paris in February and March 1999. Negotiation was minimal. NATO effectively presented the parties a document to sign.[67] While annoying the Kosovars by not granting them immediate independence, the draft angered Belgrade by calling for political autonomy for Kosovo—a region important to Serbian nationalists and to which Milošević as a Yugoslav politician had devoted much attention.[68] When Milošević balked at signing NATO's document and Serbian security forces launched new attacks in Kosovo, NATO's first war was on.[69]

The United States (and NATO) went to war to save Kosovar Albanian civilian lives; restoring the autonomy Milošević revoked in 1989 was a secondary goal.[70] That is, the civilian casualty avoidance norm drove NATO to attack Yugoslavia, not a desire to reverse Milošević's tyranny or to achieve material gain. Its advocates claimed it was the first war fought solely for humanitarian reasons. British Prime Minister Tony Blair on April 22, 1999, opined:

> This is a just war, based not on any territorial ambitions but on values. We cannot let the evil of ethnic cleansing stand. We must not rest until it is reversed. We have learned twice before in this century that appeasement does not work. If we let an evil dictator range unchallenged, we will have to spill infinitely more blood and treasure to stop it.[71]

But the noble deed was to be done under highly restrictive conditions. Before the bombing started, Clinton expressed willingness to contribute a peacekeeping contingent but said, "I do not intend to put our troops in Kosovo to fight a war."[72] Defense Department spokesman Kenneth Bacon reiterated the point, saying, "The United States has no intention of sending troops to fight in Kosovo, and the Department of Defense is not doing any planning that would enable such a deployment."[73] Even after the initial bombing failed to persuade Milošević to concede, there was no planning for an invasion even though ground troops deployed to Albania and Macedonia.[74] Clinton says he thought air strikes would produce fewer civilian casualties than a ground invasion while also saving American lives.[75] NATO Secretary General Javier Solana in March nevertheless told General Clark to develop an "assessment" of ground options; Clark and Blair argued for ground war planning but Clinton and the Pentagon resisted.[76]

Clinton thus quickly had a big problem. He had assumed that Milošević would concede after a few air strikes—as he did during the Bosnian war when the West intermittently showed determination.[77] But Milošević hung tough over core issues like Kosovo's role in Serbian nationalism and threats to Yugoslav sovereignty—and undoubtedly also because he knew NATO planned no major war.[78] Once the short-war assumption was disproved, NATO had no military plan and no plan to aid Kosovar refugees.[79] Clinton says his fallback goal was to "seriously damage the Serbs' military capacity"—similar to the vague goal of U.S. raids on Iraq the previous December—while Albright expected failure of coercive air strikes to lead to a "mess."[80] Clinton's strategy was incoherent, but Albright's analysis was accurate.

Because Clinton would not risk American lives, Clark had to make do with a makeshift, air-only campaign in violation of U.S. military doctrine that calls for "joint" operations—use of the broad range of abilities of all military services, which complement each other. To keep crews safe, planes were to fly above 15,000 feet in altitude and at night under most circumstances.[81] To keep civilian casualties to a minimum, ROE were restrictive. Pilots were to positively identify targets before releasing ordnance and were not authorized to attack Yugoslav military vehicles intermingled with civilian vehicles.[82]

Clark dealt regularly with civilian leaders and knew the alliance's unanimity rule made it essential to keep member governments informed and content, if not happy, with the conduct of the war. He knew France, Germany, Italy, and Greece had major misgivings about the war while others also had concerns, meaning he had to keep them mollified enough to refrain from vetoing key operations or Operation Allied Force as a whole.[83] NATO governments recognized the dangers of disunity and excessive oversight and delegated primary responsibility for managing the war to five states—the United States, United Kingdom, France, Germany, and Italy—which monitored operations and approved target proposals as expeditiously as politically possible. Target lists nevertheless required as many as nine separate approvals.[84] Lawyers on the Pentagon's review team vetoed one proposed target due to risks of collateral damage.[85] Political restrictions were so tight that Clark sometimes had no new targets to attack.[86]

NATO conducted phased operations. First, air defenses and command-and-control bunkers were hit, with the aim of securing the skies for allied flyers.[87] Second, ground military targets—army, police, and paramilitary units—below the 44th parallel (including Kosovo) were attacked. Third, and later, leadership targets like government buildings and economic infrastructure targets like power stations were struck. The first priority had operational value—air superiority enables aircraft to fly lower and be more effective against small, mobile targets like infantrymen—but it mainly was a force protection measure. The third contradicted NATO's claim that it was not targeting the Yugoslav people.[88]

The air war started slowly in terms of numbers of sorties flown and targets struck—consistent with both sides' short-war expectations. The Yugoslavs, having studied how U.S. air forces suppressed enemy air defenses in previous wars, used their antiaircraft weapons sparingly and moved them often.[89] As a result, the allies never neutralized Yugoslav air defenses. Throughout the war, NATO respected residual Yugoslav air defenses and kept its planes flying high—with negative consequences for collateral damage and its ability to attack Yugoslav ground forces and Serbian security personnel. Several NATO members, including France, argued that aircraft should fly lower in order to improve the effectiveness of air strikes but were overruled by U.S. commanders concerned primarily about force protection.[90]

The force protection focus of Phase I strikes meant that Serb security forces in Kosovo initially were untouched. When Milošević unleashed Operation Horseshoe, a large-scale effort to expel ethnic Albanians from the province begun shortly before the bombing started and accelerated after bombing began, NATO aircraft were nowhere to be seen.[91] Clark knew before the war started that there was little NATO could do to stop Serbian ethnic cleansing in Kosovo.[92] The lack of a ground presence, force protection–motivated air operational practices, and targeting policies ensured that the key war aim—protection of Kosovar Albanian civilians—would fail at the outset.[93] NATO attacked Yugoslav military forces in Serbia proper while Serbs attacked Kosovar Albanians in Kosovo, effectively delinking NATO's military actions and political goals.[94]

The absence of NATO ground attacks meant that Yugoslav forces could disperse without fear of piecemeal defeat by stronger ground units—just what they wanted to do anyway in order to expel the scattered Kosovars and to avoid attack from above. Yugoslav troops mingled with civilians, making them human shields that provided a degree of invulnerability, given NATO ROE, not defensibility. NATO pilots flying high could see Yugoslav ground forces only when they concentrated to fight large KLA units. Dispersion, vacant civilian structures to hide in, and a lack of detailed target data from reconnaissance troops on the ground, who the Pentagon barred from deployment out of force protection concerns, aided Yugoslav concealment and deception operations. With poor tactical intelligence NATO destroyed many "artillery pieces" made of logs and old truck wheels and "bridges" made of plastic sheet. NATO inspections after the war found little damaged Yugoslav military gear.[95]

Yugoslav personnel losses remain unclear, but at war's end Milošević said 462 soldiers and 114 policemen were killed in action.[96] Military analyst Anthony Cordesman estimates 1,000 to 1,500 were killed. Presumably the KLA killed many of them.[97] Whatever the toll, Yugoslav military and Serbian security forces remained coherent organizations throughout the war—quite a feat for a third-rate power fighting the world's premier military alliance, but one aided considerably by NATO's casualty aversion-driven policies.[98]

The exclusive use of airpower meant that General Short, who was the air component commander of NATO's southern flank, was effectively the operational commander of the campaign. Short and Clark differed sharply over issues at the intersection of the casualty avoidance norms, NATO targeting policies, and U.S. air doctrine.[99] Short advocated large-scale bombing of enemy political and military "centers of gravity" to achieve rapid victory.[100] Reflecting U.S. airmen's conventional wisdom, Short criticized NATO's European civilian leaders for meddling in the military's business; charged that Clark as a soldier did not understand air doctrine, introducing interservice rivalry to the policy debate; worried that ROE excessively endangered his flyers; and claimed that heavy blows to Yugoslavia's political leadership, by "going downtown" to Belgrade to attack "the head of the snake on the first night" would end the war quickly without having to do "tank plinking" in Kosovo.[101] In short, Short did not want to attack Serbian forces in Kosovo who were conducting atrocities of the sort that led Clinton to go to war, and who Clark wanted to attack. Short argued instead that attacks on targets close to Milošević's heart, which Short effectively claimed he knew, would cause Yugoslavia to quit.[102] U.S. strategic bombing doctrine, while recognizing the political problems civilian casualties cause, considered societal pain useful if it leads to decisive pressure on leaders.[103] This view differed from European NATO states' civilian leaders' hopes to spare Yugoslav civilians even inconvenience.[104]

The 15,000-foot rule contributed to collateral damage incidents. In one especially graphic episode, on April 12 a U.S. pilot launched a television-guided missile at a railroad bridge from 20 miles away, only to see (and record for public replay) a passenger train begin to cross the bridge just before impact.[105] Ten civilians died. Had the pilot flown closer, it is hard to imagine that he would not have seen an object as large as a train as it approached the bridge.

The Clinton administration, like its predecessors in 1965–1968 and 1991, altered bombing targets and ROE in response to civilian deaths. After a civilian convoy was bombed on April 14, Short reduced the minimum altitude for strike aircraft to 8,000 feet—still above most antiaircraft guns' range and too high for good visual identification of ground targets—and 5,000 feet for air controllers.[106] The White House forbade use of cluster bombs after they killed 15 civilians in Nis on May 7.[107] Attacks on Belgrade were suspended for a time after bombs struck the Chinese embassy, also on May 7.[108] Short complained that changes in ROE that restricted strikes on bridges to a few hours per day, late at night when few civilians were traveling, endangered his men.[109]

In May, as the air war floundered and the number of civilian casualties grew, NATO officials and several European states pressed harder for planning for a ground invasion. As Milošević continued to resist and popular support for the war in NATO states declined, Clinton moved toward the ground war option that surely would have been unpopular in the United States and led to U.S. casualties. But as tensions within the alliance grew, preserving NATO became the major goal of the war. Clinton belatedly said the political costs of losing the war were greater than those of suffering casualties and on May 18 indicated that a ground option was open.[110] European states also compromised to ensure alliance unity.[111] At roughly the same time, senior Russian diplomats and Finland's President Martti Ahtisaari accelerated the international effort to convince Yugoslavia to yield.

Milošević capitulated in early June—after receiving better terms than he rejected in March—for reasons that remain unclear.[112] Airpower partisans claim the bombing finally worked.[113] But others more persuasively emphasize loss of Russian diplomatic support, fears of domestic unrest, fears for his political position, growing KLA strength, NATO resolve and increasing pressure for a ground war from some NATO quarters, especially London, which volunteered 50,000 troops—half the British army.[114] If the latter is true, Clinton's force protection–motivated rejection of the ground option, which reduced NATO's coercive credibility through May, may have extended the war or even persuaded Milošević to fight in the first place.

Assessing the Operational Outcome

Given the American focus of this work, I mainly assess Operation Allied Force according to three U.S.-centric criteria:

1. The extent to which U.S. objectives were met. For Operation Allied Force, these goals are found in the statements of President Clinton, in NATO objectives, and in the draft Rambouillet agreement that Yugoslavia rejected, triggering the war. General Clark's operational goals embedded those of his civilian bosses.
2. Time. Strong forces usually do better when they win quickly. Guerrilla theorists and others argue that weak forces win simply by surviving.[115] U.S. military doctrine and culture called for achieving victory rapidly.
3. Cost defined as NATO military and Yugoslav civilian fatalities.

At two levels, Operation Allied Force was a success. NATO got most of what it demanded at Rambouillet and the alliance held. But for a variety of reasons, no major commentator gives NATO high marks for planning or executing Operation Allied Force—except in the arena of force protection.[116] The convoluted target approval practices violated U.S. Air Force doctrine.[117] U.S. Air Force Colonel Charlie Lyon says that operational restrictions designed to protect aviators reduced operational effectiveness, lengthened the campaign, and increased long-term risks to aircrews.[118] American airmen were again frustrated by what they considered to be inept meddling by civilian leaders.[119]

This modest performance is unsurprising because the norms driving the war—civilian casualty avoidance and force protection—were contradictory when implemented using 1999-vintage military technology.[120] To effectively control ground and protect civilians, one had to use ground troops who were vulnerable to attack. To hit small, mobile targets like infantrymen from the air without the ground spotters the Pentagon refused to deploy, pilots had to fly low. But flying low meant increased risk of losing pilots. NATO resolved the dilemma by choosing a force structure and operational practices that favored force protection over protecting civilian lives—Serbian and Kosovar Albanian—and damaging Yugoslav military and Serbian security forces. The air-only war plan driven by force protection imperatives meant Serbian security forces could more easily kill ethnic Albanian civilians throughout the war.

Yugoslav attempts to break NATO unity by publicizing NATO violations of the civilian casualty avoidance norm, thereby challenging members' identities as states that fight only just wars, failed.[121] Undoubtedly, Belgrade's crude propaganda and its lack of credibility during the Bosnian and Croatian wars contributed appreciably. Some analysts argue plausibly that Milošević's biggest strategic error was Operation Horseshoe—the murders, rapes, and expulsions of ethnic Albanians.[122] Lurid stories of atrocities and pictures of pathetic refugees at border camps convinced many Westerners that Yugoslavia was a major perpetrator of human rights violations, not just a victim, and that Clinton and Blair were right—Milošević had to be stopped. Yugoslavia thereby forfeited the moral high ground that NATO's collateral damage incidents gave it, simultaneously bolstering NATO resolve.[123] The lesson is not that the civilian casualty avoidance norm was irrelevant, but rather that each side had normative complaints against the other and the West's public relations efforts largely offset Yugoslavia's.

Military Goal Achievement

General Clark had five goals, which he called "measures of merit": (1) minimizing loss of friendly aircraft; (2) "impacting" Yugoslav military and Serbian security forces in Kosovo; (3) minimizing collateral damage; (4) achieving the first three in order to hold NATO together; and (5) protecting allied ground forces in neighboring Bosnia, especially, from Yugoslav raids.[124] Goals 1 and 5, aspects of force protection, were accomplished successfully. No NATO troops were killed in action. Yugoslavia did not conduct ground raids abroad. By carefully choosing targets and adjusting ROE, alliance unity held. Despite NATO leaders' worries that excessive civilian casualties could endanger Operation Allied Force and even NATO itself, they made no moves to improve bombing effectiveness in ways that might compromise aircrew safety—as by telling Short to fly his aircraft low enough to accomplish the mission regardless of losses. Because it did not immediately attack Yugoslav ground forces, flew high, and generated poor tactical intelligence to guide its precision munitions, NATO poorly achieved objective 2. And, because force protection trumped civilian casualty avoidance, NATO poorly (against high expectations) achieved objective 3, although not badly enough to break the alliance.

Time

The bombing, which most of American civilian and military leaders originally expected to end in a few days, lasted 78 days. Belatedly effective coercive diplomacy led to an agreement that created a UN civil administration and a NATO-led Kosovo Force (KFOR) with open-ended mandates; both remained in Kosovo even after Kosovo declared independence in early 2008. U.S. military imperatives per the Weinberger/Powell doctrine for an *exit strategy* and a decisive outcome therefore were not achieved. Yugoslavia's resistance strengthened Milošević's hand during negotiations in May and June. He got more than NATO offered at Rambouillet, arguably because NATO feared that his resolve might outlast its own. Milošević resisted long enough to successfully defend Serbian honor and to generate grievances available for recall in future conflicts. By the time criterion, NATO did poorly.

Cost

NATO caused significant collateral damage, although it was tiny compared to the millions of people killed during World War II when civilians were targets, the tens of thousands killed in Vietnam when they were not but when U.S. bombing was comparatively inaccurate, and 2,278 in Iraq in 1991.[125] Human Rights Watch counted 90 NATO attacks on civilians and confirmed 488 civilian deaths, while saying as many as 527 died; other reputable Western sources estimate civilian fatalities at 1,200 to 2,000 (Table 5.1).[126] NATO initially reported 20–30 incidents and about 500 civilian deaths, and Washington later admitted 38 incidents.[127] Several sources, including the U.S. Department of State, estimate that Serbs killed about 10,000 Kosovar Albanian civilians *during* Operation Allied Force—many times the number of civilian deaths that prompted Clinton to launch the war.[128] About 863,000 Kosovars were forced out of Kosovo and another 590,000 were internally displaced at the peak of the Serbs' expulsion campaign, meaning that about 90 percent of Kosovar Albanians were uprooted during the war.[129] Moreover, in Kosovo under KFOR "control," ethnic Albanians murdered Serbs chronically after 1999.

Against these numbers, NATO lost two aircraft but suffered no combat fatalities. NATO thus generated a *kill ratio* of about 500:0—and about 12,000:0 using broader accounting—when comparing operational results of the casualty avoidance norms. While the fog of war is real and some civilian casualties were to be expected under the best of circumstances, the numbers indicate that NATO's military commanders looked out for their

Table 5.1
Military and Civilian Deaths During Operation Allied Force

	United States	NATO	Human Rights Watch	Yugoslavia
U.S. Combat Fatalities	0			
NATO Combat Fatalities		0		
Yugoslav Military/ Police Deaths				576
Collateral Damage incidents	38	20–30	90	
Civilian Collateral Damage Deaths	500	500	~500	1,200–5,700
Kosovar Civilians Killed by Serbs	10,000			

Source: Author summary of sources cited in text.

own at the expense of civilians.[130] Humanitarian concerns about civilian suffering were at the root of NATO's decision to go to war, but military commanders conducted the war—with the approval of civilian leaders—in ways that served their values and interests first. General Short, for all his complaints about civilian interference and objectionable ROE, achieved his primary goal completely.[131]

OPERATION ALLIED FORCE AND MATERIALIST THEORY

Material superiority played little role in achieving NATO's victory. The war featured massive material disparities between grossly mismatched adversaries—powerful NATO and a rump Yugoslavia stressed by dissolution, eight years of Balkan conflict, and associated Western sanctions. Yet Yugoslavia accepted war with an alliance whose military assets it well understood. It attempted to use ideational means to neutralize, not defeat, NATO's military *capabilities*. Lack of credible NATO political will initially convinced Milošević to fight, and the belated emergence of credible NATO determination seems to have led Belgrade to concede—all with essentially the same NATO military force structure in place. The vulnerability created by NATO's unanimity rule meant Yugoslavia could neutralize NATO's conventional military power if it could use normative appeals to win a single vote in Brussels.

Material military assets, especially the technologies of PGMs, enabled the casualty aversion norms that largely determined the war's conduct. PGMs and Pentagon public relations hype about their accuracy generated unrealistic demands for "immaculate" warfare that both enhanced NATO's political vulnerability to collateral damage incidents and enabled the Pentagon to assemble a force structure that ensured force protection.[132] NATO could not have kept its pilots out of harm's way without technologies that permitted standoff bombing and could not have avoided a ground war if air and missile forces did not offer a plausible alternative. In turn, NATO's unbalanced force structure gave Yugoslavia an opportunity to adjust asymmetrically to reduce its vulnerabilities to bombing.

This conflict well illustrates the limitations of the technological focus of offense-defense theory.[133] Targeting restrictions, restrictive ROE, and force protection policies shaped operational practices sufficiently to make the actual operational offensive and defensive characteristics of manned aircraft very different from their technological potential. The main focus of Short's lament that Operation Allied Force violated air doctrine was that civilian casualty avoidance-driven ROE and gradual target approvals squandered airpower's allegedly inherent offensive advantages—the same point airmen made in Vietnam. The offensive/coercive effectiveness of NATO's air-only force structure was sharply degraded by Yugoslav air defense forces that simply survived.

Skill again largely determined whether and how material military assets were used. The KLA manipulated Clinton—meaning his skill as an intelligence analyst and policymaker was very poor and theirs was good. Clinton's perceptions of Milošević's susceptibility to mild coercion were wrong. Clinton's initial public disavowal of the ground option, like President Johnson's promise not to invade North Vietnam, gave the Yugoslavs a big advantage by reducing their vulnerabilities while simultaneously limiting NATO's policy and military options.

The heavy political pressure coming from Washington and Brussels led to operational violations of U.S. air doctrine—which focused on producing decisive force in conventional wars, not coercive diplomacy—and irritated airmen.[134] The JCS for months used unilateral U.S. command channels to prohibit General Clark from preparing even contingency plans for a land attack—reflecting a lack of desire to fight in Kosovo that Clinton shared—which violated the military principle of unity of command and significantly complicated Clark's job as NATO's military commander.

Poor collective NATO operational skills, generated largely by force protection–motivated operational restrictions like the 15,000-foot rule, and interoperability problems, offset most of NATO's material advantages. Incompetence at strategic political and military levels in Washington and Brussels nearly cost NATO the war. Tactical military competence was little better. Force protection–mandated lack of a ground presence meant tactical intelligence was weak—meaning PGM accuracy and effectiveness was poor because even the best PGMs can only hit targets they are told to hit. At age 50, NATO still did not have secure radios interoperable between national air forces, meaning air crews had to talk in the clear, enabling the Yugoslavs to react, or not communicate; new American equipment the European militaries had not adopted was mainly to blame.[135]

Such problems account for U.S. military personnel's chagrin at their collective performance in 1999.[136] The problems, many of them American generated, also had longer-term significance: they convinced many U.S. military personnel that coalition warfare was ineffective and encouraged them to fight Iraq in 2003 with few coalition partners.[137]

OPERATION ALLIED FORCE AND STRATEGIC INTERACTION THEORY

In a summary assessment of this war, Arreguín-Toft concludes that the strong actor won a same-approach interaction.[138] Using Arreguín-Toft's framework, I posit three strategic interactions within Operation Allied Force. First, NATO attempted to attack military, government, and selected industrial targets designed to persuade Milošević to capitulate. Second, NATO attacked economic infrastructure to try to induce a civilian morale-driven capitulation by Milošević. Third, NATO posed a credible threat of land invasion. Yugoslavia defended conventionally in the first two.

First, NATO's original conventional air attacks on military and government targets amounted to a conventional-conventional conflict. NATO attacked with conventional aircraft and missiles according to prevailing air doctrine. Yugoslavia initially defended conventionally with aircraft and ground-based air defense weapons, but quickly grounded aircraft it did not lose. The Yugoslavs did not use their unconventional warfare capabilities against NATO personnel and materiel in Bosnia and elsewhere—as General Clark feared. Hence, they did not use GWS. The Yugoslavs won because they kept their air defense system intact, denying NATO the air supremacy it desired and leading NATO to keep its aircraft high and thus ineffective at bombing Yugoslav military and Serbian security forces. Hence NATO's same-approach attack failed—contrary to Arreguín-Toft's expectations.

Second, NATO briefly tried to ratchet up popular pressure against Milošević by bombing electric power and oil installations, railroad lines, and bridges—a mild effort to influence Yugoslav civilians by inconveniencing them. Many Europeans regarded these attacks as barbaric, although they did not qualify as *barbarism* by Arreguín-Toft's definition.[139] But even this mild effort was quickly abandoned at the insistence of European governments reflecting popular domestic sentiment. Because some European governments regarded NATO policy as barbaric, and Arreguín-Toft considers government intent a crucial determinant of the presence of barbarism, I rate NATO's actions as reflecting barbarism.

Therefore, with Yugoslavia still defending conventionally, NATO lost an opposite-approach interaction—consistent with Arreguín-Toft's expectations.

Third, a decisive military interaction posed by a ground invasion threat that Western leaders only implied appears to have been effective by belatedly convincing Milošević they were serious, thereby generating conditions favorable for coercive diplomacy. NATO's increasingly credible ground threat in May posed a potential conventional-conventional ground fight the Yugoslavs knew they could not win. They therefore conceded a same-approach interaction before it began—consistent with Arreguín-Toft's expectations.

Once again, the weak belligerent employed barbarism—inconsistent with Arreguín-Toft's theoretical framework even though Yugoslavia was the materially dominant power in its bilateral war with the KLA. Atrocities by Yugoslav military and Serbian security forces, and some semiofficial Serbian paramilitary groups, as planned parts of Operation Horseshoe fairly clearly satisfy Arreguín-Toft's criteria for barbarism. While these actions helped Belgrade win temporary success in a same-approach barbarism-GWS interaction with the KLA, the barbarism outraged Western electorates, encouraged Westerners to overlook NATO collateral damage incidents, and offset Yugoslav propaganda about NATO's collateral damage incidents.

In sum, strategic interaction theory accounts for two of the three interactions I identify. My assessment is generally consistent with Arreguín-Toft's conclusion that the NATO/Yugoslav war as a whole was a same-approach interaction the materially dominant actor won. But strategic interaction theory cannot address Serbian barbarism, which played an important role in the politics that produced Operation Allied Force and influenced its conduct. It also does not address third party roles, meaning it has nothing to say about the KLA's ability to instigate the war and achieve its goal—effective immediate autonomy and independence in 2008—by manipulatively mobilizing a powerful coalition "partner."

DIMENSIONAL ANALYSIS

This three-sided war featured differential successes among the actors in identifying, exploiting, and ameliorating vulnerabilities. The KLA did best.

NATO

Led by the United States, NATO had modest success in using military operations to attack Yugoslavia's will to resist, while little ameliorating its own significant vulnerabilities. This poor performance accounts for NATO's inability to get even what it initially demanded of Belgrade and the flawed military operations that almost broke the alliance.

NATO focused its air strikes on Yugoslav military and civilian government facilities it thought were meaningful to Milošević, meaning it mainly attacked Yugoslavia in the *leadership* dimension. This made some sense because Milošević was the key Yugoslav/Serbian decision-maker. But Clinton erred in thinking another small military threat would cow Milošević over strategic national issues far more important to him than tactical military considerations in Bosnia, meaning the initial, modest attacks on facilities were ineffectual. As in Vietnam, Americans underestimated the will of their adversary and overestimated their own military capabilities and credibility.

After the coercive failure of the initial bombing, NATO expanded its target list to industrial plants like oil and electric power generation facilities, bridges, and railroads, which were chosen to attack *national will* with an ultimate aim of influencing Milošević.[140] These efforts, too, had little success in the short time they were in effect, but they roused opposition from European NATO electorates who wanted no military pressure on Yugoslav civilians. The bombing may have had a delayed effect as voters turned Milošević out of office in 2000 and the new government eventually sent him to the Hague to stand trial for war crimes, but it seems to have little affected Milošević's decision to settle the conflict.[141]

Finally, after two months of ineffectual bombing and at Prime Minister Blair's prominent suggestion, Clinton and the Pentagon approved planning for a ground invasion—a prospective action in the *operational execution* dimension. Yugoslavia could not have defeated NATO in a ground war and Milošević surely knew it. NATO imposed modest economic sanctions but made no major efforts to attack Yugoslavia in the *resource mobilization*, *resource conversion*, or *force generation* dimensions.

Clinton launched what he expected to be limited air strikes over the objections of many in Congress and at the Pentagon—reflecting domestic opposition to taking risks with service members' lives and thus U.S. weaknesses in the *national will* and *force generation* dimensions of American warmaking powers, respectively. Clinton's sensitivity to those concerns and to civilian casualties produced additional huge U.S. vulnerabilities in the *leadership* dimension. These in turn contributed to poor U.S. military performance in the *operational execution* dimension.

Clinton and the U.S. military (and thence NATO) were surprised by, and were unprepared for, the real war that resulted from Milošević's refusal to capitulate after a few air strikes, leading to ad hoc, reactive policies that yielded neither effective coercion nor speedy and decisive military results. The White House reviewed target lists for risks to civilians and agreed to the Pentagon's preference for an air-only force structure. When norm incompatibility arose, Clinton deferred to the military's preferences for operational procedures that emphasized force protection, thereby also precluding a Vietnam-like, casualty-driven decline in popular support for the operation. Clinton never made a case for a ground war to Congress, although Daalder and O'Hanlon argue that he might have been able to persuade Congress to accept one.[142] As a result, NATO never attempted a serious attack on the Yugoslav military. Therefore, from planning to the poor execution caused by the narrow air-focus of attacks, U.S. military operational performance was weak even though NATO won.

European states' *national will* constrained NATO substantially by working through domestic political channels to influence national *leadership*. Yugoslav civilian casualties generated widespread popular indignation in Europe, which rapidly generated pressure on European politicians and thence on NATO. European concerns influenced U.S. military and civilian leaders through alliance, not domestic American, politics. Senior American officials viewed NATO's center of gravity—its major vulnerability—as alliance unity, not congressional or domestic American popular opinion.[143] Washington had to respect the unanimity rules of NATO and adjusted ROE and targeting to ameliorate a key institutional vulnerability and preserve the alliance—leading to the operational practices that angered Short.[144] Thus, European popular will also negatively affected NATO *force generation* capabilities—keeping NATO's deployed force structure far less capable that the sum of its material military asset holdings. These factors led some U.S. military officers, chagrined by their weak operational performance, to blame NATO as a whole, not NATO's American leaders, for norm-related violations of U.S. military doctrine.[145]

Yugoslavia

Yugoslavia primarily attacked NATO in the *national will* dimension by making collateral damage the centerpiece of its political campaign.[146] NATO claimed to be waging a just war to protect human life, but if Yugoslavia could show that NATO was the bully and killer of civilians, Belgrade hoped to persuade NATO electorates, and thence their leaders, that the best way to protect human life was for NATO to end its attacks—thereby giving Yugoslavia victory. Yugoslavia called NATO the "Nazi American Terrorist Organization" and accused Western leaders of war crimes. Amnesty International made similar allegations.[147] (Long after the war ended, ICTY declined to press charges.) Some human shields were killed in air strikes, boosting the civilian toll and Belgrade's political case. The Yugoslavs brought Western journalists to sites of collateral damage incidents. State television showed damage to civilian property and burning civilian corpses; Anthony Cordesman and others concluded that much of the footage was crudely fabricated.[148]

Nevertheless, the facts NATO admitted, reputable journalists' accounts, and the imagery outraged European electorates and led European politicians to plead for fewer civilian casualties after NATO aircraft killed 87 civilians near Korica on May 13.[149] Solana personally called a Spanish television station to complain that it aired Yugoslav footage of collateral damage, and Clark later said he thought NATO lost the propaganda war.[150] After gradually increasing the severity of bombing to pressure Milošević, by late May NATO was under intense political pressure from European member states to reduce collateral damage, and it reduced attacks in response.[151] That is, Yugoslavia activated the civilian casualty aversion norm in European states, which generated domestic political pressures on government *leadership* to demand changes in NATO operations. Clinton, the Pentagon, Solana, and Clark bowed to the pressure because popular unhappiness in Europe over collateral damage threatened European governments' support for the war and, for a time, perhaps even the alliance.

Belgrade focused its efforts on European electorates, who the Yugoslavs apparently saw as more vulnerable than Americans. Yugoslavia surely knew it would defeat NATO if even a single alliance member state opposed the war. While Washington could have gone on alone, Clinton's reactive, risk-averse behavior suggested that the United States, too, might abandon the war if Europeans quit.

At the same time, Yugoslavia operated in the *force generation* and *leadership* dimensions by ensuring that its air defense capabilities remained intact—thereby denying NATO the air dominance U.S. airmen craved and maintaining a small threat that Yugoslavia could kill a few flyers. By maintaining credible air defense capabilities that kept NATO planes high and thence NATO's bombing inaccurate, Yugoslavia also protected its ground units and increased the likelihood of politically valuable collateral damage incidents. In short, it both well identified and exploited NATO political-military vulnerabilities consistent with its limited material capabilities and its own ideational baggage.

Against the KLA, the Yugoslavs conducted both a fairly brutal counterinsurgency campaign and ethnic cleansing. NATO military actions little hindered these successful efforts, either directly or by coordinating actions with the KLA.

The KLA

The KLA understood Western humanitarian concerns and, given Clinton's obvious anti-Serb attitudes, saw civilian casualty aversion as a vulnerability in Clinton personally that it could exploit to entice the United States and NATO into its war of independence. That is, it sought to manipulate NATO by exploiting normative vulnerabilities in NATO's *leadership* and *national will* dimensions—not to defeat or even to fight NATO. By provoking the Serbs into atrocities, in part by committing enough of its own against Serbian civilians to lead U.S. special envoy Robert Gelbard to call the KLA a terrorist group, the KLA generated favorable reactions by Clinton, Albright, and European electorates.[152] U.S. intelligence warned that the KLA "intended to draw NATO into its fight for independence by provoking Serb forces into further atrocities."[153] Holbrooke later said:

> The KLA was arming, the KLA was taking very provocative steps in an effort to draw the west into the crisis. That was ultimately successful, the Serbs were playing right into the KLA's hands by committing atrocity after atrocity, way overreacting, wiping out entire villages, outrageous actions that had to be responded to.[154]

KLA leaders confirmed the assessments after the war.[155]

The KLA's strategy was not unique. It emulated a successful model in nearby Bosnia.[156] UNPROFOR's Bosnia commander in 1994, British Army Lieutenant General Michael Rose, says Bosnian Muslims repeatedly fired on their own people and UN troops in calculated efforts to blame atrocities on Bosnian Serbs and thereby garner international assistance; consistent with Rose, eight Western intelligence services concluded that Muslims fired the mortar round that in February 1994 killed 68 civilians in Sarajevo, which throughout the war Muslims controlled and Bosnian Serbs besieged.[157] French UNPROFOR troops believed immediately, and Western intelligence agencies later confirmed, that Muslims fired the mortar round that killed 38 civilians in Sarajevo on August 28, 1995, which at the time was widely blamed on the Serbs and which provoked U.S.-led NATO air strikes on Bosnian Serbs soon thereafter.[158] UN and Russian officials identified a third Muslim mortar attack on Sarajevo.[159] One might suppose that in 1999 the Clinton administration, well aware of the situation, used tragic events as an excuse to attack a long-standing enemy, as President Johnson in 1965 used attacks on U.S. troops at Pleiku to publicly justify air strikes on North Vietnam that he had decided earlier to launch. However, if Rose and the French are right about the motives, sources, and effects of some of the Sarajevo shelling, Clinton was manipulated repeatedly in both Bosnia and Kosovo.

This kind of norm-based manipulation of third parties is qualitatively different than the long-standing guerrilla practice of placing civilians in harm's way to generate politically useful blame or to hinder enemies' operational effectiveness—as when Yugoslavia used Kosovar civilians as human shields, as Somali leader Aidid in 1993 mixed his fighters with civilians to protect them from U.S. soldiers trained to avoid shooting civilians, and as Vietnamese communists did in several ways.[160] But whereas Vietnamese communists primarily focused on U.S. public opinion, Bosnian Muslims and the KLA directly attacked U.S. leadership in the form of Clinton's personal normative vulnerabilities.

The KLA's leaders also recognized the importance of casualty avoidance to NATO's European electorates and civilian leaders, betting on the basis of Bosnian Muslim experience with NATO member states operating under UNPROFOR that if they could bait Serbian security forces enough, they would react in ways the KLA could exploit to bring NATO's European members to their aid.[161] But the key exploitable vulnerability was American leadership. Only Clinton, given America's unilateral military strength and it role in NATO, alone could wage war on Yugoslavia.

Summary of Actor Interactions

On balance, the United States and NATO poorly identified enemy vulnerabilities, poorly handled their own, and ended up with a technologically impressive but operationally flawed force that nearly lost the war. NATO was particularly weak in the national will, force generation, and leadership dimensions, which led directly to poor operational execution.

Both the KLA and Yugoslavia, like Bosnian Muslims, primarily wanted to manipulate the West through integrated strategic political and tactical military means. Civilian decision makers were their best avenues to influence NATO's force structure, its operational behavior, and ultimately the military operational outcome. Yugoslavia knew it could not beat NATO militarily and the KLA had no desire to try. The KLA more successfully used Western norms to achieve its goals than did Milošević, but both crafted reasonable strategies to effect such manipulation. Both caused civilian deaths—directly and indirectly—specifically to influence NATO politically, meaning that Western casualty aversion perversely gave both Yugoslavia and the KLA reasons to kill civilians in pursuit of conflicting political goals.

U.S. LEARNING AND ADAPTATION

The United States in general poorly identified, exploited, and ameliorated vulnerabilities. Only a late policy change that entertained the possibility of a ground invasion amounted to an adaptation in strategy that sealed the "ugly" coercive success of Operation Allied Force.

Clinton, the U.S. military, and NATO poorly understood Milošević's vulnerabilities to coercive diplomacy over Kosovo, despite years of personal dealings with him. They learned little about Milošević's priorities and did not understand that Kosovo was sufficiently important to him to risk a real war. It is unclear whether NATO members' intelligence services alerted their militaries to Yugoslavia's strategy of using its air defense capabilities sparingly to protect them and maintain a credible air defense threat, but the Pentagon seems independently to have been strongly committed to its air-only force structure for reasons of force protection. While intelligence agencies and senior diplomat Holbrooke understood the KLA's plan to alter facts on the ground to manipulate NATO into war on its behalf, such awareness seems not to have influenced Clinton and Albright.

The United States and NATO extremely poorly adapted to exploit Yugoslavia's major military weaknesses, mainly as results of important internal policies, institutional practices, and military preferences. First, Clinton and the Pentagon did not want a real war, initially sharply limiting Clark's planning options and destroying the credibility of NATO's threats. Second, NATO at U.S. insistence opted for an air-only force structure whose restrictive use provided limited offensive power, gave Yugoslavia many means to avoid NATO's strengths, and limited Clark's options. Third, force-protection motivated decisions (like the 15,000-foot rule) further reduced operational options, diminished NATO's bombing accuracy, and thence damaged NATO's political credibility and diplomatic options. Fourth, restricted bombing accuracy led to civilian casualties, despite tight targeting and restrictive ROE, which led European NATO states to demand yet more restrictions on bombing in general. Fifth, NATO's unanimity rules massively complicated and slowed political decision making, exacerbated NATO's fragmented military command-and-control relationships, and limited adaptive potential.

NATO performed even worse in identifying and ameliorating its vulnerabilities. The United States as matters of policy designed for other reasons, especially force protection imperatives, created many of NATO's operational and political vulnerabilities. By using only air forces after touting the technical characteristics of precision bombing systems under ideal conditions, the U.S. military generated unreasonable popular expectations of *immaculate warfare* that became a major NATO political vulnerability. The KLA and Yugoslavia both played on NATO sensitivities to civilian casualties, which generated operational dysfunctions that were mitigated only when breaking the alliance became a bigger threat to Clinton than having to explain some American deaths. The alliance worked to ameliorate the weaknesses, and thence vulnerabilities, its unanimity rule caused, but normative vulnerabilities were never successfully addressed.

In short, NATO under America's leadership displayed many dysfunctions that reflected many learning and adaptation failures. The ultimate causes were virtually entirely internal U.S. and NATO policies and institutions designed for other purposes. Although NATO eventually prevailed, massive U.S. and other Western political and military vulnerabilities remained unresolved through the end of the war. They reappeared in the U.S. war in Iraq after April 2003 and caused greater problems in Afghanistan.

SUMMARY

Operation Allied Force illustrates well how complex norm- and military culture-motivated ideational and institutional vulnerabilities interact with material factors in contemporary Western military operations. The civilian casualty avoidance norm drove NATO to go to war. Once fighting began, civilian leaders directed their armed forces to achieve largely humanitarian-oriented objectives, but only in the context of powerful force protection imperatives that damaged operational effectiveness. U.S. military leaders incorporated civilian casualty avoidance concerns in their targeting plans and ROE when they were consistent with force protection and airpower doctrine, but when the civilian casualty avoidance norm conflicted with force protection, force protection imperatives prevailed.

Were Western civilian leaders hypocritical? Perhaps. But they, like Clinton, may not have recognized the incompatibility of the norms or anticipated the extent and effectiveness of new, instrumental uses of both norms, from several directions, against them.

The case demonstrates again that estimates of relational military power must reflect negative as well as positive values. NATO's most egregious collateral damage incidents—like the Chinese embassy bombing—generated much negative political utility and nearly broke the alliance.[162] But Operation Horseshoe, Belgrade's most dysfunctional use of force, reduced the net cost of NATO's errors and helped preserve the alliance. Normatively inappropriate uses of force generated offsetting political negatives, hardly the logic of materialist notions of military power.[163]

There is no question that material military assets were operationally important. Combined finally with political will, they convinced Milošević to negotiate. But technology designed to enhance material military power became the key enabler of two casualty avoidance norms that dominated the conduct of the war by limiting the use of force. A key lesson therefore is that material-based analyses of *capabilities*—and capabilities-based explanations of military operational outcomes—must consider norm-based restrictions on force structure and focus on normatively *usable* forces.

The case has implications for the theories of Shimshoni, Biddle, Merom, and Mack. The KLA's strategic success in manipulating the United States, on the one hand, and multiple examples of American ineptitude on the other, show again why Shimshoni places such emphasis on leadership. NATO chose for normative reasons not to use its considerable skills at conventional warfare against Yugoslavia in 1999. The ugliness of the win in this exercise in coercive diplomacy, compared to the convincing wins in 1991 and March–April 2003 in medium-intensity conventional warfare, illustrates one set of tactical and strategic consequences of a U.S. decision not to use its version of Biddle's *modern system* when it had not mastered alternative political-military skills.

This war illustrates Merom's point that modern Western democracies are willing to bear low levels of human cost in limited wars, and it eclipses Mack's emphasis on U.S. unwillingness to bear costs in a long and limited war. In this case, Clinton launched a war of choice driven by normative concerns but sought to keep its human costs tiny. The U.S. military and Congress nearly derailed launch of the war over concerns about only the possibility of injury to U.S. military personnel, while Europeans nearly collapsed NATO because a very small number of civilians—by the European standards of 1939–1945 at least—died in accidents of war that NATO clearly worked hard to prevent.

CHAPTER 6

The U.S. War in Iraq (2003–2011)

On March 19, 2003, the United States and its partners invaded Iraq for purposes of deposing President Saddam Hussein, thereby eliminating the perceived threat he posed to U.S. interests, finding and destroying Iraq's WMD programs, and establishing a new Iraqi government that President George W. Bush expected to be democratic, peaceful, and Western-oriented.[1] Bush hoped the new Iraq would be a model for political reform in the Muslim world and would not threaten Israel. Achievement of the first goal required only part of what was widely described as the finest conventional military force ever created. In a television spectacle, U.S. troops helped Iraqis topple a statute of Saddam in Baghdad on April 9. Although mopping up operations continued for several more days, the "war" was won within a month. The second goal evaporated when it became clear that Iraq destroyed its WMD stockpiles before 2003 while retaining residual WMD production capabilities.

The third, far more ambitious, objective required building a new Iraqi state. But it was not one that much concerned U.S. planners before April 2003. Senior civilian officials and military officers believed most Iraqis would welcome U.S. troops, that a coherent Iraqi government would nearly spontaneously reemerge, and there would be no internal Iraqi discord that would threaten U.S. troops or the nascent Iraqi democracy.

It was not to be. U.S. planning assumptions proved to be nearly wholly wrong.[2] Despite its initial successes, the United States soon became bogged down in a counterinsurgency (COIN) war. After April 2003, the military task required skills the U.S. military previously denigrated and which Bush, like several presidents before him, showed little interest in developing. The now-unavoidable COIN effort prompted a major military effort to relearn the lessons of earlier *small wars*—including the army's war in the Philippines, U.S. Marine missions in Latin America before World War II, and Vietnam.

The United States ran this war of choice. Unhappy with the operational consequences of working within the NATO institutional framework in 1999, the Department of Defense (DoD) wanted to control this operation more directly. Bush's case for regime change in Iraq failed to persuade many countries, including France and Germany, leading to assembly of a modestly sized group of states willing to participate meaningfully in the war. While Washington cajoled 48 states into at least token support for the war and postwar peacekeeping and reconstruction tasks, the bulk of the coalition's combat power consisted of American and British forces; Australia, Poland, and a few other states contributed smaller combat forces.[3] Unlike in the Gulf War, no Muslim state contributed troops, and only Kuwait offered logistical support. From the start, the dedicated core of what the Bush administration called the "coalition of the willing" was small, and the commitment of the larger coalition was shaky.

Senior American civilian leaders' views of modern warfare dramatically shaped the conventional war and its aftermath. Most critically, Secretary of Defense Donald Rumsfeld strongly influenced the initial strategy and force structure the United States used. While regional U.S. military commanders draft war plans for approval by civilian leaders, Rumsfeld played an unusually active role in shaping CENTCOM's operational plan (OPLAN), which was designated 1003V and implemented as Operation Iraqi Freedom. He also changed it in mid-execution, causing problems for theater commanders.[4]

An advocate of defense "transformation" enamored with technology, speed, and agility as war-winning characteristics, Rumsfeld rejected the overwhelming force model used in the Gulf War of 1991 in favor of a model closer to what seemed to have worked in Afghanistan in 2001—aircraft-delivered precision munitions guided by U.S. Special Operations Forces (SOF) in support of a conventional Afghan ground force (see chapter 7).[5] He rejected use of the joint service logistical program, which helps plan deployments of standard force packages and produces a deployment plan known as the Time Phased Force Deployment List (TPFDL), in favor of ad hoc decisions about deployments that left at home some artillery and support units; Rumsfeld's directives violated the military's principle of "train as you fight" by making last-minute changes in the deployed force structure.[6] Deputy Secretary of Defense Paul Wolfowitz declared that a purpose of the war plan was to break the post-Vietnam military paradigm (that is, the Weinberger-Powell doctrine) that held that more material assets and force mean less risk to troops and missions.[7]

Rumsfeld's insistence on speed and technology and Iraq's many military weaknesses led CENTCOM commander General Tommy Franks to use a powerful but small and unbalanced force and a campaign plan that emphasized speed.[8] Operation Iraqi Freedom therefore differed appreciably from Operation Desert Storm in 1991, but its differences were evolutionary in nature and it was consistent with Desert Storm–era U.S. military practice in many respects—especially aversion to U.S. casualties and concerns about collateral damage.[9]

Franks's primary invasion force consisted of the army's V Corps, consisting of the 3rd Infantry, 82nd Airborne, and 101st Airborne (Air Assault) Divisions; the I Marine Expeditionary Force, consisting of the 1st Marine Division and its related aviation units augmented by additional infantry regiments; and, 26,000 troops of the British 1st Armored Division. Some 8,000 American, British, Australian, and Polish special operations troops comprised the largest special operations force in history. U.S. and UK air and naval forces supported coalition ground forces. In all, some 290,000 coalition troops from 23 countries participated in some way; nearly half were American.[10] Some 50,000 to 60,000 Americans actually fought.[11]

Iraqi troop strength considerably exceeded that of the attacking coalition force—contrary to the normal 3:1 or so numerical superiority of attacking forces commonly thought necessary to win conventional fights. Franks thought his force's qualitative superiority in speed, firepower, tactical intelligence, and reaction time obviated a need for even quantitative parity.[12] Indeed, Franks said before the war he thought the force ratio would be 1:6.[13]

Plans called for great efforts to avoid collateral damage in order to keep key economic infrastructure operating, facilitate reconstruction, avoid the recriminations that civilian casualties produce, and avoid the political costs of damaging Muslim holy sites. Target selection procedures and rules of engagement again were designed to minimize collateral damage. As in 1999, U.S. military planners identified civilian casualty minimization as a war goal, not just a measure of operational effectiveness or a constraint. However, Wolfowitz pointedly declared that no Iraqi civilian was worth the life of any American soldier—a view consistent in practice if not word with U.S. policy during the war against Yugoslavia.[14] U.S. military leaders recognized that this war was important and did not, as in 1999, insist on force structures and operations designed to preclude U.S. casualties. When U.S. casualties occurred, it was antiwar civilians, not military personnel, who complained about them.

Rumsfeld's invasion force was successful in the conventional phase of the war, but the reductions in numbers of support troops needed to keep the overall force small, operating assumptions, and the military's Vietnam legacy of visceral dislike of counterinsurgency and nation-building operations left U.S. forces, and therefore the coalition as a whole, ill-prepared for stability operations after April 2003. Rumsfeld and Franks focused on taking Baghdad, not the more difficult task of administering a sullen, troubled country. The decision was a strategic blunder.[15]

U.S. failure to end the war speedily and successfully, along with mounting American casualties, generated considerable unhappiness in the United States.[16] In the congressional elections of 2006, Democrats running on an antiwar platform took both houses of Congress. Democrats immediately began to try to impose restrictions on, and to end, American involvement in the war—much like their predecessors did in 1970–1975. Bush forestalled congressional efforts to legislate a precipitous U.S. withdrawal, however. The war also was a major issue in the 2008 elections, which saw the Democrats increase their congressional majorities and Democrat Barack Obama win the presidency after vowing to withdraw from Iraq. But operational successes beginning in 2007 led to much reduced violence, leading Bush in December 2008 to sign two agreements with Iraq calling for withdrawal of U.S. combat troops by the end of 2011 and long-term U.S.-Iraq cooperation.[17] The progress enabled Obama to ignore his campaign promise with impunity.

After many adjustments, the U.S. government at the time of this writing in mid-2011 is conducting variously effective political, social, economic, and specialized military operations against residual insurgent groups. Obama in mid-2010 celebrated the end of conventional U.S. combat operations and the transition of the focus of the much reduced U.S. military force to training Iraqi units by renaming the mission Operation New Dawn. International support for a fragile Iraqi government still involves many U.S. government agencies, thousands of U.S. government contractors, intergovernmental organizations (IGOs) including the United Nations, and nongovernmental organizations (NGOs).

Improved U.S. performance beginning in 2007—along with better Iraqi governance and no small amount of luck—appears increasingly likely to lead to an outcome palatable to the United States and therefore a win per strategic interaction theory. But whatever the ultimate outcome, Bush's initial goal of rapidly establishing a functioning, democratic, and Western-oriented Iraqi state was not quickly or efficiently achieved, meaning that by many other criteria the war was unsuccessful from the U.S. perspective.[18]

OVERVIEW OF THE WAR

The war is easily divided into two phases: (1) a conventional phase during March–April 2003 in which coalition forces soundly defeated the conventional Iraqi military, deposed Saddam Hussein, and occupied Iraq; and, (2) an unconventional phase thereafter in which similar international forces and eventually growing Iraqi security forces fought domestic Iraqi insurgent groups and foreign fighters. The insurgents conducted unconventional warfare using traditional guerrilla tactics along with some newer methods—most notably suicide bombings. Through 2006, insurgents were numerous, were effective in the field, and posed serious challenges to the coalition and the Iraqi government. Thereafter, most Iraqi Sunni insurgents turned on their former al-Qaeda allies and began to work with the Americans and a finally stabilizing Iraqi government, Shi'a militias also largely ceased fighting, and the Americans adopted better COIN techniques. These developments belatedly enabled the coalition and improving Iraqi government forces to contain the insurgent threat.

The Conventional Phase

CENTCOM's war plan called for operations on five fronts, including (1) a southern effort launched from Kuwait involving capture of Basra and movement northward toward Baghdad; (2) occupation of Baghdad; (3) rapid movement of SOF units into southern and western Iraq to interdict highways and hunt mobile SCUD missile launchers that potentially threatened Israel; (4) SOF deployment to northern Iraq to join Kurdish militiamen and tie down Iraqi units in the north; and (5) an intelligence/information campaign including efforts to disrupt Iraqi command-and-control activities, a CIA effort to bribe Iraqi military units into avoiding combat, and deception operations.[19] The plan called for capture of Iraq's oil fields before Saddam loyalists could torch them and rapid movement toward Baghdad.[20]

In the conventional phase of the war, the United States and Britain initially faced some 389,000 Iraqi troops, including 350,000 soldiers in 17 regular army divisions, seven Republican Guard divisions, and five Special Republican Guard brigades.[21] The air force had some 20,000 personnel, and the navy had 2,000 personnel.[22] Military regulars were augmented by several paramilitary forces, including special units from security and intelligence agencies; militia units of Ba'ath Party loyalists; volunteers known as the "Lions of Saddam"; *Saddam Fedayeen* ("Saddam's Martyrs"), men from Saddam's Tikriti tribe under the command of his son Qusay, who were variously estimated to number from 15,000 to 40,000; and fighters from other Muslim countries, who began arriving in Iraq shortly before the war started.[23] Most of these foreign fighters, who eventually numbered about 4,000, were not trained soldiers and were not affiliated with al-Qaeda.[24] Collectively, the irregulars were known as *fedayeen*.

Iraqi equipment holdings were considerable. The army had some 2,200 to 2,500 tanks; 3,700 other armored vehicles, including BMP infantry fighting vehicles and armored personnel carriers; and 2,400 major artillery pieces.[25] Iraqi air defenses included 210 surface-to-air missile launchers, 150 early warning radars, air defense fighter aircraft including modern MiG-29s, and many antiaircraft guns.[26] Iraq's air force had some 320 aircraft, including fighters, bombers, reconnaissance aircraft, transports, and helicopters.[27] Iraq's small navy was operationally negligible.[28] Iraq also dispersed throughout the country large quantities of arms, ammunition, and explosives, which would play a significant role in the unconventional phase of the war.

Despite its material assets, Iraq's armed forces were operationally weak.[29] Western sanctions and low-level U.S. and UK air attacks since 1991 prevented rebuilding and modernizing the force badly damaged in 1991, meaning the Iraqi military was technologically obsolescent compared to American and British forces. The regular army consisted largely of conscripts who had little desire to be in uniform. Conscript morale was poor, degrading the effectiveness of army units and leading to large-scale desertions when the fighting started. Training was limited, and the army had not conducted a major field exercise since 1994.[30] Maintenance was poor, meaning much equipment was inoperable. Leadership was heavily politicized; Saddam appointed some 11,000 generals, largely to foster domestic political loyalty, whose military skills were minimal.[31] Command-and-control networks were fragile. The war revealed severe deficiencies in doctrine and simple ineptitude in making and executing war plans. In total, Iraq's conventional military deficiencies were serious.

To surprise Iraq, Franks ordered a rapid advance without the extended, preliminary air campaign the United States used in 1991. Franks saw "maneuver speed" as his most important asset; it would enable him to keep the initiative.[32] The coalition's intelligence capabilities and its ability to quickly place precision munitions on targets its sensors found would produce "effective" firepower. The Americans would "shock and awe" the Iraqis into capitulation according to a new variant of airmen's old morale bombing theory.[33] Rapid movement and surprise would complement the coalition's technological superiority; fast-moving U.S. columns would outflank Iraqi forces, defeat them by fire and maneuver tactics, or force them to move, making them visible to U.S. sensors and shortly thereafter to devastating attacks by PGMs delivered by coalition aircraft and artillery.

Focused on Baghdad and to keep movement rapid, Franks told his troops to destroy Iraqi forces on open ground but bypass pockets of resistance in towns. V Corps also bypassed many Iraqi military bases and supply depots by traveling in the lightly populated desert west of the Euphrates River. Marines initially followed army troops but then veered east to drive on Baghdad from the southeast through the area between the Euphrates and Tigris Rivers and from east of the Tigris. The British took Basra and much of southern Iraq.

While Franks and his staff were confident of victory before the fighting began, they worried about possible Iraqi use of chemical and biological weapons. WMD could kill unprotected coalition troops or force them into protective "MOPP" (Mission-Oriented Protective Posture) full-body suits that were hot, restricted visibility, slowed movement, and increased logistical requirements for replacement suits.[34] Therefore, even alerts that forced the wearing of protective gear degraded coalition operational performance. Finds of Iraqi protective suits early in the fighting heightened Franks's concern that the Iraqis would use WMD, which proved unfounded in the fact.

The Americans also worried that Saddam would withdraw his forces into the cities and fight urban warfare that is notoriously tough, especially when defenders prepare fortified, camouflaged fighting positions.[35] Franks wanted to avoid a "Fortress Baghdad" or a "Saddamgrad"—a reference to the bloody German/Soviet battle at Stalingrad in the winter of 1942–1943—in which Iraqis might inflict significant casualties on attacking American troops.[36] Fighting in built-up areas also would negate many of the coalition's sensor capabilities and cause it to curtail air strikes for fear of collateral damage, leading one senior U.S. targeting expert to describe the urban warfare scenario as like "trying to have a fist fight in a really dark room."[37]

To avoid a major battle in Baghdad, CENTCOM devised a multi-axis advance designed to destroy three Republican Guard divisions positioned in open country south of Baghdad before they could withdraw into the city. But Saddam never intended an urban fight and never trained his troops for urban warfare; he apparently feared they might use such skills against his regime. Franks also was concerned about possible Iraqi use of key terrain features like the Karbala Gap and water barriers like lakes, the Tigris and Euphrates Rivers and canals between them, and marshes, but the Iraqis poorly used these potential advantages, too.[38]

In American military parlance, the coalition conducted "effects-based operations" (EBO).[39] The EBO concept referred to choosing targets to achieve specific functional results—like disruption of command-and-control-related activities—rather than physical destruction of targets. Modern EBO, enabled by precision munitions with a circular error probable of as little as two meters, aided coalition aircrews and ground troops in 2003 to dismantle the Iraqi military while destroying little of Iraq's economic infrastructure. Some 70 percent of the bombs the coalition dropped in 2003 were guided—compared to about 30 percent against Yugoslavia in 1999, less than 10 percent during the Gulf War, and a few dropped on North Vietnam.[40]

The coalition tried to "get inside the Iraqis' OODA loop."[41] The OODA (Observe, Orient, Decide, Act) concept refers to the time it takes military organizations to recognize salient characteristics of their environment, decide on new courses of action, effectively act on the decisions, and then start the process again. Getting inside Iraq's OODA loop meant U.S. troops consistently acted faster and more effectively than the Iraqis.[42] PGMs destroyed Iraqi sensors and communications nodes, degrading Iraq's ability to observe coalition actions and limiting communication to decision makers of what its deployed forces learned, thereby hampering Iraqi analysis of battlefield conditions and coalition actions. Communications problems also degraded Iraqi commanders' ability to direct their units, reducing the effectiveness of Iraqi operations. Coalition ground forces' mobility and air superiority meant they could maneuver faster than the Iraqis could, further complicating Iraqi decision making. U.S. sensor technology generated asymmetric advantages in situational awareness, especially at night and in bad weather, which markedly enhanced overall coalition superiority.

These advantages meant that the coalition could press the Iraqis almost constantly, while avoiding or easily defeating most Iraqi counter thrusts. Said V Corps commander Lieutenant General Scott Wallace, "That means doing as many things simultaneously to the enemy as you can, horizontally in terms of hitting him from multiple directions on the ground and vertically in terms of the joint fight with Air Force and Navy air power . . . the point is to make this a multi-dimensional fight that overwhelms the enemy's command and control and keeps him off balance."[43]

The rapid advance was designed to keep the Iraqis from discovering and exploiting tactical vulnerabilities produced by the small size of the coalition force.[44] CENTCOM planners thought rapid movement was the best way to keep Iraq from using the WMD they then believed it had. But speed also stretched U.S. LOCs; supply shortages forced a short pause before the final assault on Baghdad.[45] Aggressive *fedayeen* raiding parties also threatened LOCs the lean coalition force could not adequately guard. CENTCOM before the war identified some 170 "sensitive sites" along its LOCs from Kuwait to Baghdad.[46]

The American gamble was a good one because the Iraqi military performed poorly.[47] U.S. ground forces rapidly advanced on Baghdad as air and ground forces pummeled Iraqi troop concentrations. The drive on Baghdad was the fastest-moving major offensive in history.[48] Even sandstorms did not blind U.S. sensors or prevent use of precision munitions. Iraq's regular army divisions fought little, and poorly when they did fight; often they just melted away. The coalition took only about 20,000 prisoners.[49] Republican Guard divisions fought harder but not well and were overwhelmed. No major conventional force awaited U.S. troops in Baghdad, and no WMD were used. Saddam's air force and navy effectively did not fight at all.

Fedayeen fighters posed a threat U.S. troops did not expect.[50] They were organized and trained as guerrilla fighters just before the war for a variety of reported missions, including defense against a possible Shi'a revolt, because Saddam feared a repeat of the uprisings of 1991 by Shi'a and Kurds after the United States called for Iraqis to overthrow him, and irregular warfare after losing the conventional fight with coalition forces.[51] The *fedayeen* fought hard but not wisely, mostly in frontal attacks on U.S. armored columns using rocket-propelled grenades (RPGs) and truck-mounted machine guns.[52] The cannon and machine guns of U.S. tanks and Bradley infantry fighting vehicles killed many of them.[53] When U.S. sensors discovered that *fedayeen* commonly staged their attacks from Ba'ath Party headquarters buildings in towns, surveillance of the buildings enabled coalition aircraft to bomb them when *fedayeen* fighters congregated, killing many more.[54] Hence, the *fedayeen* in March and April 2003 minimally disrupted coalition operations, but their attacks were harbingers of the unconventional war to come.

Iraqi efforts to stop the coalition advance failed. The 3rd Infantry Division reached Baghdad in early April and conducted a doctrinally nonstandard "Thunder Run"—a charge by an armored column into and then out of the city that killed, by some estimates, at least 2,000 Iraqi irregulars and foreign fighters while suffering few casualties.[55] After a second Thunder Run, U.S. troops stayed in the city, Saddam's government fled, and the Iraqi army quit fighting, ending the conventional phase of the war. It lasted from March 19 to April 16.

The conventional campaign went largely as American planners expected. U.S. forces crushed Iraq's military and occupied Baghdad well ahead of OPLAN 1003V's timeline. On May 1, 2003, at General Franks's request, President Bush declared an end to major fighting.[56] Prospects then seemed good, to wishful thinkers in Washington at least, for rapid establishment of a democratic Iraq and a rapid return of U.S. troops to America.[57]

Most independent observers initially gave coalition forces high marks for executing a good plan for what then seemed to be the entire war.[58] They recognized Iraqi weaknesses and command errors but credited superior U.S. technologies, effective leadership, good doctrine and training, and well-coordinated joint (U.S. multi-service) and combined (multinational) operations as keys to success. Andrew Krepinevich, a retired U.S. Army officer and military analyst critical of the army's operations in Vietnam, said the U.S. performance was so impressive as to encourage all prospective U.S. opponents to confront the United States only asymmetrically.[59]

The Unconventional Phase

The unconventional phase of the war began before the end of conventional fighting with *fedayeen* attacks and a few suicide bombings but accelerated markedly in August 2003. After the fall of Baghdad, Iraqi soldiers and paramilitary personnel, some of whom fought as *fedayeen*, shifted to fairly standard guerrilla warfare. They used the regime's huge caches of weapons and ammunition and some $4 billion that Saddam's government held abroad.[60] Within weeks, Iraqi nationalists, Sunni Arab tribal groups, Sunni Islamists who resented conquest of Iraq by foreign infidels, and some criminal groups began to take up arms.[61] By March 2004, some Shi'a militiamen, most notably the *Jaish al-Mahdi*, or Mahdi Army, led by Moqtada al-Sadr, son of a prominent Shi'a cleric Saddam's agents killed in 1999 and a leader of Iraq's Shi'a poor, also began to fight coalition troops.[62]

Initially, these groups fought poorly. When they stood and fought, U.S. troops usually defeated them easily. But they learned to raid and shell coalition bases and to ambush coalition convoys, at first mainly in the "Sunni Triangle" north and west of Baghdad.[63] Survivors quickly learned to fight better. By 2004, most U.S. troops and intelligence officers considered their enemies to be dedicated and resourceful adversaries.[64]

Insurgents manufactured roadside bombs to attack coalition convoys.[65] These "improvised explosive devices" (IEDs), in American military jargon, initially were made mainly from the artillery and RPG ammunition, and explosives, cached before the war and from what the insurgents scavenged from former Iraqi army supply depots. They, too, became more sophisticated and deadly over time.[66] Exploiting U.S. troops' propensity to consume large quantities of supplies, their use of mounted patrols, and troop concentrations in large bases that necessitated road movement to operational areas, insurgents regularly detonated IEDs on, under, and near roads as coalition convoys passed. Sometimes they repeatedly placed bombs on the same spot. These generated steady U.S. casualties that soon considerably exceeded those of the conventional fighting.

The arrival of appreciable numbers of al-Qaeda fighters from other countries in late 2003 and 2004 reinforced Iraqi Sunni Islamist groups and changed collective tactics and strategic goals of the insurgency. Foreigners flocked to Iraq to attack U.S. troops, undaunted by U.S. dominance in the conventional war. The foreign fighters were not foolish; they planned to fight in ways that gave them, not the Americans, asymmetric advantages.

Thus, by early 2004, the "insurgency" was fully formed. While the insurgent groups' goals, strategies, and tactics varied, and some groups fought each other, the groups' common goal was to expel the foreign occupiers.[67] Iraqi insurgents called their actions *muqawamah,* or resistance.[68] Motives included nationalism, religion, and a desire for revenge against those who harmed or humiliated individual Iraqis, sectarian and tribal groups, and Iraq as a whole.[69] Some Islamist groups, including al-Qaeda, also fought to increase their political control of the Muslim world.[70] Insurgent groups' sharply different hopes for postwar Iraq produced vulnerabilities the coalition finally exploited in 2007.

The insurgents attacked several U.S. weaknesses. They shared Saddam's view that U.S. military casualties were a key American political vulnerability and sought to bleed U.S. forces even if they had no hope of defeating them militarily.[71] They aimed to (1) break the coalition by attacking U.S. and other coalition troops because they saw Western casualties as the coalition's center of gravity; (2) shatter the international effort to stabilize Iraq by attacking foreign national civilians working for IGOs and NGOs; and (3) prevent the reestablishment of normalcy by killing Iraqis working with the coalition and destroying critical infrastructure to prevent the coalition and the new Iraqi government from restoring law, order, and basic utility services—thereby degrading the credibility and legitimacy of both.[72] Bombings of oil-related facilities, including oil export pipelines, reduced Iraqi government revenues and thereby negatively affected the government's ability to provide services.[73] Attacks on other infrastructure facilities kept gasoline, electricity, and clean water in chronically short supply.[74]

The insurgents used fairly standard guerrilla techniques, which long have included expedient bombs. But they innovated based on local conditions and perceived Western political vulnerabilities. Insurgents extensively used mosques to protect supplies and combatants and to produce propaganda when coalition troops attacked mosques that lost immunity under international law because insurgent combatants used them; after the second battle for Fallujah in November 2004, for example, U.S. troops found that all 77 mosques in the city had been fighting positions and/or weapons storage depots.[75] Saddam's cash stash and foreign contributions made some insurgent groups relatively wealthy, enabling them to offer attractive salaries for fighters and bounties for killing Americans, adding a mercenary aspect to insurgent recruiting. Initially $300, the value of an American death rose to $2,500 in May and to $5,000 by October 2003—large sums for jobless men with families to support.[76] And, Saddam appears to have facilitated the looting and chaos of April–May 2003 by releasing some 100,000 common criminals from prison before the war and widely distributing assault rifles, adding significant criminal violence to the challenges the coalition faced.[77]

The IED war quickly became a struggle of technological innovation and counterinnovation.[78] The term *improvised* indicates that the bombs were not products of defense industrial factories. But engineers trained by Saddam's security apparatus built many of them, and they reflected considerable technological sophistication, including use of passive infrared and radio command detonation features, a variety of buried and roadside bomb types, elaborate camouflage, and a variety of explosives and armor penetrating projectiles. Insurgents also reportedly received training and bomb-making materials from Iran.[79] U.S. troops modestly reduced their vulnerability by dispersing vehicles, adding armor to their vehicles, jamming IED command signals, and causing premature detonations. Insurgents reacted quickly and effectively to U.S. technological measures, generating a dynamic interaction.[80]

Al-Qaeda played a major role in shaping the insurgency. It believed the U.S. invasion of Iraq provided a great opportunity to bleed America and then drive it from Iraq, producing a humbling American defeat.[81] Ayman al-Zawahiri, Osama bin Laden's deputy and al-Qaeda's chief strategist, identified two approaches to attacking the United States.[82] First, he advocated inflicting casualties on Americans because he believed casualties are the only language Americans understand, a perspective many Iraqi insurgents soon adopted.[83] Second, he advocated use of suicide bombings because he thought they were the most cost-effective means the insurgents had.[84] The technique became popular among insurgent groups (including non-Islamic groups like Sri Lanka's Tamil Tigers) after Hezbollah used suicide vehicle bombs in the 1980s to drive the United States, France, and Israel out of Lebanon.[85] Bin Laden called for suicide bombings against U.S. troops in Iraq as early as April 8, 2003.[86] But as bin Laden and Zawahiri evidently were in Afghanistan and/or Pakistan, they were unable to directly control al-Qaeda forces in Iraq. What eventually became known as al-Qaeda in Iraq (AQI) was led by Abu Musab al-Zarqawi, a Jordanian national. AQI soon developed a personality that eventually alienated many of its Iraqi supporters.

Al-Qaeda and other insurgents attacked coalition members to break their will to support the United States, cause them to quit the coalition, and thereby increase the burden on American troops and taxpayers.[87] In late August 2003 a bomb killed 19 Italian Carabinieri.[88] An apparent al-Qaeda car bomb killed 17 Italian soldiers on November 12.[89] Insurgents killed seven Spanish soldiers on November 29, 2003, and bomb blasts struck commuter trains in Madrid on March 11, 2004, killing 191 civilians, injuring more than 1,800, and apparently influencing voters to elect a new prime minister who vowed to bring the troops home.[90] Soon thereafter, the Dominican Republic, Honduras, and Nicaragua also withdrew their few troops; by early 2005, at least 13 countries that contributed about one-third of the coalition's original non-U.S./UK force had withdrawn.[91]

Al-Qaeda considered civilian organizations and their personnel legitimate targets. Bin Laden, in a widely publicized 1998 *fatwa*, urged Muslims to kill all Americans, including civilians.[92] Zawahiri identified six largely civilian "tools" that the United States allegedly used to fight Islam: the United Nations, governments of Muslim states that work with the West, multinational corporations, physical communications systems, international news agencies, and international relief agencies.[93] Al-Qaeda repeatedly attacked them all. Al-Qaeda leaders also considered the deaths of noncombatant Muslim civilians acceptable costs of the broad war against the United States.[94]

Military, police, and civilian officials of the new Iraqi government installed in early 2005 were primary targets of assassination.[95] To discourage Iraqis from cooperating with the new government, insurgents attacked army and police units, facilities, and recruits—often as they waited in lines to enlist. In the three months ended June 27, 2005, insurgents killed 52 Iraqi officials while the number of military and police deaths tripled from January to July 2005.[96] Regular attacks on policemen and police stations continue through this writing. Gunmen periodically assassinated Iraqis they called traitors or American spies—acts of both retribution and intimation consistent with the tactics of many other insurgencies.[97]

To demonstrate the inability of the coalition and the Iraqi government to protect foreign civilians and to encourage coalition members to quit, insurgents killed or kidnapped foreign civilian commercial workers, NGO employees, and foreign diplomats. By summer 2003, al-Qaeda had begun a car/truck bomb campaign against foreign civilians that included strikes: on August 14 on the Jordanian Embassy in Baghdad, killing 17 people; on UN headquarters on August 16 and again on September 22; and on the Turkish Embassy on October 14.[98] The United Nations cut the size of its mission after the first blast wrecked its headquarters and killed 22 UN personnel, including Special Representative Sergio Vieira de Mello, and effectively shut down its operations in Iraq after the second.[99] Facilities of the International Committee of the Red Cross were bombed on October 27, 2003.[100] In 2005, al-Qaeda attacked diplomats from Arab states supportive of the new Baghdad government, kidnapping and killing envoys from Egypt and Algeria.

In January–July 2004, more than 70 foreigners from two dozen countries were kidnapped.[101] Germany, France, and Russia warned their nationals to leave the country, and some 600 Russians left.[102] Politically motivated kidnappers (as opposed to criminals seeking ransoms) typically made clear that hostages would be killed unless the target organization/state withdrew support from the Iraqi government or the coalition effort. When concessions were not forthcoming, kidnappers repeatedly executed hostages.[103] Some firms (like trucking companies whose drivers were kidnapped), NGOs, and other states withdrew or reduced their commitments in order to protect their people.[104] Three big international NGOs—CARE, World Vision, and the International Rescue Committee—declined to participate in a U.S. Agency for International Development program as early as May 2003 out of concern for the security situation as well as what the NGOs considered excessive military oversight.[105]

AQI destabilized Iraq by fostering Sunni-Sh'ia sectarian violence.[106] On August 29, 2003, Zarqawi's group killed Shi'a leader Ayatollah Muhammed Bakr al-Hakim. It also blew up several Shi'a mosques, killing hundreds of people. AQI's bombing of the Askariya mosque in Samarra, one of the most holy in Shi'a Islam, on February 22, 2006, prompted Shi'a retaliation against Sunnis that soon looked much like civil war and severely complicated coalition security and nation-building efforts.[107]

Al-Qaeda demonstrated growing political sophistication in its generation of, and communications about, civilian casualties.[108] Before 2003, al-Qaeda used the Internet to glorify its martyrs, to recruit new members, and to distribute training materials. After 2003, al-Qaeda's information operations aimed at the Muslim world and attacked propagandistically coalition member states' support for U.S. policies in Iraq.[109] AQI released videos of captives begging their national leaders to save their lives by withdrawing from Iraq and graphic imagery of executions when its demands were not met.[110] Even the style of execution conveyed a locally salient message: when beheading captives, Zarqawi wielded knives butchers use to slaughter animals rather than the long swords Saudi executioners use to dispatch criminals.[111] Western governments that failed to save their citizens often received more criticism domestically than the murderers, creating a large and obvious political vulnerability that terrorists easily continued to exploit using low levels of targeted violence and widely available communications means. Other images showed insurgent tactical military successes like shooting down U.S. aircraft and ambushing coalition vehicles with IEDs.

Despite coalition military strength levels that fluctuated around 150,000 troops, some 20,000 foreign security contractors, and gradually reconstructed Iraqi army and police forces, the insurgent groups thrived. After a slow start, their attacks increased in late 2003. Attacks on U.S. forces rose to 26,456 in 2004 and 34,131 in 2005.[112] Chairman of the Joint Chiefs of Staff General Richard Myers said on July 21, 2005, that U.S. forces were then on the receiving end of about 70 attacks per day.[113] After the United States said in mid-2004 that insurgent strength was about 3,000, the U.S. military abruptly raised its estimate to 5,000 and then to 20,000, raising questions about how well the United States understood its enemies.[114] Official estimates in mid-2005 were 16,000 to 20,000, including about 1,000 foreign fighters, most by then affiliated with al-Qaeda.[115] CENTCOM commander General John Abizaid in 2006 estimated Sunni insurgent groups to have 10,000 to 20,000 fighters and said Shi'a fighters numbered in the "low thousands."[116] Insurgent force levels remained high despite appreciable combat losses. The United States obviously was not winning this Iraqi war. American critics of the war alleged that contrary to Bush's claim that the Iraq war was an essential part of the so-called Global War on Terror, the war attracted and trained more terrorists than it neutralized.

The insurgents' integrated political/military/economic strategy from the beginning was successful. Polls quickly showed that a strong majority of Iraqis regarded the coalition as an illegitimate and incompetent occupation force; an appreciable minority, especially Sunnis, believed that living conditions were better under Saddam Hussein.[117] Anecdotal evidence indicated that many Iraqis blamed the United States for the civilian deaths the insurgents' terror attacks caused. Many people therefore thought attacks on Americans were legitimate. Growing antiwar sentiment in the West, including the United States and Britain, indicated that the insurgents were successful abroad as well.

Faced with an obviously thriving insurgency and unhappy with DoD's handling of it, President Bush ordered a change in approach. After several months of consideration, Bush in November 2006 replaced Secretary Rumsfeld with Robert Gates; kicked General George Casey, the commander in Baghdad, upstairs to be army chief of staff; appointed General David Petraeus to implement the change in strategy he announced in January 2007; and temporarily increased in troop strength in Iraq.[118] Together, these changes soon became popularly known as "the surge." U.S. troop strength peaked at about 171,000 in 2007. Petraeus altered military operation in Iraq in ways consistent with Bush's desires, traditional COIN practices, and the new army and marine corps counterinsurgency doctrine whose writing he supervised.[119] Petraeus emphasized clearing and permanently holding areas first in Baghdad and Anbar province, and then elsewhere, with small U.S. and Iraqi units, not just pursing enemy fighters. This approach improved human intelligence collection—a key to successful COIN operations.[120] It also led temporarily to more fighting, higher American casualties, and heightened criticism of the war at home.

Meanwhile, starting in roughly May 2006, insurgent Sunni tribal leaders began to turn on their al-Qaeda allies, many of them foreigners, whose brutality, religious fundamentalism, and demands to marry Iraqi women in violation of tribal norms had by then alienated many Iraqis.[121] Petraeus recognized its potential and in 2007 created the "Sons of Iraq" program, which paid Sunnis to enlist in local police forces. By 2008, some 100,000 Sunnis had joined the Americans in policing Sunni areas against AQI.[122] The Iraqi government of Prime Minister Nouri al-Maliki also aided what became known as the "Sunni Awakening."[123] Maliki's Shi'a-led government, meanwhile, cracked down on some Shi'a militias and eventually arranged cease-fires with them. Together, these developments sharply reduced the size of insurgent combat forces and greatly reduced popular support for them.

By early 2008, Baghdad was relatively peaceful. Attacks on American troops and U.S. combat fatalities declined. Anbar province, previously a hotbed of Sunni insurgent activity, was relatively quiet. Although residual sectarian violence continued, the return to relative normalcy and Bush's December 2008 agreements with Maliki enabled new President Obama in 2009 to turn his attention to the worsening situation in Afghanistan. The counterinsurgency campaign thus was won politically—not on the battlefield—after most Iraqi adversaries recognized that their best interests called for at least some mutual cooperation. Skeptics worried that enhanced cooperation was tentative and that the relative peace was unstable, but both Washington and Baghdad, for different reasons, saw the progress as rationalizing a declining U.S. military presence in Iraq.

American Gifts to the Insurgency

The insurgents started the "postwar" war with several major advantages handed to them. American ineptitude helped trigger the insurgency and sustain it.[124] Indeed, Ahmed Hashim, a professor at the U.S. Naval War College, even argues that U.S. actions largely created an insurgency that was not otherwise inevitable.[125] American military planners had assumed that their military victory would be complete and that physical reconstruction and reconstitution of an Iraqi government would be fairly easy and so swift that the United States could reduce its troop strength in Iraq to 30,000 by the end of the summer of 2003.[126] The ground war plan code-named Cobra II, a part of OPLAN 1003V, had an "end state" of regime change, meaning there was no identified need for military efforts to help rebuild a functioning Iraqi state.[127] Planning for postwar stability operations, known in the U.S. military as Phase IV operations, therefore was done largely as an afterthought.[128] The 3rd Infantry Division had no divisional Phase IV plan until after it took Baghdad.[129] Rumsfeld's small attack force meant there were not enough troops to secure the country.[130] Rumsfeld's amended TPFDL also delayed arrival of military policemen who might have helped secure the country before it descended into the chaos that fostered the insurgency.[131] Without orders to maintain law and order and no Phase IV plan, U.S. soldiers and marines in Baghdad simply watched looters dismantle most government buildings in April 2003.

As U.S. troops marched on Baghdad, they ignored thousands of weapons depots and cache sites that contained an estimated 650,000–1,000,000 tons of weapons and ammunition.[132] Many of the unsecured munitions fell into insurgent hands—sharply reducing the need for external supply that is the bane of many insurgencies. In mid-2007, the Congressional Research Service estimated that coalition and Iraqi government forces had secured only 40 percent of the munitions while Ali Allawi, a former minister in the post-Saddam Iraqi government, says the coalition could not account for 250,000 tons.[133] The Americans also largely ignored Iraq's borders for almost a year, only partly due to troop shortages, leaving them open for infiltration of foreign fighters, cash, and specialized munitions, and fostering sanctuaries in neighboring countries.[134] Both omissions were major violations of well-established tenets of counterinsurgency operations.[135]

American troops' personal conduct stimulated the insurgency. For example, Americans broke into houses of innocent Iraqis, humiliated men in front of their families, and intruded on family space that offended Iraqis' sense of women's honor.[136] Iraqi cultural norms called for such insults to be avenged. "Presence patrols" of U.S. troops in vehicles accomplished little but chronically reminded Iraqis of their new status as defeated people living under foreign occupation.[137] Repeating a well-known error of the U.S. war in Vietnam, units randomly fired artillery rounds as "harassment and interdiction" fires that were wasteful, had little effect on insurgents, and alienated Iraqis.[138] Cultural insensitivity and inability to adjust to the requirements of COIN operations drove most of these dysfunctional actions.[139] And, some American units rounded up large numbers of innocent Iraqis for questioning and alienated many further by mistreating them in crowded prisons.[140] The International Committee of the Red Cross repeatedly warned American officials of problems at Abu Ghraib prison before the scandal of U.S. troops' abuse of Iraqi detainees there broke in early 2004—without evident effect.[141]

American troops chronically killed large numbers of Iraqi civilians for at least three reasons. First, insurgents repeatedly tricked Americans into firing on civilians.[142] Second, to reduce casualties to suicide bombers and drive-by shooters, American troops regularly, consistent with force protection-motivated ROE, opened fire on vehicles that came too close to them for their comfort.[143] Hundreds of civilians died, generating grievances that further stimulated the insurgency, thereby increasing the long-term threat to U.S. troops.[144] Third, despite the limited applicability of air strikes to counterinsurgency operations, American air forces continued to bomb, regularly killing civilians.[145] Causes of collateral damage incidents included the human errors, mechanical malfunctions, and purposefully inaccurate information provided by Iraqis hoping to dupe the Americans into killing local rivals or innocent civilians.[146] Given the tribal nature of Iraqi society, American mistakes that directly affected a few individuals often had large political ramifications. For example, an air strike that killed 20 civilian members of one Sunni clan deeply affected the attitudes of the entire clan toward the Americans.[147]

Major American policy errors alienated still more people, including U.S. allies. DoD's Office of Reconstruction and Humanitarian Assistance (ORHA), the initial civilian administrative body in Iraq headed by retired army Lieutenant General Jay Garner, who was plucked from civilian life in early 2003 to head ORHA after CENTCOM planning for the war was well advanced, prepared primarily for humanitarian relief operations, not running the country.[148] The informal Garner also seemed personally ill suited to the role of proconsul. In May, retired career diplomat L. Paul Bremer replaced General Franks as head of the Coalition Provisional Authority (CPA), which Washington created to administer Iraq, simultaneously replacing Garner as the senior American civilian in Iraq; the CPA at the same time absorbed ORHA. Bremer soon issued edicts that drew widespread criticism. Most prominently, he disbanded the Iraqi army and fired many Ba'ath Party members from their government jobs, crippling the government's ability to function, eliminating salaries at a time of high unemployment, and driving many unhappy young men familiar with military weapons to joint insurgent groups.[149] In a May 2003 memo to Prime Minster Blair, the United Kingdom's special envoy in Iraq, John Sawers, described ORHA as an "unbelievable mess" that had "no leadership, no strategy, no coordination, no structure, and inaccessible to ordinary Iraqis."[150] The CPA got only marginally better reviews.[151]

Despite the importance of establishing a secure environment, the United States delayed training the new Iraqi army and police forces. Army Special Forces for decades specialized in training foreign troops for internal defense missions, but senior commanders told SOF instead to hunt insurgents; the conventional military personnel and contractors tasked to train Iraqi troops and police did a poor job.[152]

Interagency squabbling hindered the conduct of stability operations, especially in the first year of occupation.[153] Rumsfeld won a bureaucratic battle with the State Department over control of the immediate postwar civil administration in Iraq, giving the Defense Department responsibility for an occupation.[154] DoD had few of the skills needed for such work but borrowed limited numbers of State Department employees with relevant regional and functional expertise.[155] CENTCOM little supported the hapless Garner, who received few resources and little political support from anywhere.[156] Interagency cooperation improved somewhat after establishment of a U.S. Embassy in Baghdad in mid-2004 and after Bush issued in December 2005 an executive order that gave the State Department primary responsibility for stability operations and DoD concomitantly issued a directive ordering its components to cooperate. By this time the difficulties of stability operations were apparent and the military welcomed relief from an unwanted responsibility.[157]

Poor intelligence helped make bad U.S. policy making worse. U.S. intelligence agencies before 2003 had few human assets in Iraq and a weak understanding of the likely shape of postwar Iraq.[158] But Bush administration decision makers, including senior military officers, were mainly culpable for intelligence-related errors by making rosy operational planning assumptions and initially relying excessively on inaccurate, self-interested reporting from Iraqi exile Ahmad Chalabi, a favorite of senior administration officials who the CIA and the State Department viewed with skepticism.[159] While the CIA warned of a budding insurgency, it did not take a strong stand; the intelligence community as a whole did not generate an assessment in the form of a National Intelligence Estimate (NIE), although the National Intelligence Council issued two warnings about the possibility of an insurgency.[160] Presumably military commanders and congressional leaders were content without a NIE; they could have had one if they asked.

As a result, no one in substantial authority for the war effort expected the Iraqi government and law and order to collapse as they did.[161] Military planners and senior commanders badly misunderstood the *fedayeen*, underestimated their determination to fight, and, incongruously given the then-recent September 11, 2001, attacks, did not anticipate the appeal to foreign Islamists of a battleground on which they could attack even armed Americans relatively easily.[162]

The rise of the insurgency forced the coalition to rely on their weak human intelligence collection capabilities because the sensor-based collection systems that American military technophiles prefer, which accurately locate tanks in the open, cannot find dispersed guerrillas or determine people's political attitudes.[163] With U.S. situational awareness now poor, the insurgents seized the initiative the Americans had in the conventional phase and got inside the Americans' OODA loop, keeping them chronically in reactive postures.

Many of the failures reflected the U.S. military's denigration of "nation-building" operations. As discussed in chapter 4, a prominent perceived lesson of the Vietnam War was that the U.S. military should never again conduct counterinsurgency operations or do nation building. The conventional military therefore purposefully did not prepare for such operations.[164] As self-identified "warfighters," conventional military people denigrated skills useful in COIN operations and were content to leave nation building to civilian agencies and what they regarded as less capable allied military establishments.[165] The result was a U.S. military uninterested in accepting, or preparing for, occupation missions.[166] While civilian leaders badly misjudged the situation and made poor policy decisions, many U.S. military failures in Iraq were, per the American division of civil-military responsibility for military affairs, clearly the generals' responsibility. When U.S. diplomats could not round enough allied troops to keep the peace and help rebuild Iraq, U.S. troops found themselves stuck with an unwelcome mission they were ill prepared to accomplish.

The numerous U.S. government weaknesses, and policy and operational errors, dramatically and rapidly transformed the force that conducted a very successful conventional campaign into an incompetent occupation force. Said Anthony Cordesman in mid-2003:

> [T]he paradox of American power at the start of the 21st century . . . [is] the superpower's staggering prowess at winning wars and its equally remarkable ineptitude at securing the peace. Twenty-one days of shock and awe . . . followed by seventy-five days in which American civilian administrators and hapless soldiers have stumbled and bumbled around like undermotivated keystone cops.[167]

Williamson Murray and retired U.S. Army Major General Robert Scales similarly note that the precision munitions useful in the conventional phase of the war were irrelevant after April 2003 and that what was needed was "intelligent thinking."[168] They note that technologies like PGMs cannot substitute for knowledge of opponents and the nature of war and its aftermath. Failure to master the latter, they reasonably argue, produces political and military defeat later at greater cost.

While policy and planning errors were obvious by late 2003, many U.S. government officials were uncomfortable with U.S. plans even before the war started.[169] Chairman Myers and Joint Staff director for operations (J-3) marine Lieutenant General Gregory Newbold was unhappy with the war plan; Newbold retired in quiet protest soon after the war began.[170] Army Lieutenant General David McKiernan, who was the ground component commander in the conventional phase, wanted the units that Rumsfeld cut out of the TPFDL.[171] Within the Office of the Secretary of Defense, retired army Colonel Joe Collins, chief of post-hostilities operational planning, thought the planned invasion force was too small.[172] In Congress, Representative Ike Skelton and Senator Joe Biden wondered about the evident lack of a Phase IV plan.[173] State Department officials who were planning an occupation administration before Rumsfeld won his turf fight with Secretary of State Colin Powell expected tougher postwar challenges than did CENTCOM. Yet, Rumsfeld and Wolfowitz rejected as unimaginative early versions of OPLAN 1003 that held that 350,000 troops would be needed to occupy Iraq and they publicly rebuked army chief of staff General Eric Shinseki for saying the coalition might need several hundred thousand troops to secure Iraq.[174]

As during the Vietnam War, discussions within the administration reflected concerns about conditions in Iraq and the war plan. The system procedurally "worked." But the strong opinions of senior officials, especially Rumsfeld and Vice President Dick Cheney, and severe DoD parochialism vis-à-vis the State Department and the CIA, especially, precluded effective interagency cooperation and helped generate poor operational performance.[175]

Casualties

In different ways, the United States, Saddam's government, and the insurgents made casualties major parts of the war. The casualty statistics for the two phases of the war vary sharply—consistent with the duration, decisiveness, and different tactics of each side in each campaign.

In the conventional phase, coalition casualties were low. As of May 1, 2003, when Bush declared major fighting over, the United States had suffered 138 deaths (114 in combat and the rest to a variety of nonhostile causes) while the British had lost 42 troops, 19 of whom died in accidents.[176] Some of the combat deaths resulted from friendly fire.

Coalition casualties in the unconventional phase were much larger and differed considerably in cause. U.S. combat fatalities in the unconventional phase exceeded those of the conventional phase by the fall of 2003. At year-end 2003, the U.S. military death toll neared 500, and by mid-2005, the number killed as result of hostile action exceeded 1,300.[177] By mid-October 2007, the combat death toll was 3,115; another 12,622 troops were so badly wounded that they did not return to duty within three days.[178] British and other coalition partners' military deaths then were 302.[179] Coalition casualties fell sharply in 2008 and stayed low thereafter. As of April 8, 2011, some 4,449 Americans had died while engaged somehow with Operations Iraqi Freedom and New Dawn; of them, 3,493 were "killed in action."[180]

Hundreds of contractors working for the U.S. government also died in the line of duty—mainly protecting military supply convoys.[181] But the Pentagon (and the American people) cared so little about the contractors that DoD never bothered to count their deaths.[182] Only the Army Corps of Engineers kept a tally of the deaths of its own contract employees. In mid-October 2007, incomplete data indicated 158 U.S. citizens and 260 other non-Iraqi contractors had been killed while working for the U.S. government—a total of 418 deaths, or more than 13 percent of total U.S. military combat-related fatalities at the time.[183]

Coalition military-related deaths were due to many causes, but the largest single cause was IEDs. In the period May 2003 to March 2004, the first months of the unconventional war, 30.8 percent of all U.S. military fatalities in Iraq and its vicinity were due to IED blasts, more than all other hostile causes combined.[184] In January–July 2004, 759 roadside bombs detonated near American and Iraqi convoys.[185] The pace of attacks continued to grow and by September 2007 the military reported a total of some 81,000 IED attacks on American troops in Iraq, which accounted for 63 percent of U.S. combat fatalities and a larger share of serious wounds.[186]

Iraqi military, insurgent, and civilian deaths are not known with precision. Some 4,530 to 6,050 Iraqi troops are thought to have died in the conventional fighting.[187] Several thousand irregulars apparently also died. In September 2007, *USA Today* reported that a U.S. military database then counted 19,429 insurgents killed by coalition forces since June 2003.[188]

Violent Iraqi civilian deaths were much higher and due to several major causes. Estimates of civilian deaths vary substantially, reflecting methodological differences and evidently ideological positions on the war.[189] The Iraqi Ministry of Planning and Development Cooperation estimated that 18,000 to 29,000 Iraqis—combatants and noncombatants—died in fighting through May 2004.[190] The NGO Iraq Body Count (IBC) estimated that of nearly 25,000 violent civilian deaths it counted between the beginning of the war and March 19, 2005, the coalition killed 9,270 civilians (37%), insurgents killed 2,238 (9%), and criminals killed 8,951 (36%); it did not know causes of the other deaths.[191] IBC counted nearly 18,000 violent civilian deaths from May 1, 2003, to March 19, 2005, the first part of the unconventional phase of fighting. Of those for which IBC found reliable data on causes of death, coalition forces killed 2,388 persons (14%), insurgents killed 5,237 (30%), and criminals killed 8,894 (51%); another 1,047 people died by "shared causation" such as cross fires.[192] These figures imply that the coalition killed 6,882 civilians during the conventional phase of the war and another 3,000 or so in the first two years of the counterinsurgency campaign—for a total of about 10,000. By mid-October 2007, the total death toll for Iraqi civilians had risen to an estimated 74,982 to 81,709.[193] In February 2011, IBC put the toll at 99,393 to 108, 514.[194]

Accurate counts do not seem to exist for foreign civilian employees of foreign commercial firms unaffiliated with the coalition, IGOs, and international NGOs. By mid-2007, journalists counted 119 of their own killed in Iraq since March 2003, mainly Iraqis working for foreign news agencies.[195]

THE IRAQ WAR AND MATERIALIST THEORY

This Iraq war dramatically demonstrates the irrelevance of materialist measures of military power. The numerically inferior coalition quickly and decisively defeated a heavily (if not well) equipped conventional Iraqi army in 2003 but for years thereafter was unsuccessful against numerically much smaller and materially much weaker irregular forces. The war also illustrates evolution in the operational significance of military technology.

A superficial glance at the conventional phase of the war suggests that technology accounts for the operational outcome: superior combat-related technologies of PGMs, mobility, and armored protection produced a rout. Offense had the advantage. Airpower partisans for a time reveled in a "shock and awe" campaign that reconfirmed their faith in offensive airpower.[196]

But a closer look reveals that the outcome of the conventional phase was not determined by mass or technology. The Iraqis were inept for a variety of nonmaterial institutional reasons, and the coalition had unusual advantages. The Gulf War defeat, economic sanctions that included a partially effective arms embargo, and steady if modest American and British bombing as enforcement of the UN-mandated no-fly zones over northern and southern Iraq substantially degraded Iraq's residual military forces. A decade of attention by British and American intelligence agencies gave the coalition good understanding of Iraq's air defense system, conventional military force structure, and military facilities infrastructure. At the beginning of the war, CENTCOM had a bombing encyclopedia of some 4,000 targets.[197] Allied aviators for a decade practiced attacking Iraqi targets; British fliers called the chronic attacks that increased in volume in 2002 in preparation for the invasion "recreational bombing."[198] These intrusions gave American and British air forces advantages not available to states that fight fully sovereign adversaries. Force size was largely irrelevant as the coalition's qualitative superiority offset Iraq's quantitative advantages. Anthony Cordesman, in assessing the conventional campaign, notes that old rules of thumb—like the need of attackers to have the 3:1 numerical advantage commonly cited in military and IR literatures—never were very reasonable but now are thoroughly invalid.[199]

Both phases of the war illustrate how operational asymmetry matters regardless of relative material holdings, making linear, arithmetic summations of military "capabilities" impossible. In the conventional phase the coalition little cared that its ground forces were numerically smaller than Iraq's because U.S. and UK ground units were qualitatively better and could count on massive, reliable fire and other support from their dominant air and naval forces, PGMs, space-borne communications, and sensors. Saddam declined even to try to redress the imbalance by conducting urban operations, which might have given him other, appreciable asymmetric advantages.

But coalition advantages dissolved almost instantly when insurgents shifted to irregular warfare. Large parts of the U.S. force structure were irrelevant to counterinsurgency warfare and many conventionally minded commanders initially operated sub-optimally, even foolishly. Lightly equipped Special Forces and civil affairs personnel suddenly became more valuable because they were among the few American troops who had any idea how to conduct counterinsurgency operations.

Technologies and material mass played very different roles in the war's two phases. In the conventional phase, PGMs and technical intelligence collection systems that produce target data devastated Iraq's combat formations but did little collateral damage. But in the unconventional phase, excessive and inappropriate use of airpower, suddenly poor intelligence, and purposeful Iraqi disinformation fed to gullible Americans led to frequent misuse of firepower that alienated many Iraqis and fostered the insurgency. As in Vietnam and Yugoslavia, U.S. firepower produced much negative operational value. Ostensible military assets again often became political liabilities.

CENTCOM initially used mobility to thwart Iraqi situational awareness to generate force protection as well as to move to Baghdad. But in the unconventional phase the Americans' presence patrols were ineffective and irritated Iraqis. Insurgents chronically ambushed convoys, generating the U.S. casualties that fueled domestic American antiwar sentiment. Insurgents relied on material asset-based U.S. mobility to place their targets into their ambush kill zones—the producers of most U.S. casualties.

Again, Americans were enthralled with technology that failed to fulfill technophiles' promises. The officer in charge of anti-IED efforts in 2003 vainly expected a technological solution to his problem within six to eight months.[200] General Abizaid talked of need for a "Manhattan Project" to defeat IEDs; marine Lieutenant General James Mattis lamented technologists' inability to solve the IED problem.[201] Through 2007, the Americans' multibillion-dollar effort to solve the IED problem ineffectually focused primarily on technological "solutions" while insurgents working with comparatively primitive laboratory and manufacturing facilities readily adapted to Americans' technological fixes and introduced new IED capabilities, including weapons that could penetrate the best armored vehicles. Tactical efforts to counter IEDs at ambush sites and more armor on vehicles and troops were weakly complemented by efforts to root out bomb makers before they could deploy weapons and to alter American operational patterns to reduce vulnerability to such attacks.[202] The IED threat receded only after 2007 when U.S. intelligence, not technology, got better.[203]

Al-Qaeda's use of television and the Internet challenges materialist theory's concept of the role of technology in warfare.[204] Traditionally, theorists view militarily relevant technologies as tools of overt military forces that enhance firepower or provide protection or mobility, for example.[205] Yet television, digital cameras, mobile phones, and the Internet are not weapons; they have no firepower and cannot protect troops or transport supplies. They move only ideas. The media are neutral; they become strategically important only when they deliver politically relevant messages to receptive audiences.[206] Al-Qaeda's and other insurgent groups' images of violence—including beheadings, IED ambushes, and aircraft shoot-downs—as well as recruitment messages and training films provided the content that made television and the Internet very valuable for information warfare.[207] Because the messages would not be as valuable without effective communications means, insurgents' information campaigns were joint products of externally procured information technologies and content they generated. Ideational and material contributions were tightly linked, making sterile any claims for the superiority of either.

At the same time, the Internet provided fairly secure intra- and interorganizational communication. While Western security agencies shut down a few servers, the engineering design of the Internet gave insurgent groups protection and security measures like encryption made it largely invulnerable to attack. When Gabriel Wiemann, a professor at the University of Haifa, began tracking terrorist-related websites in 2001, he found 12; by mid-2005, he monitored more than 4,500 sites.[208]

The Americans displayed fine conventional war-making skills in March–April 2003 but were largely inept at counterinsurgency operations immediately thereafter. Skills useful in conventional warfare were counterproductive when used excessively or inappropriately in COIN operations. The problems started with poor leadership. Military commanders poorly understood the political/military challenges implicit in the mission they received from President Bush, poorly crafted war plans, and poorly produced a force to accomplish the mission.[209] Andrew Krepinevich, who praised U.S. performance in the conventional phase, called the failure to plan for stability operations "inexplicable."[210] U.S. Army Lieutenant General Ricardo Sanchez, commander in Iraq in 2003–2004, clashed personally with CPA chief Bremer and was widely seen as unable to comprehend his mission.[211] Andrew Bacevich called him the Westmoreland of Iraq.[212]

But in 2004 and later, some learning occurred. Special Forces and civil affairs officers criticized U.S. operations as dysfunctional.[213] General Casey, upon replacing Sanchez, established in Iraq a "COIN Academy" that taught long-standing principles of counterinsurgency.[214] After the Abu Ghraib prison scandal broke, the military issued new orders on the treatment of detainees.[215] General Petraeus, who led the 101st Airborne Division into Iraq in 2003, then supervised training of the Iraqi army and headed the U.S. Army Combined Arms Center as a three-star, and was the four-star commander in Iraq in 2007–2009 and then CENTCOM commander, embraced a much more sophisticated approach to the war that reflected understanding of counterinsurgency principles. His successor in Baghdad, General Ray Odierno, did the same. The joint army-marine corps counterinsurgency doctrine published in December 2006 received generally good reviews.[216] Army chief of staff General Peter Schoomaker in 2005 provided command emphasis by writing an introduction to a rerelease of a book by Lieutenant Colonel John Nagl on lessons of Vietnam for the U.S. military.[217] Aided by such command emphasis, it became fashionable again for military personnel to read classic books on counterinsurgency like those of David Galula and Robert Thompson.[218]

Although some learning and adaptation occurred in 2007 and afterward, aspects of U.S. military operations in Iraq remained dysfunctional. Force protection policies that allowed troops to use deadly force against Iraqi civilians if they merely felt threatened remained largely intact.[219] And, like in Vietnam, efforts to keep up troop morale by providing creature comforts meant most troops still lived in big bases separated from the populace.[220] Short tours, similarly designed to keep troops happy, continued to hinder development of local expertise, including cultural awareness and language skills, and thence the HUMINT that is essential in COIN operations.[221] Both faults were violations of the basic tenet of counterinsurgency doctrine that calls for close military association with local people.

U.S. officers chose not to learn vicariously even from their British allies. Based largely on its long and difficult experience in Northern Ireland, the British army developed patrolling and other counterinsurgency techniques that emphasized maintaining low profiles, operational flexibility, and tactical HUMINT collection. They employed the techniques in close proximity to U.S. troops in Bosnia starting in late 1995 and in Kosovo after mid-1999. British officers criticized U.S. force protection policies and excessive reliance on technology as shortsighted; they, like Anthony Cordesman, suggested tightening ROE for troops who felt threatened as a way to improve force protection over the long term by reducing Iraqi civilian deaths—and resultant alienation of the rest of the populace.[222] None seems to have had appreciable effect, perhaps partly because the British made mistakes in the Basra area that tainted their reputation for counterinsurgency expertise.

In contrast to American institutional learning problems, many U.S. officers in Iraq and external analysts of the war saw Iraqi and foreign insurgent groups as innovative and competent.[223] Retired Lieutenant General J. T. Scott, former head of the Army Special Operations Command, said of al-Qaeda's capacity to adapt and to learn:

> What separates this guy from all his one trick predecessors is, he really has developed a learning organization . . . They learn from their successes and failures, as well as ours. Each time, we've prepared to defend ourselves against his last operation, and each time he's discovered a new, asymmetric way to get at us.[224]

Bruce Hoffman concurs, saying that al-Qaeda is a "remarkably nimble, flexible and adaptive entity"—and that one of its three special capabilities is an ability to identify "a key vulnerability or gap in the defenses of its principal enemy—the United States—that could be ruthlessly exploited."[225] Hoffman reiterated the point in September 2007, saying "al-Qaeda has shown itself to be remarkably nimble and adaptive—able to compensate for and even obviate some of our [U.S.] most effective countermeasures."[226] Domestic Iraqi groups similarly developed learning skills and showed operational flexibility. The Americans were lucky that AQI was, uncharacteristically for al-Qaeda affiliates, inflexible enough theologically to alienate Iraqi Sunnis in strategically relevant ways.[227]

THE IRAQ WAR AND STRATEGIC INTERACTION THEORY

This war features two major strategic interactions: the conventional phase of March–April 2003 and the unconventional phase thereafter. First, the United States waged a conventional war against Iraq's conventional forces. Despite significant supporting roles for coalition special operations forces and Iraqi *fedayeen*, most of the troops employed by both sides were conventional. They mainly fought conventional fights. U.S. commanders worked hard to reduce Iraqi civilian casualties, meaning the United States tried to avoid even the appearance of barbarism. Hence, the interaction was a conventional-conventional, same-approach interaction the materially dominant United States and its coalition partners won handily—consistent with Arreguín-Toft's expectations.[228]

In the unconventional interaction, remnants of Saddam's regime now stripped of most of their major material military assets, and sectarian, tribal, and foreign groups used fairly standard unconventional warfare techniques against a coalition force that remained primarily conventional in structure and operations despite a considerable effort by special operations troops. Poor U.S. intelligence, operational errors, and force protection policies sometimes generated unintended Iraqi civilian casualties, but when individuals and small groups of junior American troops broke military law by harming Iraqis, violators were prosecuted. U.S. policy did not qualify as barbarism by Arreguín-Toft's definition. Iraqi insurgents' actions were much closer to Arreguín-Toft's GWS than to conventional defense. Although the insurgents for several years stymied America's initial ambitions, the changes Bush ordered and Petraeus implemented in 2007, along with the Sunni Awakening and Iraqi government actions, helped produce what as of this writing seems likely to be a largely strategically successful, if drawn-out and costly, operation. Hence, the United States eventually "won" a mostly conventional-GWS, opposite-approach interaction—inconsistent with Arreguín-Toft's logic.

Operational details again differ appreciably from Arreguín-Toft's model. Most obviously, none of the insurgent groups closely resembled the people's war actors whose Marxian ideology and tactics he embeds in his GWS. Instead, they included residual military and security force elements of a defeated tyrannical regime, Sunni tribal groups offended by their loss of status and economic perquisites, Islamists of various sorts, and fighters motivated primarily by money.[229] And, as in other cases, insurgent groups' policies of killing and kidnapping of thousands of noncombatant civilians in support of political goals easily qualify as barbarism.

DIMENSIONAL ANALYSIS

The several belligerents conducted very different wars in the conventional and unconventional phases of the war, and achieved very different operational results. The performance of essentially the same U.S. force that easily won the conventional campaign deteriorated markedly and rapidly in the unconventional phase as its opponents shifted operational dimensions in ways the Americans found hard to counter.

The U.S. Conventional War

The conventional war against Saddam's regime was fought and decided almost entirely in Iraq's *leadership* and *operational execution* dimensions. The coalition did not attack Iraqi *national will* because U.S. planners and policy makers assumed that coalition troops would be welcomed as liberators. The coalition did not attack Iraq's *resource mobilization* and *resource conversion* capabilities because it planned to remake the country using those resources. And, Rumsfeld initially meant to use the core of Iraq's defeated military to make a new Iraqi army, meaning there was little reason to attack violently in the *force generation* dimension.[230] The only significant U.S. actions in this dimension were psychological and intelligence operations designed to keep Iraqi regular army units from fighting.

The United States was successful in both of its chosen dimensional operations, although an attempt to kill Saddam on the first day of the war failed. The coalition quickly smashed Iraq's conventional armed forces and captured Baghdad, leading to the collapse of Saddam's government. Focusing on decapitating the Ba'ath regime, the United States widely distributed images of senior officials on playing cards to coalition troops. When American soldiers captured Saddam Hussein in December 2003, most of the faces on the cards were in custody and his sons Uday and Qusay were dead. But neutralizing Saddam and his senior subordinates had little effect on the insurgency as effective new leaders emerged.[231]

The United States did, however, ameliorate its own vulnerabilities in the *national will* dimension. U.S. commanders, keenly aware of their political vulnerabilities to casualties, made force protection and collateral damage minimization two of their three planning imperatives.[232] Expected civilian casualties were a primary determinant of whether a potential target would be approved for attack. Before the war, the U.S. Air Force identified 437 prospective targets; of those, a military model identified 22 as "High CD" (high collateral damage) targets that, if attacked, probably would generate 30 or more civilian casualties.[233] Rumsfeld insisted on personally approving all attacks on High CD targets. Good planning and a short war kept the number of civilian casualties relatively low and thereby irrelevant to the outcome of the conventional phase of the war. But senior decision makers confirmed publicly yet again that Americans are highly sensitive to both civilian and friendly military casualties.[234]

Saddam's Conventional War

Iraq recognized the importance of finding enemy vulnerabilities but failed to exploit any significant coalition vulnerabilities and showed little awareness of its own vulnerabilities.[235] Saddam's conventional military tried to fight the coalition symmetrically in fairly open terrain—using weapons and other equipment sometimes decades older than the newest coalition gear—and lost badly in the *operational execution* dimension to enemies who exploited their many tactical military vulnerabilities with asymmetric strengths based on technology, well-trained and well-equipped forces, and well-coordinated joint and combined operations.

Iraqi military operations targeted three prospective U.S. vulnerabilities in two other dimensions—*force generation* and *national will*. First, a limited Iraqi military effort to exploit an acknowledged U.S. technological weakness—by jamming the signals of the GPS satellites that guide many precision munitions—failed when F-117s destroyed the jammers with GPS-guided bombs; Iraq tried futilely to jam GPS signals in 1991 as well.[236] Had the Iraqis been able to jam GPS signals effectively, they might have neutralized a considerable portion of America's air and missile arsenal.

Second, Iraq tried to exploit American and other coalition member states' aversion to civilian casualties and their commitments to international conventions that protect schools, mosques, hospitals, and archaeological sites.[237] U.S. ROE again made such sites politically invulnerable so long as they were not used in combat operations. But as they had since 1991, the Iraqis deployed personnel and equipment in and near protected sites and *fedayeen* used human shields, in effect trying both to protect their assets and to lure coalition forces into politically exploitable collateral damage incidents.[238]

The Iraqis employed information operations to exploit civilian casualties caused by coalition air and missile strikes, especially, and they blamed the coalition for some civilian casualties apparently caused by errant Iraqi antiaircraft artillery rounds and missiles.[239] But their propaganda was again crude.[240] These efforts altered some tactical-level coalition military operations but failed as strategic war-winning political devices. Nevertheless, prominent U.S. efforts to minimize political vulnerability to collateral damage incidents prompted retired General Bernard Trainor to quip that the world had entered a period of "Lawfare"—replacing traditional warfare.[241] This comment amounts to a recognition of, and lament about, the expansion of the scope of war into legal and ethical aspects of the *national will* dimension.

Iraqi efforts to exploit coalition aversion to civilian casualties in the conventional phase were largely passive. That is, the Iraqis mainly positioned themselves and civilians in ways that could create political embarrassment for the coalition if coalition troops made mistakes that killed civilians. They generally did not directly cause civilian casualties, although Iraqi forces murdered some civilians in the Basra area in an apparent attempt to blame British troops for the deaths.[242]

Third, after Saddam Hussein at first doubted U.S. willingness to bear casualties and thus U.S. resolve to attack him, he and his government believed that inflicting coalition military casualties would erode domestic political backing for the war in America and Britain, thereby causing the coalition to break and its member states to quit Iraq, independent of the effect of the losses on coalition military capabilities. Noting America's aversion to casualties, Saddam called America a "paper tiger" before the 1991 Gulf War and apparently continued to see inflicting U.S. casualties as a war-winning strategy.[243]

In sum, Iraq accurately identified civilian and military casualties as key U.S. political vulnerabilities but could not exploit them. Coalition dominance was so great that U.S. fatalities in March and April 2003 were lower than during the Gulf War. The war was so short, friendly military casualties so low, collateral damage incidents so few, and Saddam's propaganda apparatus so ineffective that American casualty aversion played little role in the conventional campaign. Moreover, the United States for a time seemed really to have won the war. As a result, modest U.S. domestic opposition to the war through May 2003 was strategically irrelevant.

The U.S. Unconventional War

In April 2003, the U.S. military found itself fighting growing numbers of irregular enemies and helping coalition member governments' civilian agencies and private actors working under U.S. government contracts to build and stabilize a new Iraq. As part of a multifaceted U.S. government and international effort, it also tried to conduct competent counterinsurgency operations.

Counterinsurgency operations by nature are complex. As army and civilian administrators demonstrated successfully in the Philippines and the U.S. government as a whole, despite generally good work by CORDS after 1968, demonstrated by its failures in Vietnam, successful counterinsurgency operations cross all six dimensions. Rumsfeld's Defense Department initially failed to conduct coherent COIN operations in any dimension.

The Americans at first denied they faced an insurgency. When the insurgency had grown to the point that it could not be denied, the Americans failed to adapt well to conduct counterinsurgency operations. Only in mid-2004 did General Casey, after replacing Sanchez as the ground commander in Iraq, write a campaign plan for counterinsurgency operations. Even after Petraeus's changes in 2007, many U.S. troops still said they believed their job was to find and kill enemies—not secure the peace. American troops therefore continued to conduct themselves in ways almost purposefully designed to alienate Iraqis—thereby losing the political battle that is critical in counterinsurgency operations.

Effective counterinsurgency campaigns normally subordinate combat operations to winning the "hearts and minds," or at least the acquiescence, of most people—that is, operating successfully in the *national will* dimension.[244] Paraphrasing Mao, the fish of Islamist, tribally based, and al-Qaeda-affiliated insurgent groups swam in the sea of the Iraqi populace.[245] But for the same reasons the United States avoided military attacks in the *national will* and *resource* dimensions in the conventional phase, it made little effort before the unconventional phase to ensure that activities in these dimensions would augment its efforts after the fighting stopped. By failing initially to influence Iraqi political and economic life, the Americans abandoned opportunities to prevent the insurgency from claiming the loyalty of the people, thereby both creating and failing to ameliorate then latent but eventually major threats to coalition political and military objectives. Only years into the war did conventional troops in large numbers begin to internalize the new counterinsurgency doctrine.[246]

The Americans' unwillingness for about a year to try meaningfully to close Iraq's borders to inflows of foreign fighters from Iran, Syria, Jordan, and Saudi Arabia, or to secure Saddam's vast stocks of munitions, were major failures in the *resource mobilization* dimension. The coalition faced disadvantages already in this dimension given the weapons caches and cash hoard the Ba'athists prepared before the war. Counterproductive U.S. troop behavior and policies also aided insurgents' efforts to keep popular support and to recruit fighters.

Trained security personnel and the explosives Saddam cached before the war gave the insurgents big advantages in developing IED manufacturing capacities. But because the Americans focused on developing tactical devices to protect their vehicles and to defeating the devices after they were planted, they inadequately concentrated on generating the intelligence needed to attack key bomb makers, their laboratories and bomb-manufacturing sites, and operatives—and thereby to destroy IED manufacturing and emplacement capacity.[247] The Americans thus failed until 2007 to effectively attack key insurgent capabilities in a strategically important arena of the *resource conversion* dimension.

Insurgent groups sprang from many sources—including elements of Saddam's regime, tribal groups, Islamists, and foreign groups. Coalition actions initially angered many Iraqis enough to induce them to take up arms, fostering insurgent *force generation*. The Americans began to neutralize insurgent groups effectively only in 2007 with more sophisticated military, political, and economic approaches that redressed some grievances and exploited tensions among insurgent groups. AQI's brutality, which produced the Sunni Awakening, helped considerably.

Poor coalition HUMINT hampered efforts to round up the *leadership* of the insurgency by killing or capturing individual leaders. Insurgent groups, some of which were based on existing tribal and former regime organizations with established command relationships, also readily replaced their losses.[248]

And, while U.S. troops fought well when they could find their enemies, they did not do so often.[249] The insurgents continued to initiate many engagements, especially ambushes, meaning U.S. performance in the combat portion of the *operational execution* dimension was mixed. Moreover, in counterinsurgency wars the most critical operations often are nonviolent pacification efforts; American troops too often focused excessively on combat operations and either ignored or poorly performed other important counterinsurgency-related tasks.

The Insurgents' Unconventional War

The unconventional phase of the war prominently featured insurgent guerrilla attacks designed strategically to influence U.S. and other foreigners' *national will*. Degraded popular will would encourage states and nonstate entities to withdraw their people from Iraq rather than suffer more casualties, thereby negatively affecting external *resource mobilization* for the United States and *force generation* for the military component of the coalition as a whole. The insurgents clearly could not defeat the coalition directly in combat. The insurgents instead generated coalition military casualties to help create the external strategic political conditions for insurgent victory without winning tactical battlefield victories. They thereby used a variant of the model North Vietnam used successfully against America. Attacks on aid agencies and Iraqi civilians damaged the new Iraqi government's ability to mobilize internal resources, get meaningful help from abroad, meet its promises, and achieve legitimacy.

Using fairly standard techniques of unconventional warfare, the insurgents initially sharply altered to their own advantage the pattern of asymmetric strengths and vulnerabilities that previously favored the coalition. Saddam's Iraq—a weak, rogue state with limited conventional military capabilities and little external support—had vulnerabilities the U.S. military spent years finding ways to exploit. But the insurgent groups, while having modest offensive and defensive capabilities, were secretive irregular organizations with only a few vulnerabilities susceptible to identification or exploitation by narrowly military means. After April 2003, anti-U.S. forces had the invulnerability to conventional U.S. military firepower afforded by a lack of a territorial base they felt obliged to defend.[250] They thereby rendered U.S. air and missile forces largely useless and sharply degraded the technological and other advantages of U.S. ground forces designed for conventional combat.

Senior Americans understood that the IED was the primary means by which insurgents attacked U.S. *national will*. Retired army General Montgomery Meigs, head of the Joint IED Defeat Organization (JIEDDO), the Pentagon task force charged with neutralizing the IED threat, aimed to "defeat IEDs as weapons of strategic influence," which insurgents used "against our structural weaknesses."[251] CENTCOM commander General Abizaid worried, "This thing could unravel on us by wearing down the American public with these IEDs."[252]

Al-Qaeda's two major goals—to increase U.S. military casualties and to raise the financial cost of the war to the United States—amounted to attacks in the *national will* and *resource mobilization* dimensions, respectively.[253] Bin Laden cited the high level of damage to the U.S. economy caused by the September 11, 2001 attacks as an indicator of the success of that mission; he counted real and opportunity costs of at least $1 trillion due to physical damage, lost work in the immediate aftermath of the attacks, reduced stock market valuations, and layoffs.[254] Against these costs, the 9/11 Commission estimated that the hijackers spent $400,000 to $500,000 on the operation—a benefit-cost ratio of about 2,000,000:1.[255] Even allowing for large estimation errors on both numbers, al-Qaeda on 9/11 and again in Iraq conducted operations that were very cost-effective.

As Iraqi insurgents both stymied U.S. political and military goals and increased the human cost of the war, antiwar sentiment in the United States again demonstrated U.S. vulnerability to American casualties. In May 2004, Senator John McCain expressed concern that U.S. resolve was wavering and that the United States risked confirming the lessons of Mogadishu: pain inflicted on the United States would cause Washington to quit.[256] U.S. military and civilian officials in Baghdad feared the loss of domestic support for the war would doom the effort and focused their public relations efforts on shoring up American support.[257] General Franks stated the point bluntly:

> The lessons terrorists took away from these earlier attacks on America was simple: *If you kill Americans, they will quit* [emphasis in original].[258]

Antiwar Democrats' gains in the 2006 congressional elections after only about 2,500 American combat fatalities reconfirmed the point. And, in 2007, Admiral William Fallon, who replaced Abizaid as CENTCOM commander, said he focused on only two metrics of the status of the war: violent incidents and U.S. combat fatalities.[259]

The insurgents' main asset was popular Iraqi opposition to the occupation—a reflection of *national will*, which in turn provided a substantial *resource mobilization* capability. Of the four motives that Hashim argues led Iraqi Sunnis to join the insurgency—nationalism, honor, revenge, and pride—the latter three reflected the humiliation of defeat and the occupation and the loss of social status and economic perquisites the fall of Saddam's regime entailed.[260] Many insurgents' focus thus was mainly negative—a desire to expel the foreigners. Positive goals were much more diffuse. There was no single, positive collective vision of Iraq's future and Sunni insurgents vaguely promised that life surely would be better when the Americans leave Iraq.[261] Disparate ambitions suggest that the internal strength of the insurgency was modest, and that it was vulnerable to coalition and Iraqi government actions that addressed honor, revenge, and pride motives, at least. When Sunni tribal leaders attacked AQI, partly in exchange for U.S. financial payments, local autonomy, and respect, foreign Sunni Islamists' destruction of their welcome produced the vulnerability U.S. commanders effectively exploited in 2007 and later.

U.S. LEARNING AND ADAPTATION

The United States showed sharply different learning and adapting characteristics in the two phases of this war. The same people displayed excellent learning abilities in the conventional phase and often very poor ones in the unconventional phase soon thereafter—indicating that no consistent, systematic dearth of cognitive or organizational capabilities, information, or material resources accounted for the performance problems. Rather, the differences in U.S. military performance mainly reflected different types of focus, operational preferences, and training.

The United States displayed far better understanding of the vulnerabilities of Saddam's conventional military than of the much broader set of factors relevant to counterinsurgency operations. The United States built technical intelligence collection capabilities heavily focused on supporting conventional warfare, mainly with targeting data, to the purposeful exclusion of learning about, and preparation for, counterinsurgency wars the military eschewed. The Americans easily won a conventional war against Iraq in 1991, seemingly confirming the value of a conventional focus. Years of attention to Iraq since 1991 gave the U.S. military excellent knowledge of Saddam's remaining conventional military force structure, but technical intelligence collection systems could not provide information or understanding about Iraqis' attitudes, concerns, and other factors relevant for counterinsurgency operations. Poor HUMINT, abetted by congressional unwillingness to risk American case officers in operations against Saddam's police state, especially limited America's knowledge of, and ability to learn about, Iraq as a whole.

The Bush administration decided to attack Iraq in the absence of a warning from the intelligence community of an imminent threat. Instead, officials reporting to Under Secretary of Defense for Policy Douglas Feith reportedly compiled fragments of intelligence material into a case that supported the invasion decision.[262] Feith denies the charges.[263] Such "cooking the books," if it occurred, is a legitimate if unwise prerogative of policy makers, but not of intelligence agencies, and was not directly a fault of cognitive errors.

In the unconventional phase, having belatedly recognized that it had no choice but to fight a counterinsurgency war, the U.S. military and intelligence community increased HUMINT collection efforts and achieved much better success in 2007.[264] But short tours aimed purposefully to maintain troop, government civilian, and contractor morale, and deficient language and cultural training hampered collective U.S. government abilities to learn about Iraq throughout the war. In addition, as in the bombing campaign and in Vietnam, U.S. troops little heeded relevant lessons of others. The British in Iraq offered sound advice on how to better run counterinsurgency operations, but the Americans seem to have reacted negatively for at least two reasons: they again knew better, and the advice impinged on bureaucratically powerful force protection imperatives.

The Americans similarly showed very different abilities to exploit Iraqi vulnerabilities in the war's two phases. The U.S. military's *OODA loop* concept highlights the importance of rapid and accurate learning and adaptation relative to adversaries and the U.S. military very well achieved it in the conventional phase. But in the unconventional phase, the still largely conventionally focused U.S. military—in force structure, doctrine, technology, and cultural outlook—was slow, ineffective, and often counterproductive in its adaptations. Insurgents got inside the Americans' OODA loop with predictably negative operational consequences for coalition forces. Only the Bush/Petraeus initiatives of 2007 enabled the United States to adapt enough to influence the course of the war—and its apparent outcome.

Americans' awareness and amelioration of their vulnerabilities also varied sharply by phase. In the conventional phase, CENTCOM planners and General Franks accurately understood the tactical risks of a rapid advance on Baghdad by a relatively small force and took successful gambles. They accurately identified civilian and U.S. military casualties as strategic political vulnerabilities and largely avoided them. In the unconventional phase, U.S. military casualties remained a major vulnerability, but the military and the Bush administration showed little ability to ameliorate them until the intelligence-related successes of 2007. The major effort to find technological solutions to the IED challenge itself reflected an enduring failure to learn the limits of military technology and to avoid dysfunctional overreliance on it. Even after 2007, the military continued to stoke the insurgency by maintaining force-protection-motivated ROE that gave troops wide latitude to shoot Iraqi civilian noncombatants; military commanders still chose to accept more civilian casualties in order to keep down their own. This again was a policy choice, not a learning or adaptation flaw—at least in its narrowest sense. And the air force, as ever pursuing organizational interests, continued to bomb, generating a steady flow of civilian casualties.

SUMMARY

This case study offers a compelling rebuttal of the standard materialist conception of military power as a linear sum of actors' material asset holdings. Virtually the same force that crushed Iraq's conventional military stumbled badly in counterinsurgency and stability operations against materially much weaker opposition immediately thereafter. This near simultaneity of drastically different operational performance refutes Kenneth Waltz's assertion that power is fungible for great powers.[265] It is not. It also refutes Charles Glaser's version of materialist theory, which holds that what matters is capabilities to perform a mission, which he defines as COW-like power modified by the offense-defense balance.[266] Instead, the case shows that contingent, dyadic, directionally and dimensionally different interactions determine operational success and failure—and thence the possession of military power.

The conventional phase of Operation Iraqi Freedom and its aftermath is an excellent example of sharply different operational performances of a purported military superpower based virtually exclusively on the changed strategies and tactics of its opponents. Compared to Saddam's approach, insurgents altered the relevant dimensions of conflict, created different dyadic asymmetries vis-à-vis the coalition, and changed the operational significance of each side's vulnerabilities. Saddam tried to match the superpower in conventional combat and lost badly. Much weaker insurgents later chose to be different and had much greater success.

The two sides displayed sharply different levels and types of institutional learning. While American institutional learning in Iraq after a slow start arguably was better than in Vietnam, it was still very slow compared to that of insurgent groups and slow compared to the army's experience in the Philippines in 1899–1902. Effective Sunni insurgent groups' leaders and many fighters once worked for Saddam's Ba'ath regime—a government with a long record of conventional military incompetence—indicating that fighters displayed vastly different propensities for learning and adapting when their institutional circumstances changed. The strategic error of al-Qaeda in Iraq—alienating its initial Sunni allies—seems more prominently to reflect strident theological beliefs than learning deficiencies.

This case also illustrates that Shimshoni's emphasis on leadership is well placed. American military leaders built a force that arguably was the best the world has ever seen at medium-intensity conventional combat—especially against a weak opponent. But the narrowness of perspective, driven by institutional lessons of the defeat in Vietnam and cavalier assumptions, led to creation of a force with major weaknesses that a produced horrible Phase IV plans, fostered an insurgency, and for years badly conducted counterinsurgency operations. When the insurgency broke out, military institutional constraints and leaders' commitment to conventional methods kept U.S. military learning and adaptation slow in comparison to their far more nimble adversaries.

Biddle's theory works well for the conventional phase of the war but his *modern system* theory is applicable only to conventional warfare. The unconventional phase featured an even more modern form of warfare in which unconventional tactics focus on strategic political and other ideational vulnerabilities, not the traditional military objectives of winning battles and taking ground.[267]

The theories of Merom and Mack both apply. Americans' patience with a lack of immediate success and acceptance of human costs were small in a war in which immediate American interests were low—especially after the Bush administration's WMD rationale for the war seemed to evaporate when no WMD stockpiles were found. Consistent with both Merom and Mack, American voters gave Congress in November 2006 to Democrats who campaigned against casualties that, while surely tragedies for the deceased and their friends and families, were trivial losses for a genuine superpower focused on a mission—the scenario materialist-oriented IR theories typically presume. Unlike in Mack's Vietnam case, many Iraqi insurgents were motivated by concerns other than nationalism, but their commitment levels were high—providing the determination to hold at bay for years a much weaker-willed United States.

CHAPTER 7

The U.S. War in Afghanistan (2001–?)

Al-Qaeda's attacks on the United States on September 11, 2001, provoked an American war in Afghanistan.[1] When it became clear that al-Qaeda conducted the attacks, and that its leader Osama bin Laden was in Afghanistan, President George W. Bush demanded that Afghanistan surrender bin Laden to American authorities. But the Afghan government, headed by Mullah Mohammed Omar, leader of a group known as the Taliban whose most prominent feature was adherence to a peculiarly Afghan, and especially Pashtun tribal, version of the Deobandi school of Islamic thought, refused.[2] Bush in turn ordered military action to capture or kill al-Qaeda leaders and depose the Taliban regime. The American people, stung by the deaths of 2,976 people at the hands of 19 suicide attackers, strongly backed Bush. Bin Laden, a veteran of the Soviet defeat in Afghanistan in 1989, said he hoped the 9/11 attacks would entice the United States into a war in Afghanistan that would have consequences for America similar to those of the Soviet Union, which collapsed soon after its defeat there.

Bush soon after 9/11 asked the Pentagon for a plan, but the military did not have contingency plans for war in Afghanistan and could not promise a quick response.[3] Director of Central Intelligence George Tenet then proposed a CIA-led campaign that would exploit the agency's ties to the major Afghan group fighting the Taliban—a group of mainly ethnic Tajik, Uzbek, and Hazara fighters called the United Islamic Front for the Salvation of Afghanistan, better known as the Northern Alliance, whose leaders were veterans of the U.S.-supported war against the Soviet Union in the 1980s.[4] Tenet's plan satisfied Bush, who was aware of problems that invading powers from Alexander to the Soviet Union had in "the graveyard of empires" and correspondingly decided to keep the U.S. military "footprint" in Afghanistan "light."[5] Quick success at deposing the Taliban in late 2001 led to a modest effort thereafter designed to conduct counterterrorist (CT) operations focused on al-Qaeda.[6]

Even this small effort was soon overshadowed, however, by Bush's decision in 2002 to authorize planning for the invasion of Iraq. Afghanistan became what the U.S. military calls an "economy of force" operation—one designed to achieve modest objectives while freeing resources for the main effort elsewhere. As the war effort in Iraq floundered in 2003–2006, operations in Afghanistan suffered from some of the same policy and military operational errors discussed in chapter 6, from unique Afghanistan-focused misunderstandings and mistakes, and from senior policy makers' neglect, which David Kilcullen, a counterinsurgency specialist and advisor to senior U.S. civilian and military officials, calls "episodic strategic inattention."[7] U.S. challenges also included management of a large and diffuse multinational effort that reflected widespread international belief that this war, unlike the war in Iraq, was justified by the September 2001 attacks. Indeed, NATO on September 12, 2001, for the first time invoked Article 5 of its founding treaty, which obligates members to assist other members under attack. British troops joined the Americans immediately; other national contingents arrived later.

With the international effort in Afghanistan modest, and American attention focused elsewhere, the Taliban and al-Qaeda regrouped and launched by 2003 an insurgency against the new Afghan government led by President Hamid Karzai. Initially modestly capable, by 2008 they posed an existential threat to Karzai's government. In response, the United States and its allies boosted their aid and troop strength somewhat. A Bush administration policy review in late 2008 indicated a deteriorating situation but Bush in his last days in office could not make a major course change of the sort he ordered in 2006 for Iraq. At the request of the incoming Obama administration, Bush withheld public release of the review.[8] The United States and its coalition partners continued to muddle along.

New President Barack Obama embraced the struggle in Afghanistan as a just war and began a series of policy reviews that featured internal administration wrangling about war goals and strategy that lasted nearly a year as the situation in Afghanistan continued to deteriorate.[9] In December 2009, Obama finally decided to deploy another 30,000 troops to Afghanistan, tripling the force there since he assumed office to nearly 100,000 by late 2010, and to focus on defeating al-Qaeda while countering somewhat the Taliban and helping to stabilize nuclear-armed Pakistan.[10] But he refused to embrace an open-ended COIN strategy aimed at either neutralizing the Taliban threat to Karzai or stabilizing the Afghan government and he announced that the United States would begin to "transfer" security responsibilities to the Afghan government and start withdrawing U.S. troops in July 2011—a very short window of time to achieve meaningful, enduring success in the context of typically long-running counterinsurgency wars and the growing Taliban insurgency.[11] He repeatedly told his staff that he primarily wanted an "exit strategy" designed to limit U.S. casualties.[12]

Obama initially gave the military mission to General Stanley McChrystal, a career Special Forces officer. In June 2010, General David Petraeus took a nominal demotion from command of CENTCOM to tackle Obama's major military challenge. Obama's early withdrawal date generated acrimony in Washington about the wisdom of U.S. policy and unease in the U.S. military. It also reconfirmed popular opinion in South Asia that the Americans, who abandoned Afghanistan soon after the 1989 Soviet defeat there as the country descended further into civil war, again were in Afghanistan for the short haul.[13] As of this writing in mid-2011, Afghan government and U.S.-led international efforts to build Afghan governance capacities are floundering despite some local successes, the Taliban continues to gain strength, and Afghan government efforts to achieve a political solution to the conflict remain futile. While U.S. defeat is not inevitable, the trends are not favorable.[14] Whatever eventually happens, as in Iraq initial U.S. goals were not achieved, meaning the war has been unsuccessful from the U.S. perspective.

OVERVIEW OF THE WAR

The war can be divided into two phases—a conventional phase between October and December 2001 and an unconventional phase thereafter.[15] Both phases are unusual, however, and the long unconventional phase has shown significant shifts in U.S. operations that largely reflect evolving relative priorities for CT- and COIN-focused strategies—as well as long periods of strategic drift.

The Conventional Phase

Tenet's plan to overthrow the Taliban was unusually imaginative and emphasized American strengths.[16] The CIA in late September 2001 inserted into Afghanistan small teams led by officers who quickly reestablished ties with Afghan leaders they personally knew, including Mohammed Qasim Fahem, a Tajik who assumed command of the Northern Alliance after al-Qaeda agents assassinated its longtime leader Ahmed Shah Massoud on September 9, 2001, and General Abdul Rashid Dostum, an Uzbek. CIA personnel literally bore suitcases of cash and promised American airpower to support a Northern Alliance offensive against the Taliban and al-Qaeda. Bombing began on October 7. After U.S. air forces destroyed the small number of fixed military targets in Afghanistan and the Taliban's modest air defenses, they attacked Taliban fighters in the field. After the Pentagon completed elaborate logistical, search and rescue, and air basing arrangements predominantly for force protection reasons, the first Army Special Forces team arrived on October 19 and linked up with Northern

Alliance and other anti-Taliban forces.[17] Other teams arrived soon thereafter. The teams communicated to aircraft overhead the locations of groups of the 30,000 to 50,000 Taliban and al-Qaeda troops who concentrated, mainly near Kabul and other northern cities, to battle some 20,000 North Alliance fighters in largely conventional battles.[18]

The combination of Northern Alliance ground attacks and U.S. air strikes soon defeated major Taliban formations, many of which broke and scattered.[19] Al-Qaeda fighters fought more tenaciously but soon also were defeated. Northern Alliance troops seized Kabul on November 12. Mullah Omar retreated first to Kandahar, then to Helmand province and eventually to Pakistan. Al-Qaeda's senior leaders mainly went first to the mountains of eastern Afghanistan, particularly the Tora Bora area near Jalalabad, and then into Pakistan. By the end of 2001, Taliban influence in Afghanistan was minimal. By one estimate, the Taliban and their foreign allies suffered 8,000 to 12,000 men killed, twice as many wounded, and some 7,000 fighters taken prisoner.[20] Although hurt badly, Taliban and foreign forces were not destroyed and most senior leaders escaped. The Americans continued futilely to search for bin Laden until April 2002 when they tired of the chase, began to draw down troop strength that had reached about 7,000, and recruited Afghan militias to continue to hunt "terrorists." For Washington, for the moment, the Afghan war was largely over.

The conventional phase symbolically ended on December 5, 2001, when the United Nations sponsored an assembly in Bonn, Germany, of senior leaders from all major Afghan groups who agreed to form an interim government led by Karzai, a member of a prominent family of the small Popalzai Pashtun tribe. Tajik leaders of the Northern Alliance received important defense, police, intelligence, and foreign affairs portfolios. The Bonn meeting established a series of political goals but was not a peace conference, and the United Nations did not invite Taliban representatives to attend—a decision that discarded an opportunity to reconcile the defeated to the new government in Kabul.[21] In January 2002 in Tokyo, international donors pledged $4.5 billion in economic aid to help Afghanistan recover from the ravages of two decades of war.[22] UN Security Council Resolution 1401, passed on March 28, 2002, established the UN Assistance Mission to Afghanistan (UNAMA) to monitor implementation of the Bonn agreement. In June 2002, major Afghan groups conducted an emergency *Loya Jirga,* or grand council, that formally elected Karzai head of state and reconfirmed his senior Tajik ministers. Karzai was elected president decisively in a popular election in 2004 and reelected in a more closely fought, fraud-plagued contest in 2009.

The Unconventional Phase

Karzai's administration from the start was weak and international support, while strong in principle, was initially and stayed materially modest and largely uncoordinated. Karzai and his foreign backers could not convert their battlefield successes into stable and effective governance. Poor governance generated widespread popular disappointment and grievances the residual Taliban and other groups soon exploited to regenerate politically, to build substantial military forces, and eventually to seriously threaten the government in Kabul. Some knowledgeable observers argue that international aid and political activism ostensibly designed to improve Afghan governance actually damaged it because Western promises of security, policy initiatives, and tight control of aid funds insulated the Karzai government from a need to make its own policies and take responsibility for them—key aspects of political legitimacy. The Taliban learned lessons from their mistakes of 1996–2001, and their goals and, to some extent, strategies and tactics differed from those of the 1994–1996 civil war that had a much smaller international component.

The "insurgency" was comprised mainly of six groups that, while differing on some issues, generally cooperated at tactical, operational, and strategic levels of war.[23] Kilcullen calls the collective insurgency the "Taliban confederation."[24] First, the resurgent Taliban, sometimes called the "neo-Taliban" to reflect its infusion of new members, like the old Taliban was led by Mullah Omar and was based mainly in southern and eastern Afghanistan, especially in Kandahar province.[25] Second, Hizb-i-Islami (HiG) led by Gulbuddin Hekmatyar, a Ghilzai Pashtun and a leader of the anti-Soviet war, operated mainly in northeastern and eastern provinces of Konar, Nangarhar, Nurestan, Paktika, and Paktia.[26] Third, the so-called Haqqani network, or faction, led by Jalaluddin Haqqani and his son Sirajuddin Haqqani, who are Jadrani Pashtun tribesmen, operated in eastern Afghanistan, including Paktika and Paktia provinces.[27] Fourth, the rejuvenated al-Qaeda, based in Pakistan, employed forces in eastern Afghanistan comprised largely of Afghans but led mainly by expatriate Islamists from Uzbekistan, Chechnya, and elsewhere.[28] Fifth, Pashtun tribal militias operated in both Afghanistan and Pakistan. And finally, criminal groups associated with opium production and trade and "mafias" that specialized in smuggling and illegal forest cutting, notably in Kunar province, dealt opportunistically with the other groups.[29]

Pashtuns in Pakistan, eventually organized loosely under the umbrella group Tehreek-e-Taliban Pakistan (TTP), focused largely on domestic targets like the Pakistani army and predominantly operated in the so-called Federally Administered Tribal Areas (FATA) along the Afghanistan-Pakistan border, where in fact the Pakistani government had little presence, and in Baluchistan province.[30] They gave Afghan insurgent groups bases in these areas; the Haqqani network, for example, had a strong presence in North Waziristan, Hekmatyar operated in other areas of FATA, and Mullah Omar's headquarters was in Quetta, Baluchistan.[31] The Taliban confederation's extensive use of Pakistan as a sanctuary, its ability to draw human and material resources from Pakistani *madrassas*, or Islamic religious schools, and its longtime ties to elements of Pakistani government agencies, especially the Inter-Service Intelligence Directorate of the Ministry of Defense, made the war much more complicated from the coalition perspective.[32]

All of these groups except al-Qaeda were composed primarily of ethnic Pashtuns, some 30 million mainly Sunni Muslims who comprise the largest ethnic group in the world without a national homeland. The insurgents displayed varying degrees of commitment to the Pashtun social code called *Pashtunwali,* Islamist fundamentalism, and international *jihad.*[33] *Pashtunwali* is an unusually strong social code that emphasizes independence, honor, revenge, and hospitality. The latter norm prohibited Mullah Omar from complying with President Bush's demand in September 2001 that the Taliban deliver bin Laden to the United States. *Pashtunwali* and Pashtun nationalism long have contributed to Pashtun resistance to foreign invaders and to any Afghan government that Pashtuns have not led—a rare occurrence since the establishment of the modern Afghan state in 1747. The brutality and policies of Tajik warlords in 1992–1996 contributed to the original rise of the Pashtun-based Taliban, and the prominent inclusion of some them in Karzai's initial administration contributed to the resurrection of the Taliban.[34] But while Pashtuns traditionally have been united against external groups, they long have also been deeply divided among themselves along tribal lines.

While the insurgent groups had somewhat different perspectives and objectives, they featured combinations of Sunni Islamic and Pashtun tribal imperatives; there was no significant Iraq-style sectarian gulf between Sunni and Shi'a insurgent groups. Most Pashtun ignored the Durand Line, established in 1893 as the border between British India and Afghanistan that continues, as the international border between Afghanistan and Pakistan, to divide the Pashtun nation. Many Pashtun wanted a unified, independent Greater Pashtunistan, which, if created, would imperil both Afghanistan and Pakistan. Many Taliban and al-Qaeda people wanted the new state to have transnational Islamist ambitions.[35]

The insurgents established a coherent strategy and employed tactics with potential to achieve their goals.[36] Kilcullen argues that the insurgents adopted a three-point strategy, which he calls an "exhaustion strategy," designed to convince Western governments to tire and quit Afghanistan, leaving the weak government in Kabul easy pickings.[37] First, their information operations effectively communicated a platform that was faithful to Islamist principles, promised governance that would meet people's needs and be better than Karzai provided, emphasized the corruption and incompetence of Karzai's administration, and demanded an end to the foreign presence in Afghanistan. Second, the Taliban conducted effective guerrilla operations that killed coalition troops, thereby supporting strategically focused information operations. Third, they conducted "asymmetric attacks" in forms like terrorist bombings, assassinations, and threatening "night letters" delivered to government supporters' homes.

They killed foreign relief workers—starting with the murder of an expatriate employee of the International Committee of the Red Cross in January 2003 and including the unusually large-scale killing of eight American and European medical personnel in August 2010.[38] The attacks successfully drove out some NGOs.[39] The Taliban recognized that international organizations and NGOs, however apolitically neutral their ostensibly humanitarian goals appear at first glance, in practice generally aided the Karzai government, and in 2009 issued a *fatwa* that ordered death to infidels and to Muslims who help the Afghan government.[40] Some NGOs in response unilaterally sought accommodation with the Taliban in order to continue to operate in Afghanistan. Obviously minimal Western will to persevere enabled this strategy and obviated a need to defeat the coalition militarily.[41]

To achieve their goals, the Taliban built an effective organization that Kilcullen calls the "Taliban operating system," which included aspects of standard guerrilla tactics, Maoist people's war operations, some al-Qaeda techniques, some unique Afghan characteristics, and extensive use of sanctuaries in Pakistan.[42] In Afghanistan, the Taliban employed core groups of full-time fighters operating as foot-mobile columns typically composed mainly of Afghan and Pakistani Pashtuns with some foreign leaders, part-time Afghan fighters who supported the core fighting groups in their locales, and village underground infrastructures that performed shadow government functions, gathered intelligence, and supported and coordinated the operations of the fighting groups. Local leaders sometimes were members of the former Taliban government. The main activities of the core fighting groups, which had significant authority over local groups, were political and religious indoctrination, armed propaganda, intimidation, and killing people who backed the government or disagreed with their interpretation of Shari'a law, including their view that suicide bombing was theologically acceptable. In Pakistan, the Taliban ran training and logistical support operations, maintained supply depots, rested fighters, provided medical treatment to sick and wounded fighters, recruited new fighters, and maintained links with external supporters who were private donors and governments, including elements of the Pakistani government.

Non-Pashtun Afghans largely eschewed this insurgency. Fighting initially was confined mainly to Pashtun-majority areas, and polling data indicated substantial support for both Karzai and the international effort in Afghanistan by non-Pashtuns, especially.[43] The strongly Pashtun character of the insurgency both aided the Taliban initially and limited its appeal to other Afghans. Nevertheless, as the Karzai government and the coalition stumbled, the Taliban by 2011 expanded their operations into areas, especially in the north, where Pashtuns were local minority groups.

Confronting the insurgents were the Afghan government and a staggering array of foreign governments, international organizations, and private groups, including the United States and other states unilaterally;

UNAMA and numerous standing UN agencies like the World Food Programme; the European Union; NATO; the International Security Assistance Force (ISAF), a UN-authorized military organization of some 46 members that initially was charged with peacekeeping duties in the immediate vicinity of Kabul but later assumed countrywide security responsibilities; and some 3,000 NGOs.[44] The foreign entities chronically were hamstrung variously by major deficiencies, including an inability to decide whether the international mission (and their own) was peacekeeping, counterinsurgency, counterterrorism, governance capacity building, economic development aid, reconstruction, some other activity, or some combination of the above; a lack of basic understanding of the nature of the problem in Afghanistan; unwillingness to provide much assistance and then inability or unwillingness to deliver the aid amounts promised in a timely fashion; various but usually tight rules of engagement, force protection priorities, and mission-related "national caveats" that precluded many national military contingents from fighting except in self-defense and sometimes from doing much work at all; deployments of military and civilian government personnel for such short periods (often as little three months) that some never learned much about their operational environments and were little more than official tourists; and unwillingness to work together in even a loosely coordinated effort, let alone one unified by command by a single individual.[45] The NGOs and international organizations jealously guarded their independence despite their predominant funding by national governments; nevertheless, they worked in ways generally consistent with national governments' objectives and demanded that coalition military forces provide physically secure operating environments.[46]

Militarily, the small but effective Special Forces mission of 2001 became a more conventional U.S. military operation in 2002, adding layers of structure that inhibited rapid initiative by small-unit commanders. NATO operated its own headquarters; it took over ISAF in 2003.[47] Afghan security forces operated either independently or under U.S. control. American troops operated both under unilateral U.S. control and as part of ISAF. The result was a cumbersome coalition command structure that compounded the effects of other major constraints like national caveats and national reporting chains.[48] For example, Provincial Reconstruction Teams (PRTs), which were multinational civil-military organizations headed by single coalition members that focused on development and COIN, often reported to their lead nations' capitals, not ISAF.[49] In addition, the CIA operated a fleet of drones that launched missiles at "high-value targets," or insurgent/terrorist leaders, often in Pakistan, and it built and operated a 3,000-man paramilitary force of "Counterterrorism Pursuit Teams" (CTPT).[50] While details are scarce, the CTPT concept seems to resemble somewhat the PRU the CIA used effectively to attack VC cadre (chapter 4).

On the civilian side, UNAMA, apparently at U.S. insistence and evidently without anticipation that an insurgency would soon emerge, initially parceled out responsibility for major aspects of "security sector reform" to five countries: the United States (the Afghan National Army [ANA] and the National Directorate of Security, the intelligence service); Germany (the Afghan National Police [ANP]); Britain (counter narcotics); Italy (the legal system); and Japan (the "Disarmament, Demobilization, and Reintegration" [DDR] of fighters).[51] These activities initially were purposefully uncoordinated and in some cases little addressed Afghanistan's needs.

For example, the international community over the years made at least seven attempts to help build capable Afghan police forces—all without significant success.[52] The modest initial German effort focused on producing small numbers of able police supervisors when the glaring Afghan need was large numbers of moderately competent but honest street policemen and police units that could withstand small-scale insurgent attacks. When the deficiencies of the German approach became obvious, the European Union provided additional resources.[53] The U.S. Department of State then took responsibility, contracting with the American firm Dyncorp for police trainers, who by all accounts performed poorly. Former U.S. Ambassador to Afghanistan Ronald Neumann calls State's contracting procedures "a nightmare."[54] In 2006, the U.S. military took over, but continued to focus mainly on training the army and also did not create a capable, honest police force.[55] The Taliban, in contrast, recognized that able police produce security, intelligence, and government legitimacy—all undesirable from their prospective. The Taliban therefore attacked the police chronically, inflicting more ANP than ANA casualties during 2002–2010 and appreciably hindering the development of capable national police forces.[56]

The Japanese DDR program collected some 36,000 weapons and successfully demobilized some fighters.[57] But many of the some 2,000 illegal armed groups resisted demobilization, and the Taliban later rearmed some groups.

The coalition's counter-drug effort was wholly ineffective.[58] The initial British counter-narcotics effort was minimal and was ignored by many coalition governments that questioned its importance despite its corrupting effects on the Afghan government and the financial clout, and thence recruiting and logistical support, that drugs gave the Taliban. Opium poppy production unsurprisingly grew rapidly after 2001, especially in Taliban-controlled areas.[59] In 2005, the UN Office of Drugs and Crime estimated that Afghanistan produced 87 percent of the world's opium with a local value of $2.76 billion—about one-third of Afghanistan's gross domestic product; Afghanistan's share of world opium production was still an estimated 85 percent in 2010.[60] The Americans, who facilitated the Afghan drug trade in the 1980s because they thought its corrupting effects would help defeat Soviet military forces in Afghanistan, later joined the counter-drug effort, also ineffectively, in part because American aid promised to Afghan farmers to grow alternative crops often did not materialize.[61] Diplomat Richard Holbrooke called the U.S. counter-drug effort "the most wasteful and ineffective program I have seen in 40 years in and out of government."[62] The coalition as a whole did not commit troops to counter-narcotics efforts until 2008, long after the negative effects of resurgent opium production were obvious.[63]

The international aid effort initially was haphazard as some 60 countries provided assistance that was not coordinated in any formal way. Amounts of funds provided were well below those both needed and promised, and states spent on projects that were popular at home but often neither met Afghan needs nor effectively complemented the Afghan National Development Strategy (ANDS), the country's economic development plan.[64] National governments unsurprisingly insisted on overseeing closely the spending of their aid funds given the extent of Afghan government corruption, which hindered Afghan control of national economic development policy and damaged Karzai politically by disrupting traditional patron-client relationships, which were in part a legacy of Afghanistan's role as a buffer state that long has received large foreign subsidies—from Britain in the 19th century and the USSR and United States during the Cold War.[65] The international community, recognizing the problems caused by disjointed aid efforts, created the Joint Coordination and Monetary Board in January 2006, which helped some.[66] The U.S. Office of Management and Budget still insisted on using standard budget review processes despite the war and highly non-American circumstances in Afghanistan.[67]

Within the American effort, bureaucratic fiefdoms as usual jockeyed for power.[68] In 2010, the commander of a U.S. Marine unit decided to do things his way—against the wishes of his ostensible commander, General McChrystal, confirming the continuing existence of interservice rivalries and rank insubordination that again negatively influenced military operational performance.[69] American ambassadors and their military counterparts cooperated with different levels of effectiveness that reflected the personalities of the pairs. Lieutenant General David Barno and Ambassador Zalmay Khalilzad evidently worked effectively together in pursuit of COIN-focused operations in 2003–2005, only to see the U.S. effort shift back toward CT just as the insurgency was accelerating.[70] Other pairs were less successful; Generals McChrystal and Petraeus, for example, reportedly disagreed with Ambassador Karl Eikenberry, who as an army lieutenant general serving in Afghanistan got along poorly with Ambassador Neumann.[71] Obama in 2009 further disjointed the U.S. effort by making the notoriously crusty Holbrooke his special representative to both Afghanistan and Pakistan, especially complicating State Department activities. After Holbrooke's death in late 2010, Vice President Joe Biden inherited the role. And, the left wing of Obama's Democratic Party, often led by Biden, sniped at his policies and the Karzai government and demanded immediate U.S. troop withdrawal. Numerous observers note that such domestic American criticism of the war effort had many negative consequences, including encouraging Afghans to look out for themselves via corruption before an imminent American withdrawal and giving the Taliban more reason to believe that they need only hold out a little longer in order to prevail. Indeed, Taliban commanders were often quoted as saying variants of "the Americans have all the clocks, we have all the time."[72]

Despite its many performance problems, the international community did some things well. For example, U.S. Special Forces and some coalition military contingents were operationally effective in both CT and COIN roles. Generals McChrystal and Petraeus evidently improved coalition military operations as a whole. The innovative civil-military PRTs helped markedly in some areas and some aid projects were well conceived and executed.[73] Aid flows contributed to growth in the national economy and some of the advisory missions to government ministries achieved some belated success. But some of the successes caused other problems or were so narrowly focused that they made little strategic difference. The CIA's drone strikes, for example, killed some insurgents but did not kill many senior leaders or produce leadership paralysis—and they alienated many people in Pakistan, especially.[74]

More importantly, for years the Karzai government failed to establish honest, competent government, thereby damaging itself in what was, as in most COIN contests, the central political battle for legitimacy. In the Afghan context, especially important determinants of state legitimacy were the ability to provide effective and fair dispute resolution mechanisms, security, and a chance for people to earn a living lawfully.[75] Yet government corruption was initially, and remains, endemic. Increased opium poppy cultivation merely enhanced already ample opportunities for official corruption.[76] The ANA with some American help gradually absorbed some militias and improved its operational performance but the national police initially, and nearly a decade later remained, underpaid and largely corrupt, under equipped, and inept.[77] Wealthy foreign-based NGOs hired many of Afghanistan's best minds, hampering cash-poor government ministries' efforts to develop staffs of competent professionals. Karzai, a member of a small tribe, had a limited personal power base and so employed traditional Afghan methods to produce political loyalty, including dispensing government jobs, meaning that he often appointed tribally acceptable officials who were incompetent and/or corrupt. Such efforts gradually redressed the ethnic imbalance of his initial administration, but Taliban personnel often persuasively courted losers in appointment squabbles by promising better positions under a new Taliban regime.[78] In many areas, tribal groups out of favor in provincial capitals or Kabul comprised big parts of local Taliban forces. Weak governance prompted analysts to argue frequently that the biggest threat to the Kabul government was its own incompetence and corruption.[79]

The result was an insurgency that steadily became bigger and more violent. The insurgents conducted few military operations in 2002 but by 2006 were strong enough to mount steadily increasing numbers and types of attacks on coalition and Afghan forces. The U.S. military estimated that the Taliban had 9,000 to 10,000 fighters in 2006.[80] Estimates of main force Taliban fighters in 2008 ranged from 5,000 to 20,000, and up to 25,000 in 2009.[81] Kilcullen estimated Taliban troop strength in 2008 at 32,000 to 40,000—about 90 percent of whom were part-time fighters who were potentially reconcilable to a competent government with a judicious amnesty program.[82] Insurgent troop strength grew sufficiently to enable company-sized attacks by 2005 and battalion-sized operations by 2008.[83]

Increased troop strength enabled more insurgent attacks, which rose by some 400 percent from 2002 to 2006.[84] The insurgents used standard guerrilla warfare tactics, progressively better IEDs, and some new techniques, including suicide bombings—an al-Qaeda favorite that Afghan fighters never used against the Soviets.[85] The Taliban's first suicide attacks were in 2004; in 2006, they conducted 141 such attacks that killed 1,166 people.[86] Learning from other insurgencies, including the one in Iraq in which some Afghans and third-national al-Qaeda fighters participated, the insurgents attacked a wide variety of Afghan and foreign interests.[87] Domestically, insurgents assassinated government officials, school teachers, and policemen; in 2006, the Taliban killed 85 school teachers and more than 600 policemen.[88] In April 2011, after many attempts they finally assassinated the unusually effective police chief of Kandahar province.[89] They killed clerics and tribal leaders who cooperated with the government. They also bombed and burned government facilities, including schools. In 2005–2007, they burned down 1,089 schools.[90] In targeting teachers and schools, the insurgents attacked a modernizing influence as well as the government's presence.[91] And, they innovatively incorporated large-scale (a dozen or so) suicide bombings into tactical military operations.[92] Western observers generally agree that the Taliban's collective fighting skills improved over time.[93]

Al-Qaeda helped the Taliban improve their information operations, key to their political program.[94] Taliban propagandists regularly exploited collateral damage incidents, which over time killed thousands of Afghan civilians, and cases of American insensitivity to Afghan cultural norms such as a 2005 incident in which U.S. soldiers burned the bodies of dead insurgent fighters—a violation of Islamic burial practice.[95] The Taliban produced and distributed propaganda CDs and DVDs, operated mobile FM radio stations in Afghanistan, conducted Internet and e-mail operations, and assigned spokesmen to talk with journalists in Pakistan.[96] These fairly sophisticated media operations contrast markedly with those of the reticent Taliban of the 1990s.[97]

Even as the rejuvenated Taliban increasingly obviously was conducting an insurgency, many of the Western governments that sent troops to Afghanistan in late 2001 mainly to help rebuild the country retained "national caveats" that prohibited their troops from conducting counterinsurgency operations. Of some 46 ISAF members that provided a total of about 130,000 troops in late 2010, only the United States, Britain, Canada, and the Netherlands both provided appreciable contingents and authorized their troops to fight aggressively. The Taliban in turn targeted these troops and less aggressive but politically prominent contingents from Germany, Italy, and Spain, consistent with its exhaustion strategy, in a successful effort to induce their governments to quit.[98] By mid-2010, as the fighting increased and coalition casualties mounted, all four of the most aggressive states had announced plans to begin withdrawals or to leave altogether. The Netherlands began to withdraw its 2,000 troops on August 1, 2010, after suffering total war casualties of 24 troops killed in action and 140 wounded.[99] Canada's parliament decided in 2008 that Canadian forces would cease operations in Afghanistan by mid-2011 and withdraw by the end of 2011, when Germany also planned to have withdrawn its relatively large force.[100] As of June 26, 2011, 157 Canadian military personnel and four Canadian civilians had been killed in Afghanistan since 2002.[101] British Prime Minister David Cameron in 2010 said that British troops would be out of Afghanistan by the next general election, due to be held in 2015 or before, but did not announce a firm withdrawal timetable.[102] As of February 4, 2011, 351 British troops had died in Afghanistan since October 2001, 309 of them as a result of hostile action.[103]

The United States showed a similar pattern. With polls showing declining popular support for the war effort at home, as of this writing Obama still plans to begin to withdraw U.S. troops in July 2011.[104] As of April 21, 2011, DoD reported that 1,174 U.S. military personnel and one DoD civilian had been killed in action.[105] DoD does not count 11 CIA personnel and other government and contract personnel killed in Afghanistan.

By one count, as of September 24, 2011, a total of 2,738 coalition personnel from 27 countries had been killed in Afghanistan since October 2001.[106] While fighting and casualties have accelerated in recent years, these figures amount to an annual average rate of only about 260 deaths. Given casualty tolls very modest by the standards of major wars of the past but obvious war weariness in many NATO countries, it is no wonder the Taliban evidently plan to continue to produce modest numbers of coalition casualties and simply wait out the West.[107] NATO formally decided in late 2010 to conclude all security duties by the end of 2014—making the necessary wait rather short.

Civilian casualties have been much higher and seem to have increased over time, although hard numbers are few.[108] UNAMA began to count civilian casualties only in 2007. In the first half of 2011, it counted 1,462 civilian deaths—about 80 percent of which were caused by insurgents.[109]

The Taliban, like their counterparts in Iraq, effectively exploited Western and Afghan aversion to civilian casualties by using disinformation to induce collateral damage incidents and to publicize incidents when they occurred.[110] They became adept at using civilians as human shields to provide both protection and politically useful collateral damage incidents, mainly produced by the air strikes that coalition forces frequently called to compensate in part for the modest number of ground troops they deployed over large areas.[111] Errant air strikes caused large numbers of shocking deaths and the coalition often compounded the errors by initially denying culpability. Polls chronically indicated widespread Afghan outrage at civilian deaths, explaining the repeated, public and private complaints from Karzai about coalition military operations.[112]

Most Western observers therefore see a deteriorating situation that, while not hopeless, will require much more effective and enduring Afghan government and international efforts. Kilcullen in 2009 thought the task would take at least 5 to 10 years—a time period no Western coalition member state ever seemed, or now seems, willing to accept. Afghans generally recognize that fact—which presumably accounts largely for the ineffectiveness of Karzai's efforts to engage Taliban leaders meaningfully in reconciliation talks and to reach out to minimally committed Taliban through his National Reconciliation Plan, or amnesty program, despite a history of tribally based reconciliation means employed locally since 2001 and repeatedly in the Afghan wars since 1979.[113]

THE AFGHANISTAN WAR AND MATERIALIST THEORY

This war also refutes materialist theory. At first glance, U.S. military power appears to have defeated Taliban forces quickly and seemingly completely in late 2001. But the story is much more complex. American airpower supported Northern Alliance ground formations large enough to force the Taliban and their allies to concentrate, making them easy targets for aircraft. Without major ground opposition the Taliban and al-Qaeda fighters, like Yugoslav security forces in 1999, would have been able to disperse, rendering them much less vulnerable to air strikes.

And, the Taliban were militarily fairly weak. They had large numbers of young, inexperienced fighters, weak command and control, and very limited combat support capacities, meaning they could not sustain heavy fighting for long.[114] Moreover, the Taliban in 2001 had in their five years in power alienated many Afghans with their religious fundamentalism by banning music, requiring men to grow beards, closing girls' schools, and prohibiting women from working, for example; major acts of brutality like massacring several thousand ethnic Hazaras, who are Shi'a; and their inability to effectively run the national government.[115] After winning in 1996 in part by promising honest government, they showed themselves also to be corrupt. As a result, their legitimacy sharply eroded. When Northern Alliance and U.S. troops initially defeated Taliban field forces in 2001, they had no reservoir of popular support upon which to call. Given Afghans' preference for backing winners, working for the Taliban no longer looked to many people like a good personal strategy. As a result, the movement collapsed quickly.

After 2001, U.S. and NATO military might had no significant influence on improving the Afghan government legitimacy necessary to win this war, despite military involvement in many nation-building as well as combat roles—including training and the economic development work of the PRTs and conventional units. And, as in U.S. wars beginning in Vietnam, excessive or misplaced U.S. military "power" produced collateral damage, sustained the Taliban's political message, and had negative, strategically important political consequences in Afghanistan, Pakistan, and in coalition member states.[116]

OPERATION ENDURING FREEDOM AND STRATEGIC INTERACTION THEORY

Although Arreguín-Toft assesses only the "Anti-Taliban War" of 2002 [sic], we can use his method to assess both phases of the war.[117] In October–December 2001, the United States supported Northern Alliance troops by conducting a conventional, air-centric war against Taliban and foreign fighters who mainly fought conventionally as light infantry, often massed in static positions. U.S. SOF personnel fought some, but they more prominently performed liaison functions with Northern Alliance leaders and spotted targets for bombers. They did not then conduct traditional Special Forces missions like raids, unconventional warfare, and foreign internal defense. The United States won this conventional-conventional, same-approach interaction, consistent with Arreguín-Toft's expectations.

In the unconventional phase, the United States variously emphasized CT- and COIN-focused strategies and did not emphasize airpower or other conventional uses of force. U.S. commanders, especially McChrystal, explicitly ordered use of minimal force to minimize civilian casualties. While numerous collateral damage incidents occurred, they were accidental from the American perspective.[118] Hence, the United States did not employ barbarism. The Taliban like other insurgent forces did, however, through widespread assassinations, summary executions, and gruesome murders by beheadings, for example. The Taliban were brutal even by the standards of most insurgencies, generating a reputation for cruelty among Afghans as a whole. Taliban activities as insurgents overall were much closer to GWS than to conventional operations. Therefore, while Arreguín-Toft's model is especially inapplicable to the CT- and COIN-focused operations the Americans predominantly conducted, the United

States is losing a roughly same-approach unconventional interaction, inconsistent with Arreguín-Toft's expectations.

DIMENSIONAL ANALYSIS

The United States conducted a narrowly conventional war against the Taliban in 2001 and used evolving strategies with rough, albeit often incoherent, CT and COIN focuses in the unconventional phase. Strategies in the unconventional phase over time were differentially ambitious and as a result involved operations in varying combinations of dimensions.

The U.S. Conventional War

The U.S. effort in the conventional war was focused in two dimensions only—*leadership* and *operational execution*. The Americans tried to (1) drive the Taliban from power and (2) kill or capture as many al-Qaeda personnel and their Afghan allies as possible. This approach was motivated by a narrow focus on CT in the aftermath of 9/11 and by reluctance to become committed in any significant way in Afghanistan. Hence, the other dimensions were out of scope. The Americans initially succeeded to a considerable degree in both objectives, although their most senior adversary leaders all escaped. But the mistaken judgment that conventional military force could end the war, and resultant inattention exacerbated by focus on the war in Iraq, provided an opening the Taliban exploited to revive.

The Taliban's Conventional War

The Taliban initially sought to fight the Northern Alliance and the Americans in the military part of the *operational execution* dimension. But the presence of U.S. air forces employing precision munitions, accurate ground-based targeting data, and vulnerable Taliban troop concentrations led to rapid military defeats that precipitated political disaster in only a few weeks. By the time the Taliban learned how American military aircraft work, it was too late. They never engaged the Americans appreciably in the other dimensions.

The U.S. (and Coalition) Unconventional War

In the unconventional phase, the United States and its coalition partners generally failed to do well in any of the six dimensions, some notable individual operational exceptions notwithstanding. The initial 2001–2003 focus on CT, and a return to that emphasis in roughly 2005–2007, meant that the United States as a matter of policy often ignored the *national will* dimension. American actions, as in Iraq but in different ways, were often counterproductive in this dimension. Determined to get bin Laden, to avoid nation building, and to get U.S. troops home as soon as possible, the Americans initially cut deals with local warlords to hunt al-Qaeda operatives. But the warlords' behavior in the 1990s alienated many Afghans, tainting the United States when the Americans employed them, even years later. American troops' behavior, as in Iraq, often alienated Afghans, while Americans' verbiage and actions convinced most Afghans that they were again short-time, unreliable partners whose security and other promises should not be believed.[119] The United States refused initially to talk with Taliban leaders, throwing away an opportunity to prevent the insurgency. And, when some Taliban surrendered early in the unconventional phase of the war, the Americans treated them harshly and sent some to the prison at Guantanamo, alienating more people and further damaging prospects for national reconciliation.[120]

The debacle of the counterdrug campaign and resultant Taliban material wealth, sanctuaries in Pakistan, and a large supply of unemployed young Afghan men willing to fight Afghan security forces and foreign troops for $400 per month meant that the coalition unsuccessfully attacked Taliban *resource mobilization*. Given relatively secure weapons supplies and training bases in Pakistan, the coalition similarly poorly attacked Taliban *resource conversion* and *force generation* abilities. Washington failed to persuade Pakistan to deny Afghan insurgents sanctuary even though Pakistan rounded up some Arab, Chechen, Uzbek, and other non-Pashtun al-Qaeda personnel; Pakistan evidently continued to see friendly Afghans as providing strategic depth in its ongoing struggle against India.[121]

Ground-based attacks on Taliban and al-Qaeda *leadership* in Afghanistan netted "hundreds" of people, but both groups evidently replaced the losses, meaning the strategic gains for the United States probably were minimal.[122] Drone attacks on leaders in Pakistan that killed fighters also killed civilians, alienating local people and damaging U.S. performance in the *national will* dimension.[123] Guided by CIA-provided intelligence, a SOF team finally killed bin Laden on May 2, 2011, in Pakistan—a big CT success.[124] But the drone strikes and raid on bin Laden also angered Pakistani *leaders* for violating Pakistani sovereignty. And, while SOF and some conventional forces fought well when they encountered insurgents, they did not defeat the insurgent fighting forces or prevent the Taliban from increasing their fighting strength and the size, regional scope, and complexity of their attacks.[125] U.S. activities in the military part of the *operational execution* dimension thereby were only partly successful.

The Insurgents' Unconventional War

Focusing on their Pashtun roots, the Taliban and related groups first rebuilt their domestic support base and armed strength and then, in 2003 and later, sharply increased their military attacks on Afghan government, coalition military, and international civilian personnel. They thereby produced a comprehensive political, military, economic, and social program that covered all six dimensions in their shadow government and war-making activities.

Internally, they rebuilt much of the appeal they generated in 1994–1996 on the road to power but squandered thereafter. They effectively performed shadow government functions, especially by providing local security and conflict resolution services. They thereby performed increasingly well in the *national will* dimension, particularly in majority-Pashtun areas. Combined with their control of most opium poppy production and trade and by maintaining external support, including cash from Saudi Arabia and other Gulf Arab states and Islamic NGOs, they generated substantial resources with which they built fighting forces. The Taliban engineered some of the collateral damage incidents that produced anger that in turn, by one estimate, generated three to five Taliban recruits per civilian death.[126] Excellent recruiting operations in the madrassas of Pakistan and elsewhere also helped the Taliban more than offset combat losses.[127] And, they retained the support of the Pakistani government, protecting their primary sanctuaries. Therefore, they performed well in the *resource mobilization, resource conversion,* and *force generation* dimensions.[128] They preserved senior *leadership* in Mullah Omar, Zawahiri, Hekmatyar, and the Haqqanis, and over the years they quickly replaced other leaders lost in battle. As of this writing, it is unclear what the long-term implications of bin Laden's death will be. Good leaders built effective information operations, formed effective shadow governments in some places, and maintained command and control of forces despite losses to coalition attacks. They conducted increasingly effective military operations using new techniques and competently linked tactical military actions to strategically important political vulnerabilities in adversaries, meaning they also performed well in many nonmilitary parts of the *operational execution* dimension.

Externally, their exhaustion strategy primarily attacked the coalition's *national will* and *leadership* by creating casualties at a level politically unsustainable in the West while attacking propagandistically the legitimacy of the Karzai government, making Westerners less willing to provide material aid to Karzai, thereby hurting Afghan government *resource mobilization*. The Taliban thus are on the verge of defeating NATO by inflicting a small number of casualties that have no material affect on the narrowly military forces of coalition member states. The exhaustion strategy, like the attrition strategies of numerous U.S. adversaries since North Vietnam used one successfully, depended on competent tactical-level military performance, not military victory—a much less demanding requirement. The Taliban, as insurgent forces, could largely control the tempo of operations and their substantial ability to recruit meant they could take significant tactical military losses in order to produce strategically politically important casualties among Western military contingents.

Summary of Actors' Actions

Karzai's failure to produce good Afghan governance was a major failing of both the government and its foreign supporters. While primary responsibility lies with Karzai and the Afghans, U.S. and coalition efforts variously were chronically misguided, uncoordinated, and/or incompetent. A RAND Corporation study of post-1945 insurgencies indicates that security and governance are essential elements of all COIN campaigns—something the Karzai government never was close to achieving and which the United States pursued haltingly and ineptly.[129]

The United States consistently was inconsistent in its strategy. Indeed, over the course of U.S. military-related involvement in Afghanistan going back to the anti-Soviet effort in the 1980s, it is hard to see any coherent U.S. strategy to advance long-term U.S. interests.[130] Even though it successfully helped defeat the USSR in the 1980s, it also helped build many of the Islamist groups it fought after 9/11. Strikingly, after nearly a decade of inconsistent and largely reactive U.S. policy concerning Afghanistan since 9/11 and often poor U.S. political/military performance, Obama opted in late 2009 for a "strategy" of yet more half measures, including merely containing Afghan insurgents who are increasingly operating as semi-independent actors with international *jihadi* ambitions—like al-Qaeda affiliates in Iraq, the Magreb, and the Arabian Peninsula.[131] The ineffectual strategies and policy half-measures of both Bush and Obama are reminiscent of President Johnson's strategic incoherence in Vietnam.[132]

For the United States and the coalition as a whole, other glaring deficiencies included a lack of understanding of their operational environment, a core requirement for successful COIN operations; operational incoherence caused by lack of elementary coordination across the various civil-military operations; lack of understanding of the basics of COIN, leading to a striking lack of attention to the ANP, narcotics, and governance capacities generally even as the ANA received many resources and much attention; and fairly obvious political pressures on Karzai and announced intentions to withdraw that discouraged Afghans from taking personal ownership of Afghan government policies. Coalition-produced incentives made it better for many Afghans as individuals to let the foreigners lead, protect themselves against recriminations after a possible Taliban victory, and/or take advantage of the Western gravy train through corruption for as long as it lasts.

U.S. LEARNING AND ADAPTATION

Once again, the United States displayed limited capacities to learn and adapt in a COIN war.[133] The roughly contemporary war in Iraq meant that American leaders and troops were learning from both to some degree simultaneously, but in arenas unique to Afghanistan the Americans again displayed collectively slow learning and adaptation—some notable exceptions like SOF again notwithstanding. A flawed reading of history led the Bush administration to believe that Afghans pathologically oppose foreign involvement in their country, helping lead to the "light footprint" approach that failed to effectively conduct either CT or COIN operations.[134] The administration believed this despite the fact that the coalition supported an initially popular Afghan government and was, polls repeatedly showed, itself welcome as the force that deposed the abusive Taliban regime.[135] Strikingly, given this misunderstanding, the Americans also confused entry into Afghanistan with victory, fostering the strategic inattention that enabled the Taliban to revive and then to thrive.[136] The Americans thereby committed what John Nagl calls "the most fundamental of all strategic errors: insufficient resources to accomplish maximalist goals."[137]

While the Bush administration made many of these mistakes, Obama failed to learn from them, then made new ones in trying to "win" in Afghanistan on the cheap or, as some Obama administration officials fear, simply to work at it long enough to appear to have tried, and then quit.[138] Obama told Senator Lindsey Graham that he would not seriously fight a COIN war for fear of losing support within his Democratic Party—an understandable personal political motive that is inexcusable in a statesman.[139] The Americans failed to lead or more than minimally coordinate the constellation of loosely allied players in Afghanistan and maintained an internally disjointed U.S. national effort despite recognizing the importance of unity of effort as one of its doctrinally enshrined "principles of war." And, military personnel policy again denigrated the sometimes critical role of advisors by making it clear that advisory duty was not career enhancing.[140]

The Americans again overemphasized force protection and through cultural insensitivity damaged relations with locals and hampered HUMINT collection, which in turn damaged learning and adaptation.[141] Indeed, in their search for terrorists, the Americans simply failed to pay attention to the broader operational environment in which they lived and worked. In a stunning admission of incompetence, the senior American intelligence officer in Afghanistan, Major General Michael Flynn, in a think-tank forum (not a government journal) in January 2010 publicly excoriated intelligence efforts in Afghanistan and demanded better future performance.[142]

The Obama administration failed to learn that exit strategies defined as troop withdrawal dates do nothing but help their adversaries. Personally scarred by the abrupt American withdrawal in 1989 that contributed to the Afghan civil strife of 1989–1996, and much other recent U.S. diplomatic history, Afghans and Pakistanis of all political persuasions are keenly aware that Western patience is limited.[143]

The United States again underestimated its adversaries, who have been resourceful and adept at both learning and adapting.[144] The Taliban changed their strategic goals and operational approaches after their defeat in 2001. They adopted new military and political tactics and strategies, and integrated them well. Al-Qaeda built a mixed force of Afghans and foreign fighters and convinced the Taliban confederation and many Pakistani Pashtuns to fight for its international *jihadi* objectives.

SUMMARY

The American war in Afghanistan has appreciable similarities to the U.S. experience in Iraq as well as important differences. In both cases, conventional Americans troops initially fought well conventionally, initially were largely clueless in counterinsurgency operations, and gradually learned to perform better but not well. In Iraq, the United States was extremely fortunate that AQI self-destructed by alienating its Iraqi Sunni allies—something the smarter leadership of al-Qaeda central shows no signs of doing in Afghanistan. Moreover, insurgents in Afghanistan have learned from the Iraq war and others, making them a formidable adversary the Karzai government and its foreign friends have not come close to thwarting.

A path for success is apparent but probably is unattainable. Internally in Afghanistan, the coalition and Afghan government need to build a competent, honest Afghan state; come up with an attractive reconciliation program for most Afghan insurgents; and kill, capture, or chase out the remaining hard core of Afghan and transnational enemies. Externally, Washington needs to convince Pakistan to deny sanctuary to Afghan groups and convince at least some coalition members to stay the course. While potentially doable, these tasks almost certainly cannot be accomplished in the short time limited Western political patience will grant. Bin Laden's death seems to have damaged U.S. relations with Pakistan and may give opponents of the war who only barely accept a CT rationale for it an excuse to quit even sooner than Obama intends.[145]

Prospects for U.S. success in Afghanistan thus appear to be worse than in Iraq after mid-2007. If they are successful, al-Qaeda and its allies around the world would be buoyed. If the United States again abandons an ally, U.S. credibility will take another major hit internationally. The implications of a first NATO defeat are unclear but surely negative for U.S. interests; they could mean the demise of NATO as an effective organization. In total, the negative implications for the West in general, and the United States in particular, of a coalition defeat in Afghanistan are likely to be significant.[146]

CHAPTER 8

Summary and Conclusions

The case studies confirm that identification and exploitation of critical vulnerabilities are far more important determinants of strategic military outcomes than possession of military-related material assets. Of six wars in which it was materially superior, the United States clearly won only three. Belligerents consistently, explicitly, looked for vulnerabilities in their adversaries and found them with varying degrees of success. Belligerents adapted to find and exploit enemy vulnerabilities, to ameliorate their own, or both. In each case, success went to actors who accurately identified and exploited a critical vulnerability in an adversary. The bombing case is the only one in which the victor did not find and exploit asymmetrically a critical vulnerability, leading by default to symmetrical, attritional warfare in which the relevant vulnerability was a relative dearth of military assets in the operational execution dimension. It is the only case consistent with materialist theory's assertion that quantities of military-related material wealth determine outcomes.

Belligerents were variously aware of their vulnerabilities and sometimes acted coherently to ameliorate them. American civilian leaders and military commanders often worried that some U.S. actions, particularly collateral damage incidents, would both offend Americans and hand propaganda advantages to their enemies. These concerns were not strategically important until the Vietnam War, however, when U.S. military casualties became politically sensitive at home. Since then, casualty-related vulnerability amelioration in the national will dimension has been a major focus of U.S. military operations—one the United States has only sometimes accomplished successfully.

U.S. enemies attempted to ameliorate their vulnerabilities with varying degrees of success. Filipino nationalists and Saddam Hussein's government were inept in strategically critical respects, leading to their defeats. But Germany, Vietnamese communists, Yugoslavia, and Iraqi and Afghan insurgents understood their vulnerabilities fairly well and worked hard to minimize them. The Vietnamese succeeded. Iraqi insurgents generally succeeded when fighting U.S. forces alone. Germany and Yugoslavia did not. The Taliban learned and effectively redressed initial mistakes, giving them good prospects for strategic success.

In 1999 the KLA, like Bosnian Muslims in the mid-1990s, exploited American political vulnerabilities sufficiently to convince the United States to launch an optional war. Israel similarly has convinced the United States to be its material benefactor and largely unquestioning political supporter during decades of Arab-Israeli conflict—at considerable financial and international political cost to the United States.[1] Some argue that President Bush invaded Iraq in 2003 in part to serve Israeli interests.[2] There is little evidence that U.S. policy makers, and Americans generally, understand the nature and costs of exploitative manipulation by putative friends.[3]

Operationally relevant dimensions varied by conflict. Table 8.1 summarizes dimensions belligerents considered critical, as evidenced by substantial effort in them, and whether or not operations in them appreciably influenced each war's outcome. The table shows the Iraq and Afghanistan wars' separate phases because the relevant operational dimensions, and actors' performance in them, changed across the phases more dramatically than in most wars and because the unconventional phase of each remains underway as I write. All six dimensions were critical dimensions in at least one case.

The *national will* dimension was decisive in two of the six cases, and may be in a third. The United States broke the will of Filipino nationalists by winning over many Filipinos with its policy of attraction. Vietnamese communists won their war by breaking American will to fight; Afghan insurgents are close to doing the same. But Allied bombing did not weaken German morale sufficiently to influence the outcome of World War II. The United States did not try to influence North Vietnamese popular will to fight and conducted an effective pacification campaign in South Vietnam only after the guerrilla war became secondary in strategic importance. Minor NATO efforts to influence Yugoslav popular will led only to dissension within the alliance. Americans did not much try to influence Afghan national will in 2001 because they focused on counterterrorist activities and they did not try to influence Iraqi will in March–April 2003 because they thought most Iraqis would welcome them as liberators. For years, they subordinated efforts to influence Iraqis and Afghans to imperatives of force protection and chasing insurgents.

America's enemies attacked U.S. will in every conflict. Since Vietnam, national will has been the dominant dimension of attack on the United States and the primary cause of U.S. defeats.

Table 8.1
Summary of Critical Operational Dimensions

War	National Will	Resource Mobilization	Resource Conversion	Force Generation	Leadership	Operational Execution	Outcome
Philippines	X B	X	X	X	X B	X B	U.S. win
Bombing campaign	O B	X	X			X B	U.S. win
Vietnam	A	O		O	O A	O A	U.S. loss
Yugoslavia	O B			A	X B Z	X A	U.S. win
Iraq conventional	B			B	X	X B	U.S. win
Iraq unconventional	O B	O B	O	X A	X A	X A	U.S. "win"
Afghanistan Conventional					X	X B	U.S. win
Afghanistan unconventional	O A	O A		A	A	X	U.S. impending loss

X=strategically successful U.S. effort;
O=strategically unsuccessful U.S. effort;
A=strategically successful U.S. adversary effort;
B=strategically unsuccessful U.S. adversary effort;
Z=strategically successful U.S. "friend" effort.
Source: Summary of author's judgments explained in the empirical chapters.

The materialist dimensions of *resource mobilization* and *resource conversion* played modest roles in determining strategic outcomes in all cases except the bombing campaign, which battered German industry and eventually cut supplies to the *Luftwaffe*, and in the Philippine War, where American troops' control of food and other supplies complemented operations in other dimensions. U.S. airmen tried unsuccessfully to make the dimensions relevant in Operation Rolling Thunder in 1965–1968. They had no role in Yugoslavia. Seizing the oil fields was a vulnerability amelioration effort in the conventional phase of the Iraq war. Coalition states failed miserably to stop the Taliban from receiving funding from drug sales and international donors. U.S. adversaries, in contrast, tried with some success to cut U.S. resources by shattering U.S.-led coalitions, not attacking the United States directly.

One explanation for lack of major impact except in the bombing campaign, consistent with Robert Pape's reasoning, is that resource mobilization and conversion are critical only in total wars in which normative constraints are minimal; the military objective is sheer destruction of the uniformed enemy; and no exploitable asymmetry is found, meaning the default strategy is attrition.[4] These conditions occur rarely. Western states have not experienced them since 1945.

American activities in the *force generation* dimension played modest roles in wars other than the Philippine War and Operation Allied Force. The U.S. naval blockade complicated Army of Liberation efforts to acquire supplies and the amnesty program destroyed Filipino guerrilla units by making surrender attractive. The United States attacked VC cadre with moderate success, but the war's decisive arenas lay elsewhere. It largely ignored this dimension in its limited, coercive war against Yugoslavia and in its conventional campaigns in Iraq and Afghanistan. American operations in this dimension during unconventional phases of wars in Afghanistan and Iraq mainly were ineffective or counterproductive.

America's adversaries in none of the wars directly attacked U.S. force generation activities in the United States, but Filipino nationalists and Vietnamese communists (and both sets of their American friends) worked to degrade the morale of U.S. troops, and thereby U.S. military units' operational effectiveness, via propaganda. Adversaries regularly sought, with varying degrees of success, to reduce the effective U.S. force structures available to fight their respective wars by attacking American national will (and the will of NATO members in 1999 and coalition partners in Iraq and Afghanistan) in ways that had two intended results: (1) limiting the size and shifting the composition of the forces the United States and its allies were willing to deploy; and (2) affecting targeting policies and ROE in ways that hindered the operational effectiveness of deployed forces.

Internally, the U.S. military, motivated by force protection concerns and a lack of desire to fight, convinced President Clinton to severely restrict NATO force generation in 1999 by committing only air units. NATO civilian leaders further limited NATO's use of force out of concern for civilian casualties.

The *leadership* dimension was frequently a dimension of U.S. attack, a recurrent drag on U.S. operational performance, and a critical producer of U.S. military power in one case. The United States targeted Filipino nationalist leaders in combat operations (like the capture of Aguinaldo) and psychological appeals (the amnesty program). The Allies did not try to kill Hitler but ineffectively tried to influence him via morale bombing. Presidents Johnson and Nixon futilely appealed to North Vietnam's leaders with unattractive diplomatic, military, and economic inducements. In 1999, U.S. targeteers tried to influence Milošević by attacking Yugoslav infrastructure they thought he liked—without major consequence. In Iraq, the United States failed to kill Saddam Hussein in the first attack of the war but successfully accomplished its goal of overthrowing him and capturing or killing him and most of his senior subordinate leaders—only to face insurgent leaders who the coalition did not effectively neutralize. In Afghanistan, coalition forces killed many fighters and midlevel leaders but did not kill senior leaders like Mullah Omar and Zawahiri, and the Americans shot bin Laden nearly a decade after 9/11.

Consistently in the latter four cases, especially, the U.S. military establishment, with civilian leaders' acquiescence, organized for wars military leaders wanted to fight—not those actually underway. In Vietnam, most obviously, the generals maintained the conventional force structure and tactics they developed to fight the Warsaw Pact and tightly controlled forces well equipped for counterinsurgency operations—like Army Special Forces. Military leaders chose not to learn from experiences in Vietnam, only to find themselves in Iraq and Afghanistan in unwanted counterinsurgency wars they again were unprepared to fight. U.S. government leaders in general, including senior military officers, in the last five cases learned and adapted more slowly than most of their adversaries—with negative consequences for the wars' strategic outcomes. Only in the Philippines did inspired theater- and local-level U.S. military and civilian leadership make a major, positive, strategically important difference. Only in Iraq did strong presidential leadership influence a war's outcome significantly and positively—but only after Bush made serious mistakes earlier.

American adversaries did not try to attack U.S. leaders directly. Instead, Filipino nationalists, Vietnamese communists, and Iraqi and Afghan insurgents focused on American national will, understanding correctly that if they could induce the American people to strongly oppose their wars, Congress could force a U.S. withdrawal over presidential objections. But the KLA very effectively manipulated President Clinton into launching a war against its Yugoslav enemies over modest domestic opposition.

As active wars, there was combat in the *operational execution* dimension in each case, but the strategic consequences of fighting varied. In the

Philippines, the bombing campaign, and the conventional phases of the wars in Afghanistan and Iraq, combat substantially damaged enemy military units and contributed appreciably to U.S. successes. The belatedly credible threat of a conventional ground war, not bombing, apparently convinced Milošević to settle. In Vietnam, Yugoslavia, and the unconventional phases in Iraq and Afghanistan, however, American combat activities, while usually tactically successful, did not determine strategic outcomes because critical dimensions lay elsewhere. In the four wars that featured significant U.S. counterinsurgency operations, only in the Philippines did the U.S. military recognize institutionally and quickly internalize the importance of tasks other than killing people and destroying things. Failure to effectively perform non-kinetic military tasks abroad and political ones domestically cost the United States the war in Vietnam and may do so in Afghanistan.

The cases illustrate an evolution in the dimensions in which strategic outcomes were determined. Most obviously, casualty aversion became critical influences in the national will, force generation, leadership, and operational execution dimensions starting with the Vietnam War. North Vietnam made U.S. military and civilian casualties a critical determinant of American *national will* to wage war through both military field operations and effective propaganda. U.S. civilian and military leaders tried from the start of Operation Rolling Thunder to minimize North Vietnamese civilian casualties but initially were willing to take appreciable American casualties as part of General Westmoreland's attrition strategy; the latter view changed considerably during the war. Since the war in Vietnam, force protection has been a significant constraint on U.S. operations in the *force generation* and *operational execution* dimensions. The two casualty aversion norms competed as drivers of U.S. political/military actions in 1999. Also in 1999 for the first time, aversion to civilian casualties motivated an American president to start a war that involved no U.S. national material interests and had little domestic political support, indicating significant normative influence in the *leadership* dimension. But the United States and NATO nearly lost the war due to negative operational consequences of force protection—especially the increased collateral damage caused in part by pilots flying too high to be effective, which upset European electorates and helped precipitate a political crisis within NATO.

The post-1945 wars differ radically from the Philippine War and the bombing campaign, in which U.S. normative concerns about military and civilian casualties played little role in determining strategic outcomes. President McKinley took an ethically principled stance toward treatment of the Filipino people, if not about purchase of their islands; he wanted to help Filipinos but did not worry nearly as much as modern American leaders about U.S. casualties. And the anti-imperialists, many of whom were of the northern abolitionist tradition, consistently denigrated in their propaganda and domestic rhetoric alleged American racism and alleged

American atrocities against Filipino civilians. They did not lament American casualties; instead, they worked propagandistically to damage U.S. troops' operational effectiveness, in essence trying to increase American casualties. Casualties were not a significant domestic political issue. In 1943–1945, Allied aviators purposely killed hundreds of thousands of German civilians and were willing to take very large losses themselves to win attritional fights.

Recognizing the growing salience of casualty aversion, U.S. opponents since the Vietnam War used a growing variety of tactical military techniques to attack America's strategic political vulnerability to casualties.[5] Suicide bombing, for example, emerged as a technique after bombings in 1982 helped drive the United States and France from Lebanon; they played major roles in the wars in Iraq and Afghanistan.[6] Other techniques included killing individual American troops, regardless of the consequences for unit effectiveness; use of human shields; exploitation of American targeting policy and ROE to protect military assets by locating them in or near protected sites; and tactical entrapment efforts designed to get American soldiers to kill local civilians. When North Vietnam put surface-to-air missiles in residential neighborhoods of Hanoi and Iraqi insurgents hid RPGs in mosques in Fallujah, each used laws of war the United States honored to their advantage. In 1999, the KLA effectively entrapped the United States by indirectly creating the Kosovar Albanian civilian casualties it accurately expected to motivate Clinton to launch a war against its enemies. Deliberate killing of civilians seems to be a growing trend, spurred by the logic of modern politics that terrorist attacks on civilians are the responsibility of governments that fail to protect them, not the killers.

The scope of conflict is widening. Technology accounts for part of this trend. In the Philippines, there were no military aircraft and few electronic communications means, while space-based communications and technical intelligence collection now are active parts of conflict. New communications technologies (like television, the Internet, and DVDs) gave "weak" actors inexpensive but effective means of attack in ideational dimensions beginning with the war in Vietnam.[7] Information warfare and killing civilians are consistent with many actors' efforts to expand the scope of warfare beyond the conventional warfare with which most national armies are comfortable. For example, Rohan Gunaratna argues that al-Qaeda is a multidimensional organization that fights on political, military, and socioeconomic fronts.[8] Zarqawi's beheadings would have been far less politically effective in the Philippines in 1900 than they were before video cameras in Iraq a century later.[9] As Antulio Echevarria argues, it is not so much that attacks on enemy will are new as that new technologies provide enhanced access to enemy will.[10]

The number and types of participating actors have grown markedly. Nonstate actors like al-Qaeda, Hamas, and Hezbollah now fight sophisticated unconventional wars of long duration against materially wealthy states. International organizations created since 1945, especially the United Nations and NATO, and NGOs began to play prominent roles in wars during the Balkan conflicts of the 1990s; they were important actors in the unconventional phases of the Iraq and Afghanistan wars. These organizations both complement state actions at state expense and have their own agendas. NGOs, especially, often resist government requests for cooperation, usually helping but also complicating the conduct of warfare and stability operations by status quo states while giving revisionist actors new strategic opportunities.[11] Commercial firms now perform many other functions the U.S. military and civilian government agencies no longer want to do but complicate further the challenges of coordinating the activities of an already fragmented U.S. government "team."[12] Operations' legitimacy often depends on building large coalitions, which hamper efficient and effective command and control. Journalists are physical and ideational targets given their roles as communicators and the growing importance of information warfare. And, criminal groups increasingly appear as politically and militarily relevant actors.[13]

Target lists are broader. While noncombatant civilians long have been victims in wars as collateral damage, as spoils, and as collective targets when belligerents' sought to destroy enemies' resource bases, civilians as individuals have become chronic targets, primarily to activate the civilian casualty aversion norm in others. Al-Qaeda's Zawahiri and others understand why the United States enlists IGOs and NGOs to participate in political, economic, and social arenas in most of its military operations and therefore consider such civilian organizations legitimate targets.[14] States that hobble themselves normatively by maintaining the increasingly inaccurate distinction between uniformed combatants and civilians, including civilians who work directly with the military in counterinsurgency operations, give revisionist adversaries good new targeting opportunities.[15]

MATERIAL MASS, TECHNOLOGY, AND SKILL

Quantities of material resources deployed played no strategically decisive role in any war except in the bombing campaign. Of the elaborations on materialist theory, military technology marginally affected most wars' outcomes and there were no systemic offensive or defensive advantages, but skill played major roles.[16]

Technology

Technology's roles varied considerably. It contributed significantly to the success of coalitions' conventional campaigns in 2001 and 2003 but was much less important in the other wars. In the Philippines it helped the Americans some, but U.S. volunteer regiments performed well early in the war using Civil War–vintage single-shot rifles against more modern

Mauser rifles; superior skill won the war in the end. In 1943–1945, Germany possessed some superior technologies relevant to air warfare—like jet fighters and aerial rockets—but was crushed by the Allies' material preponderance. In Vietnam and the unconventional phases of the Iraq and Afghan wars, technology helped U.S. forces in some respects but adversaries found many ways to avoid U.S. technological strengths and other ways to use American material assets against U.S. interests. The most important was manipulating American firepower through disinformation to generate politically useful collateral damage incidents. Moreover, Americans' conceit about their allegedly superior technology chronically fostered hubris that hindered institutional learning and adaptation in unsuccessful U.S. wars, especially.[17] American troops consistently failed to understand the limits of their technologies. When they expected too much, they repeatedly displayed a dearth of what Murray and Scales call "intelligent thinking."[18]

Technology played more roles than the "force multiplier" functions—like firepower, "protection," and mobility—featured in many IR theories.[19] For example, in 1999 technology provided the means by which the Pentagon could avoid a ground war it did not want; without PGMs the military would have had to plan to invade Yugoslavia from the beginning—perhaps producing a credible coercive threat that might have ended the diplomatic crisis over Kosovo without fighting. And, in recent conflicts, readily available information technologies like the Internet and DVDs gave weak actors inexpensive and secure means of propagandizing, recruiting and fund raising, training, and command and control—thereby affecting the national will, resource mobilization, force generation, and leadership dimensions of wars, respectively.[20] But technologies supporting information operations were only helpful if their content resonated with target audiences. Many important military-related technologies thus were tightly linked to ideational factors as enablers or carriers.

Offensive or Defensive Advantage?

No belligerent in any of the cases had a systemic, technology-based offensive or defensive advantage. U.S. airmen promised before World War II that strategic bombing could alone defeat Germany. Other airmen operating against Vietnam and Yugoslavia complained bitterly that politicians' interference with the application of air doctrine threw away the allegedly strategically decisive advantages of airpower. But airmen's claims consistently were grossly excessive. The limited technologies of 1943–1945 could not deliver a "knock-out blow." North Vietnam's largely agrarian economy and dispersed military forces were not vulnerable to air strikes. In Iraq, the battlefield successes of March–April 2003 melted away when insurgents shifted to irregular warfare. In Afghanistan, airpower usually effectively supported coalition ground forces tactically but also generated thousands of strategically important civilian casualties.

More generally, the strategic impact of airpower was always contingent upon enemies' willingness to present suitable targets—preferably from airmen's perspective ones that were big and readily observable, did not move, and did not shoot back. When enemies dispersed or hid, airpower produced little. Moreover, in recent wars bombing was consistently restricted for normative reasons in ways (like targeting policies) that limited use of the full range of munitions' engineering characteristics. And, military technology often was simply irrelevant in counterinsurgency operations in which nonviolent activities played major roles and even the identification of offensive and defensive orientations of stability operations blurred.

Skill Important

The cases illustrate the importance of *skill* as well as its complexity—and thereby show why there is no simple definition of the term or theoretical construct that explains it. A central tenet of vulnerability theory is that actors' abilities to identify and to exploit enemy vulnerabilities, and to identify and ameliorate their own, determine strategic outcomes and thence military power. This means that ongoing learning—how actors in dynamic interaction perceive their enemies and themselves—and effective adaptation are key processes of warfare. These learning and adapting attributes are skills.

The skills of national civilian and military leaders relevant to the conduct of the specific war underway were far more important than material mass and technology as drivers of the outcomes of the wars. Materially weak but competent American troops and civilian administrators in the Philippines defeated an initially large but inept nationalist movement. The Allies won their war of attrition in 1943–1945 in significant part by degrading the skill of the *Luftwaffe*'s fighter command by killing experienced fighter pilots faster than they could be replaced. But the Allies had to wage a war of attrition because they were not skillful enough to find and asymmetrically exploit any German vulnerability that might have won the war sooner or at lower cost. Rotation policies helped keep U.S. military skills in Vietnam, Iraq, and Afghanistan chronically inadequate. Even in October–December 2001 and in March–April 2003 when technology was operationally important, the teamwork required in Biddle's *modern system* was necessary to use offensive technologies to full effect.

U.S. adversaries displayed varying but sometimes much greater capacity to learn and adapt than did the United States—a critical skill given that while the United States could (and often did) muddle along with suboptimal operational effectiveness and efficiency buttressed by its material strength, weaker actors faced rapid defeat if they did not learn well and fast.[21] In the wars the United States lost, U.S. adversaries displayed far greater collective flexibility and initiative than did American forces. Iraqi insurgent groups composed largely of former employees of Saddam's government quickly developed the tactical military competence that Saddam's regime as a whole notably lacked for decades. The improved performance of Saddam loyalists came immediately after they lost most of their major military equipment holdings—bolstering further U.S. material "superiority"—as well as Saddam's stifling institutional constraints.[22] With al-Qaeda's help, the Taliban as insurgents by 2011 developed much better capacities for governance and information operations than they displayed while in power in 1996–2001.

Vietnamese communists perhaps learned best. Because they understood that the critical U.S. strategic vulnerability was normative, their tactical military operations supported an attrition strategy designed to persuade Americans to quit fighting, producing a model many other actors emulated successfully in later wars. Vietnamese communists' ideological perspectives gave them an enormous advantage in integrating political and military operations and the *Lao Dong* party's internal discipline and determination to unify Vietnam enabled General Giap to take chances and to adapt when operations went awry. Americans never demonstrated anything close to the DRV's patience, self-discipline, and collective political-military skills.

Even Saddam Hussein, whose record of strategic errors is impressively long, in 2003 distributed copies of *Black Hawk Down*, Mark Bowden's account of the Mogadishu battle that led Clinton to quit Somalia, to his senior subordinates to help them learn how to defeat the Americans, and Iraqi television just before the war showed footage of the last American helicopters leaving the U.S. Embassy in Saigon in 1975.[23] The regime evidently wanted to remind Iraqis of the unreliability of American promises to their allies—a poignant message to Kurds and Shi'a Arabs who thought the United States let them down in the early 1990s when, after the United States called on Iraqis to depose Saddam, his security forces crushed their revolts and killed thousands of people while Washington did nothing.

The consistent importance of operational competence in different types of wars confirms materialist theorists' judgments that skill is important. But other materialist theories' rationalist assumptions that resources are used efficiently—that is to say, consistently skillfully—are not. The development of skill is hard—as Aguinaldo and Saddam showed. It is hard to maintain, as marked shifts over time in U.S. military institutional proficiency in counterinsurgency warfare illustrate. Skill also is often not very fungible across military functions and operational dimensions, as the U.S. performance in the Iraq war's two phases demonstrates especially dramatically.

U.S. LEARNING AND ADAPTATION

U.S. efforts to identify, exploit, and ameliorate vulnerabilities show a pattern. In general, the United States, despite its massive wealth and technologies, poorly learned and adapted, and hence often slowly or inaccurately identified and weakly exploited adversary vulnerabilities.[24] It also poorly ameliorated its own. Even when the United States won its wars, these deficiencies produced American operations that rarely were efficient.

American military organizations and their political masters rarely understood their enemies very well before conflicts began and, except in the Philippines, learned little about them during operations. Intelligence collection and analytical capabilities therefore were primary means of vulnerability identification in all six cases—but generally were deficient. In the Philippines, American soldiers established close contacts with locals, learned their languages, and developed good tactical human intelligence capabilities through use of Filipino scouts and local police forces; they revised policies and practices effectively in response to their learning. As a result, they maintained tactical military and strategic political initiative. In contrast, the USAAF did not learn much about the German economy before or during World War II.

Mediocre performance by the large U.S. intelligence establishment created after World War II, and the poor use of intelligence products by consumers, were significant causes of U.S. policy making and military operational errors in the last four cases.[25] In Vietnam and later wars, the U.S. government prominently did not develop effective HUMINT collection capabilities. Causes included rapid troop rotations and concentrated basing in fortified camps that kept most troops from mingling with, learning about, and caring about local people; bureaucratic denigration of area expertise and foreign language skills as military career enhancing qualities; Americans' infatuation with technology, which led to acquisition of expensive technical collection systems that can monitor communications and track physically large metal objects (like tanks) but not people's motives and attitudes; fear of danger to case officers operating in unfriendly territory, the intelligence community's equivalent of force protection; and bureaucratic warfare between the military and intelligence agencies that delivered unwelcome messages. Reflecting both weak intelligence and poor training about Vietnam and Iraq, many U.S. troops in both wars admitted they little understood their insurgent enemies and, in consequence, chronically ceded tactical initiative to them. In 1999 and the conventional phases of the Iraq and Afghan wars, the U.S. military adequately relied on sensor-based intelligence collection that fed target data to aircraft and missiles only because the wars were nearly wholly conventional; such limited information was inadequate in Vietnam, in Afghanistan after December 2001, and in Iraq after April 2003. To America's advantage, such technologies did not exist in 1899–1902—forcing U.S. troops to develop more important personal skills.

Inaccurate assumptions about adversaries and operating environments—like the Iraqis will welcome us and the North Vietnamese have a "breaking point"—greatly hampered senior leaders' decision making independent of U.S. intelligence agencies' performance and contributed markedly to strategic failure. Clinton's misunderstanding of Milošević turned a failed exercise in coercive diplomacy into a war. Senior American leaders' negative predispositions about Saddam Hussein apparently colored analyses of information about Saddam's links with terrorist groups. Learning may not even have been attractive in 2002 to Bush administration officials who had already decided that Saddam was a threat who needed to be eliminated.

American military organizations repeatedly refused to try to identify enemy vulnerabilities when operational performance problems suggested a need for adjustments that posed threats to established military doctrine, military culture, and their parochial institutional interests.[26] Most obviously during the war in Vietnam, many individual American officials (like George Ball) and whole agencies (like the CIA) fairly accurately understood the lack of North Vietnamese vulnerabilities exploitable by the operational techniques the military preferred, but they were isolated after bureaucratic battles with the military services their assessments threatened. The chronic propensities of presidents, including Roosevelt, Johnson, and Clinton to accept inaccurate military assessments reflected not the micromanagement about which military officers often complain, but rather excessive deference to the military.[27] Bush alone in the six cases decisively, successfully bucked military advice by ordering the strategic shift of 2007 in Iraq.

The U.S. military showed little capacity to learn vicariously. In 1943, only heavy American casualties—not learning from the similarly heavy losses by British and German air forces in the same airspace immediately before they arrived in England—induced the 8th Air Force to abandon its faith in the invulnerability of strategic bombers, which finally led to development of long-range fighters and thence to force structure symmetry and success. U.S. troops in Vietnam refused to study the lessons of the French in 1946–1954 because they thought the French were incompetent. And, U.S. troops in the Balkans in the 1990s and in Iraq declined to accept counterinsurgency pointers from their British allies more experienced in such operations. In all cases, the Americans' attitudes reflected hubris; they knew better how to fight wars than others. The hubris helped keep Americans' collective skills inadequate.

Again with the exception of the Philippine War, the United States adapted slowly and generally inefficiently to exploit enemy vulnerabilities. Recurrent causes of slow and partially effective adaptation included poor learning, which hampered the identification of vulnerabilities and thence of development of effective strategies to exploit them; military service unwillingness to adapt if change endangered preferred force structures or doctrines; military unwillingness to adapt if another service or civilian agency gained bureaucratically, even in relative terms, from the change; military unwillingness to adapt to improve operational performance if proposed changes endangered other important organizational imperatives, including force protection and the creature comforts of troops; and the American cultural infatuation with technology, which led recurrently to unrealistic expectations about prospects for technological solutions to inherently military problems.

The United States was slow to identify its own vulnerabilities and often was unable to ameliorate the vulnerabilities it did recognize. Most obviously, President Clinton evidently never recognized that the KLA manipulated his personal normative vulnerabilities in 1999 even though officials in his administration clearly understood the purpose of the KLA's actions. President Johnson and many of his advisors recognized their Vietnamese enemies' strategy of attacking U.S. national will but were unable to reverse Americans' growing disaffection with their war policies. In Iraq, General Abizaid understood the domestic American political consequences of IED attacks, but the military could not reduce their effectiveness through technology-based countermeasures alone; they finally developed intelligence capabilities sufficiently to find many bomb makers—partly through the fortuitous timing of the Sunni Awakening. And, despite Saddam's efforts to jam GPS satellites, the military continued to use information and other technologies that some military analysts believe create potentially major U.S. operational vulnerabilities.[28]

Military and civilian leaders over time recognized their vulnerabilities to collateral damage incidents but avoided them only by conducting short and successful wars—like those against Iraq in 1991, Grenada in 1983, and Panama in 1989, which featured materially very weak and/or inept opponents. Sensitivity to U.S. casualties produced policy decisions that created other operational weaknesses and vulnerabilities—like the narrow, force protection-motivated force structure of Operation Allied Force, force protection-motivated ROE in Iraq, and collateral damage incidents generally. American civilian and military leaders variously ignored or accepted some operational problems in order to avoid the domestic political and career prospect threats, respectively, that U.S. military casualties cause.

The operationally key attributes of U.S. learning and adaptation in all cases were more importantly institutional than they were related to the availability of material resources, accurate and relevant information, individual cognitive factors, or basic organizational capabilities. This judgment is consistent with Jack Snyder's findings that institutional factors better explain the "myths of empire" than cognitive or realist theoretical explanations, and with Brian Linn's conclusion that the U.S. Army generally has learned and adapted in ways consistent with prevailing institutional interests.[29] These institutional factors, which became more pronounced with the development of a large and bureaucratically rigid American military/security complex after World War II, in some respects seem to be growing in importance. They significantly reduce the ability of the United States to flexibly and effectively use its military assets—a view foreign to much of the materialist-oriented IR literature.

Could the United States have avoided these problems? Might alternative policies or strategies have led to more favorable outcomes or more efficient operations in the last five cases? While obviously speculative, some counterfactual analysis about major American operational deficiencies suggests the answers are no. The USAAF might have spent more time in the late 1930s learning about actual German economic vulnerabilities, rather than assuming that bomber aircraft could exploit them to win the war and bureaucratic independence. In Vietnam, the services might have worked together rather than fight among themselves and looked for ways to win the war rather than defending the force structural and doctrinal status quo and attacking deliverers of unwelcome prognoses for the war. The military in 1999 might have embraced Bill Clinton's humanitarian agenda and worked to achieve it using then-standard joint doctrine and a ground component that could have effectively coerced without fighting or quickly routed the third rate military power that Yugoslavia was, rather than argue for a minimalist approach designed primarily to keep U.S. military personnel safe. The services might have tried to learn from the mistakes of Vietnam, not purposefully to forget them, and entered the 2003 war in Iraq with coherent contingency plans for an extended occupation and capable counterinsurgency forces. And, the U.S. government as a whole might have learned that fragmented, squabbling bureaucracies and an absence of unity of effort and coherent strategies consistently over many decades have lead to dysfunctional operations and suboptimal strategic outcomes.

In each alternative scenario, required changes were largely organizational and attitudinal. They would not have required additional money or alternative hardware. But because the impediments to such change were strongly institutional in nature and recurred over time, it is hard to imagine that there really was much chance for alternative practices or different operational outcomes. U.S. government history suggests that operational deficiencies caused by institutional dysfunctions were, and are, far harder to redress than those caused by even very significant material resource deficiencies.

IMPLICATIONS FOR STRATEGIC INTERACTION THEORY

Ivan Arreguín-Toft's strategic interaction theory is a commendable effort to improve on theories of military power unable to explain why materially weak actors frequently defeat apparently much stronger ones. His recognition of the importance of actor interaction is insightful. But his theoretical framework has flaws that make it largely inconsistent with the U.S. military history discussed herein and with modern wars generally.

While each of the four strategic options his model postulates is vague, his GWS and barbarism options are particularly weak. Derived from

Maoist people's war, Arreguín-Toft's GWS is most applicable to insurgencies that are nationalist or revolutionary in the sense of mid-20th-century colonial wars.[30] But his model only partly captures even Vietnamese communists' conduct of war in 1961–1975. Irregular political/military actors, including Balkan Muslims in the 1990s and more recently al-Qaeda and Iraqi and Afghan insurgents, use(d) forms of unconventional warfare that are different in important respects. Western military practitioners increasingly recognize that effective counterinsurgency operations require identification of the type of insurgency underway, meaning they think there is more than one type of GWS they may need to counter.[31] And, GWS is not a strategically relevant option for materially weaker parties in conventional wars—like Germany in World War II—or in naval and aerial warfare, meaning that when applied to conventional wars Arreguín-Toft's theory limits weak actors to a single strategic option—conventional defense—which is not a helpful theoretical or practical proposition.

Arreguín-Toft does not recognize the role terrorism plays in most insurgencies, meaning he ignores weak actors' chronic, extensive use of barbarism and incorrectly asserts that barbarism is an option only of materially dominant aggressors.[32] Indeed, insurgents typically are far more barbaric than modern Western militaries that understand the domestic and international political constraints on, and negative operational repercussions of, activities even vaguely resembling barbarism. Arreguín-Toft recognizes that modern Western states are heavily constrained by domestic politics from practicing barbarism.[33] And, he reasonably argues that barbarism is only good for winning wars, not for achieving control of losing actors, because it "inevitably backfires."[34] But gaining some such control is a common purpose of wars! The upshot, as Arreguín-Toft seems to recognize at times, is that barbarism is not a plausible choice for many strong modern actors, effectively giving them only one strategic option—conventional attack—which also is not a helpful proposition.

Arreguín-Toft comes close to confusing strategy with tactics. While he accurately distinguishes strategy from tactics and grand strategy in principle, his discussion of GWS generally and his Vietnam interactions, as discussed in chapter 4, are heavily tactical in nature, and his analysis does not always link tactics or even strategies to outcomes. Guerrilla warfare itself is a tactic, not a strategy.[35] Conventional armed forces long have used small unit guerrilla tactics against conventional opponents when they have been useful.[36] His discussions of conventional offense and defense also often focus on tactical-level activities.

Arreguín-Toft pledges allegiance to structural realism but cites a different rational than Kenneth Waltz uses to assert a tendency for competitors to move toward sameness.[37] He argues that socialization occurs regionally instead of systemically and says the development of GWS in Asia enabled its adherents to beat conventional powers from outside the region. But there are many ways to be different, as Iraqi and Afghan insurgents demonstrate. Instead, differences that lead to favorable dyadic political/

military asymmetries, not broad geographical region-based characteristics, enable the weak to beat the strong. Arreguín-Toft's and Waltz's variants of socialization theory both fail to accurately describe or explain modern war.

Strategic interaction theory's limitations aside, how well does Arreguín-Toft explain strategic outcomes? I conclude that he errs in asserting that the Philippine War in aggregate was a same-approach, dominant actor victory. U.S. successes in the early conventional fighting led Aguinaldo to revert to guerrilla warfare—not to U.S. victory. Decisive operations occurred in the longer counterinsurgency phase of the war, in which U.S. forces successfully integrated aggressive counterguerrilla operations with a benevolent policy of attraction—a strategy and result incompatible with Arreguín-Toft's theoretical framework. His conclusion that the strong actor in the Vietnam War loses/tie is waffling; the United States lost the war. His assessments that Operation Allied Force and the conventional phases of the Iraq and Afghanistan wars reflected same-approach wins for dominant parties are reasonable.

My evaluations of my six cases studies per strategic interaction theory do not well support Arreguín-Toft (Table 8.2). Strategic interaction theory accurately "predicts" the outcomes of seven of 16 interactions—an unimpressive 43.8 percent success rate near the coin-flip results of materialist theory that he and Biddle criticize.[38] This record compares to Arreguín-Toft's assessment that his theory accounts for 15 of 17 interactions in his five cases—including two cases resolved by his ad hoc "conciliation" strategy.[39] I emphasize that five of my six analyses are my efforts to identify and analyze interactions consistent with my understanding of his theory, not Arreguín-Toft's interactions. But I find disconcerting that for the Vietnam War—the only one of our case studies that overlap—my evaluations of the interactions Arreguín-Toft specifies differ markedly from his discussions in all four instances for reasons of fact and analysis.

Table 8.2
Accuracy of Forecasts Using Strategic Interaction Theory

War	Number of Interactions	Number Correct	Percentage Correct
Philippines	2	1	50
Bombing Campaign	3	2	67
Vietnam	4	0	0
Yugoslavia	3	2	67
Iraq	2	1	50
Afghanistan	2	1	50
Total	**16**	**7**	**44**

Source: Summary of author's analyses in empirical chapters.

In sum, while on the right track, Arreguín-Toft's theory has important conceptual and empirical problems. It is applicable to a narrow part of the spectrum of political-military conflict. It is at best a theory of how weak actors win insurgencies against colonial powers, not a theory of how actors in general win wars. His 2 x 2 analytical framework is too parsimonious, too static in nature, and too inconsistent with the complex interactions demonstrated repeatedly in history to have much explanatory power. It is not the general theory of military power that Arreguín-Toft claims.[40]

IMPLICATIONS FOR INTERNATIONAL RELATIONS THEORY

This analysis has implications for a substantial part of contemporary security-related IR theory and for literature that heretofore has had little influence in security studies, particularly the constructivist perspective.

Constructivism

A key finding of this study is that norms are fundamental constitutive elements of military power and that analysts cannot accurately assess causes of strategic military outcomes, especially of modern conflicts, without considering the normative foundations of, and instrumental use of norms by, opposing forces.[41] Two militarily significant norms—civilian casualty avoidance and force protection—affected the conduct and outcome of the American war in Vietnam and in every significant U.S. military operation since. They affected operations in other conflicts involving European states that share the norms. The emerging *responsibility-to-protect* norm strongly affected NATO's decision to attack Yugoslavia in 1999. It even more strongly influenced the United Nations' authorization of air strikes against Libya in 2011.

The construction, maintenance, manipulation, and exploitation of norms have become instruments of conflict that primarily affect the national will, force generation, leadership, and operational execution dimensions of military power. Per Peter Katzenstein, norms are:

> [C]ollective expectations for the proper behavior of actors with a given identity. In some situations norms operate like rules that define the identity of an actor, thus having "constitutive effects" that specify what actions will cause relevant others to recognize a particular identity. In other situations norms operate as standards that specify the proper enactment of an already defined identity. In such instances norms have "regulative" effects that specify standards of proper behavior. Norms thus either define (or constitute) identities or prescribe (regulate) behavior, or they do both.[42]

Constitutive aspects of casualty aversion norms influence modern war by defining mission goals and shaping force structures assigned to fight.

Regulative aspects affect wars' conduct through target selections, ROE, commanders' policies, and standard operating procedures. Force protection has prescriptive characteristics.[43] All of these aspects influence strategic outcomes.

Because casualty aversion norms influence military behavior and the types and circumstances of violence perpetrated by forces composed of both people and military hardware, modern military power is produced only by the interaction of ideational and material factors. While norms do not affect the physical properties of munitions, material military assets are operationally effective only when they are used in normatively acceptable ways. Conversely materials, especially the technologies of modern PGMs, sometimes enable military compliance with both norms. The physical characteristics of material military assets determine what can be used, and how assets are used, meaning that the practical effects of the norms depends upon force structures deployed. Different force structures can make one or the other norm easier to implement, and/or put them into conflict, leading to dysfunctional operations and suboptimal outcomes like those of NATO's peculiar air-only war in 1999.

Many actors have noticed the linkage of norms to force structures and their influence on American (and other states') military operations and, as discussed in chapters 4–7, increasingly use norms instrumentally against states that adopt them. "Weak" actors, especially, use normatively inappropriate consequences of the exercise of force by "strong" ones, like inadvertent bombing of civilians, as ingredients of propaganda campaigns that try to turn the destructive power of material military assets against their users by making them political liabilities.[44] By using civilians as human shields and placing military units in or near protected sites, for example, actors seek norm-based political invulnerability to attacks—not effective defenses—and/or derive norm-based propaganda value when they are legally attacked after losing their protected status. Such activities generate technological and tactical responses, leading in turn to counterresponses—a dynamic interaction driven by learning that identifies normative vulnerabilities and generates both physical and ideational adaptations to exploit and ameliorate them.

While the constructivist literature little discusses directly the subject of military power, it provides insights on other aspects of modern warfare.[45] Consistent with theory, the two norms constituted important parts of the identities of U.S. government agencies, including the military, in all six cases discussed herein; in each, senior leaders resolved to act, or to appear to act, appropriately by the normative standards of American society as a whole. In 1899, President McKinley emphasized the benevolence of his war aims. The USAAF in 1943–1945 sought both to rationalize city bombing as consistent with American values and to deflect criticism of it toward the RAF. President Johnson ineffectively tried to address the ethical concerns of Americans about their war in Vietnam. In 1999, the United

States and NATO went further—launching for identity reasons a war that reflected collective resolve to act without material interest, to oppose human rights abuses, to advance humanitarian goals, and to coerce Yugoslavia in ways that to the maximum extent possible harmed neither their own troops nor noncombatant civilians. In 2003, civilian casualty minimization and force protection were formal U.S. operational goals. In Afghanistan in 2009, civilian casualty minimization was a clearly articulated operational constraint.

American civilian leaders to varying but generally increasing degrees after World War II used both norms in regulative fashion to instruct their military commanders. In Operation Rolling Thunder in Vietnam and in every American war since, presidents and/or secretaries of defense, as well as military lawyers concerned about collateral damage, wielded veto power over target choices of American air forces. From President Nixon's directive to General Abrams to avoid any more Hamburger Hill battles, whether under direct orders or not, generals usually acted as if U.S. casualty minimization was an operational goal.

U.S. military identities typically are some variant of warrior. After the rise of the force protection norm, military officers saw force protection as both appropriate for preserving their troops' lives, especially in limited and peacekeeping operations, and instrumentally useful for advancing their own career prospects. They often saw civilian casualty avoidance as an obstacle to force protection and to the conduct of doctrinally sound air warfare, even though American airmen recognized the negative operational consequences of collateral damage. Civilian leaders' logics of appropriateness thereby contradicted airpower purists' logics of appropriateness and consequences. The different perspectives led to internally conflicting identities in recent wars—civilian humanitarian interventionists who wanted "immaculate" warfare against similarly well-meaning warriors who wanted decisive victory and as much safety as possible for their troops—showing in another way that military cultures may not be congruent with those of the societies they ostensibly serve and that they can conflict with political goals.[46] U.S. operations against Yugoslavia, and in Iraq and Afghanistan featured significant, operationally dysfunctional conflict between the two casualty-aversion norms for reasons consistent with constructivist logic.

The development and internalization of the force protection norm discussed in chapter 5 proceeded in a manner similar to that theorized by Martha Finnemore and Kathryn Sikkink.[47] Nonexistent during the Vietnam War, it was the product of norm entrepreneur General Abrams in the 1970s. In the 1980s, Secretary Weinberger and General Powell did more entrepreneurial work with American civilians and military personnel by promulgating their "doctrines." In 1990, the emerging norm helped convince President Bush and Secretary Cheney to commit the large, force protection-motivated Desert Storm force that Powell wanted—which appears to qualify as a tipping point. In the early and mid-1990s a cascade of international decisions internalized the norm in the West, at least. Norm internalization within DoD and most NATO states led to the bizarre, air-only force structure NATO used in 1999 and force protection as a military institutional imperative in Iraq and Afghanistan. Organizational platforms included the senior positions of Abrams, Weinberger, and Powell; DoD policies; military commanders' orders; and hierarchical military organizations.

Recent U.S. wars show the critical importance of enemy and third-party actions in determining how norms influence military outcomes. Vietnamese communists, the KLA and the Yugoslav government in 1999, and Iraqi and Afghan insurgents tried to manipulate Washington through processes similar to the boomerang model of Margaret Keck and Sikkink.[48] In discussing NGOs' influence on state policy, Keck and Sikkink argue that NGOs that cannot directly influence their ultimate targets—like human-rights-violating nondemocratic regimes—instead pressure third-party liberal states to influence the "bad" regimes. In this way, the immediate targets of NGOs are "good" Western governments. The modern political/military analogy is that materially weak U.S. adversaries and putative friends, unable directly to defeat the U.S. military or to control it, respectively, conduct tactical military operations designed to activate casualty avoidance norms in domestic U.S. and international audiences in ways that induce American leaders to make decisions beneficial to them. The cases show how adversaries repeatedly acted to drive America to quit (and accept defeat) or to alter its military operational conduct in favorable ways. Nominal friends of the United States in the KLA successfully manipulated America into attacking their enemies. That is, actors targeted the U.S. national will dimension (and sometimes the leadership dimension) to induce American leaders to follow the wills of their people or Congress and direct changes in their military operations.

The cases suggest three general propositions about casualty aversion norms in the post-1945 warfare of the United States and other Western democracies that generally share the norms: (1) norms and material military assets are mutually enabled in operational contexts and together produce the contemporary military power of Western states; (2) norms can conflict, generating inconsistent policy and dysfunctional operational behavior that helps produce poor strategic outcomes; and (3) norms can be used instrumentally against their holders, meaning the presence of "good" norms may help achieve "bad" goals of actors who exploit them. Instrumental use of norms dramatically increases the ability of materially weak actors to attack strong adversaries and to ameliorate vulnerabilities to attacks by the strong that the regulative features of norms limit in scope and intensity. Exploiting normative vulnerabilities in the strong is the primary means by which weak actors now conduct *asymmetric warfare*.

These lessons suggest a need for qualification of the usually implicit notion in idealist-leaning constructivist writings that the spread of norms is a wholly good thing, and of other theoretical assertions that norms are pushed by hegemons.[49] Modern political/military history indicates instead that many contemporary political actors both reject Western casualty aversion norms and see them as societal weaknesses useful for generating Western defeats. Vulnerability exploitation typically requires generating *facts on the ground* that activate the norms, which usually means killing people, directly or indirectly. Few constructivists seem to recognize norms as weapons of war or acknowledge that instrumental use of casualty aversion norms, enabled by international conventions and the states and activist groups that uphold them, is the reason many people, especially civilians, die.[50]

This reasoning indicates that analysts cannot well assess the relative military power of any belligerent pair that includes at least one modern (post-1945) Western state without carefully examining interactions of material and normative factors in each operational context. This means that explanations of military power must simultaneously address issues and methods that traditionally have belonged to realism or constructivism, but not both. Coherent explanations of contemporary military power must integrate realist and constructivist insights.[51]

Analysts of contemporary U.S. (and Western) military power must examine normative conditions under which material military assets are employed in order to explain or forecast specific military outcomes. These can be divided into three general categories. First, norms may limit the scope of appropriate, and thereby actual, military options by determining how material power is brought to bear. In Operation Allied Force most obviously, norms narrowed sharply the scope of applicable U.S. military power to that of an air and missile force structure, limited the range of acceptable targets, and led to restrictive ROE. Second, normative imperatives determine the magnitude, or threshold, of the political grievance required to justify the use of force in a limited war.[52] If normative alarm is modest, advocates of war (like Secretary Albright in 1999) will be unable to move bureaucracies subject to rest and soldiers reluctant to risk lives and careers in conflicts not vital to national interests, and there will be no military response or a tepid one like Allied Force that emphasizes force protection.[53] In the case of NATO's intervention in Libya in 2011, the threshold evidently was lower than in 1999. Because many U.S. enemies (and putative friends) have learned these lessons, understanding military power from a net assessment perspective requires a third analytical category: the ability of a belligerent to alter normatively relevant facts on the ground, manipulate adversaries' strategic goals, or influence operational restrictions like ROE through appeals to norms. This means a need to analyze actors' ability to shape and exploit their targets' normative vulnerabilities in dynamic interaction.[54]

Other IR Theories

Vulnerability theory and the practice of modern warfare have many implications for realist theories of military power, other realist theories that also are strongly materialist in nature, and some non-realist theories. Most are negative.

Power

Prominent realists assert that military power is tangible and measurable. But because some elements of military power are intangible and military outcomes depend upon both attackers' actions and the vulnerabilities of targets in multiple physical and ideational dimensions, there can never be single measures of military power even for any one military force—let alone for many actors across time. No ordinal or cardinal measure can consistently rank actors' military "capabilities" or "power" independently of operational context, meaning the traditional IR notion that there are stable groups of great, middling, and weak powers is inaccurate.[55] Rupert Smith similarly argues that military power can be assessed but cannot be measured.[56]

We therefore should reject COW-like quantitative, static measures of power based on a few crude industrial and demographic variables and be skeptical of estimates of relative dyadic military power unless characteristics of deployed forces, targets, strategies, national commitment, and operational scenarios are very well specified. Because actors virtually always try to make and exploit favorable asymmetries, power cannot be assessed unless offensive, defensive, deterrent, or other operational orientations of belligerents also are well specified.

Geoffrey Blainey, who is often quoted in the realist theoretical literature, says, "War is a dispute about the measurement of power," which he suggests is knowable in objective terms.[57] Vulnerability theory suggests instead that wars spring in part from differential expectations about the dimensions in which wars will be fought and exploitable vulnerabilities in them. Agreement among belligerents in these areas is much less likely than over counts of the material military assets of states—meaning statesmen and generals are not as dumb as Blainey implies. The only circumstance in which COW-like measures are roughly applicable is symmetrical, attritional warfare in which structurally similar conventional forces simply pound each other. Moreover, as the case studies illustrate, military material assets became political liabilities when they generate collateral damage incidents or facilitate "barbarism."[58]

Most elements of military power are differentially applicable in different dimensions and are not very fungible.[59] The two phases of the Iraq war and the differential results the same Allied force structure generated against different morale, economic, and military targets virtually simultaneously during the bombing campaign of 1943–1945 demonstrate the error of Waltz's assertion that "power, not very fungible for weak states, is very fungible for strong ones."[60] What David Baldwin calls the "paradox of unrealized power" stems from the differential applicability of military assets in different dimensions.[61] The term "asymmetric warfare" commonly used by military practitioners reflects awareness of the limited applicability of the conventional forces to unconventional operations, especially in ideational dimensions[62]

While operations in some dimensions feature material and ideational factors that are never direct substitutes, the extent to which military power is fungible depends heavily on the flexibility actors build into their military-related assets and operations as matters of force structure and doctrine. As in economics, there are efficiency and transaction cost tradeoffs between "general" and "specific" military assets, especially in the short term, and militaries display different institutional capacities to adjust to operational challenges.[63] Limited such U.S. military flexibility is the subject of a growing literature that also refutes COW-like reasoning.[64] Others note the negative operational consequences of inflexible government civilian agencies.[65] The U.S. military in most of the case studies herein displayed little interest in increasing the fungibility of deployed forces by tailoring force structures, changing doctrine, unifying command and control, or otherwise enhancing institutional flexibility.

More material assets may provide actors with advantages but also produce problems. More wealth may enable actors to fund intelligence agencies and diplomatic corps that better identity adversary vulnerabilities. Wealth offers the potential to build military forces that are big and flexible enough to provide means to exploit a range of enemy vulnerabilities. More assets provide some capacity to identify and remedy domestic vulnerabilities. They also provide a buffer against catastrophic surprise attack; even the Pearl Harbor attack of 1941 did not alone seriously endanger U.S. national security. But as the case studies demonstrate repeatedly for the United States, substantial material wealth also can generate dysfunctional characteristics like fragmented and internally competitive bureaucracies, hubris about the alleged war-winning powers of "superior" technologies, and military forces with substantial inertia of organizational, cultural, and doctrinal rest that smaller and more nimble adversaries easily outmaneuver. The cases suggest that military institutional flexibility varies inversely with the material assets military actors possess, not those operationally deployed—a disadvantage for the wealthy United States in limited operations, especially.

If materialist concepts of military power are as flawed as I suggest, why do many analysts find COW-like measures helpful? First, as COW theorists note, good measures of power are hard to derive; even crude ones may be better than none. And, the validity of macro-measures seems to have markedly deteriorated only in recent decades. Even during World

War II, conventional warfare between nation states was fairly symmetrical, limited to comparatively few dimensions by current standards, and largely attritional in nature. The efficacy of such measures declined after 1945 with the arrival of new types of political/military events like wars of national liberation in which conventional military assets often were used suboptimally (compared to their design purposes) in unconventional wars with substantial ideational components. Western adherence to international conventions promulgated after 1945 to limit civilian casualties dramatically increased the salience of the ideational national will and leadership dimensions as determinants of military outcomes, simultaneously reducing the relative importance of material military assets. And, information technologies enabled information operations that influenced strategic political/military outcomes in nontraditional, nonviolent ways. Thus, COW measures that are reasonable first approximations of the relative power of states' similarly structured conventional land armies in Europe in July 1914 are much less accurate for recent conflicts in which the scope of operations expanded in ways the COW Project's few crude material measures cannot address. This logic suggests that large-sample quantitative analyses that draw conclusions about modern warfare from patterns of conflict going back to 1816 may be systematically flawed.[66]

Balance of Power Theory

Balance of power theory is hard to operationalize if states hardly know what to balance against and can never measure adversaries' relative power, let alone systemic power balances, due to the intangible nature of many relevant factors and the uncertainties of assumptions about future conflict scenarios. Depending upon one's perspective, there always is either no balance or multiple contingent balances, explaining why seemingly dichotomous balance of power and power transition theories both have poor explanatory records.[67]

At the same time, it is possible to balance in new ways in dimensions other than the standard materialist dimensions of resource mobilization and force generation. For example, in recent decades states have been able to make the UN Security Council (UNSC) a de facto ally against aggressors, reducing security-seeking actors' need for traditional internal and external balancing. The legitimacy of UNSC enforcement actions, generated in leadership and national will dimensions, produces contingent political-military force for aggrieved victims. Weak Kuwait, for example, temporarily acquired substantial de facto military power vis-à-vis Iraq only after it was unjustifiably, by world normative standards, conquered by Iraq in August 1990. The power lasted only until Kuwait's liberation in early 1991. Ideational circumstances also provide temporary, narrowly focused de facto military power to nonstate actors perceived to be victims—like Kosovar Albanians in 1999 and Libyan rebels in 2011. Such contingent defensive, retaliatory, and occasionally aggressive political-military assets are potentially available to most states, perhaps helping to explain the puzzle of the lack of material balancing against the United States since 1991 (see below).

Offense-Defense Theory

Most variants of offense-defense theory allege enduring, asymmetric advantages of military technologies produced in what I call the resource conversion dimension. Technological asymmetries sometimes produce offensive or defensive advantages of appreciable duration before other technologies, strategies, tactics, force structures, or vulnerability amelioration actions offset them, but other factors consistently are more important determinants of strategic outcomes, explaining the poor correlation between alleged offensive or defensive advantages and outcomes.[68] Narrow theoretical reliance on a part of one dimension as a driver of outcomes is inconsistent with political/military history, including the cases of this study.

In recent decades, ideational and institutional considerations (targeting policy, ROE, force protection-motivated force structure, etc.) sometimes radically restricted actors' use of their weapons, thereby appreciably altering their operational performance compared to scenarios in which the technological characteristics of weapons systems alone dictate operational uses and effects. Actors typically seek to win by generating favorable operational asymmetries, not by developing systemic (that is, widely available) abilities that sometimes produce operational advantages.[69] Moreover, "weak" actors long have gained military advantages in insurgencies by using asymmetrically obsolescent equipment—not the "advanced" technologies implicit in offense-defense theory and explicit in the views of many American technophiles.[70] My cases and many other wars therefore refute offense-defense theory's assertion that systematic, technology-driven advantages to offensive or defensive operations determine wars' likelihood and their results.[71] For modern wars, especially, offense-defense theory is so flawed that it should be discarded.

The Security Dilemma

The security dilemma concept is more intractable than generally presented because it depends on weak offense-defense theory and because states cannot confidently judge the offensive and defensive power-related implications of conventional armed forces, let alone dual-use assets like nominally civilian television networks, for example, which both entertain innocuously and influence strategically important national will by delivering politically relevant messages.[72] Actors can increase their aggressive unconventional warfare capabilities by assembling dual use civil-military assets (like the 9/11 attackers' skills as pilots of civilian aircraft) without alerting other actors that they may be disadvantaged. Yet states' awareness of the changing military operational capabilities of other states is a requirement of security dilemma theory. At the same time, use of social, political, and economic policy tools that unambiguously alone pose no external offensive threats but ameliorate domestic vulnerabilities may improve security without generating the dysfunctions of the security dilemma, opening new security policy options for status quo states, especially.

Alliance Politics

The ability of the KLA in 1999 to exploit Western political vulnerabilities to induce NATO to start a war on its behalf suggests a need to reassess theories of alliance politics. "Chain gang" and "buck passing" do not well characterize situations in which actors recruit de facto allies by engineering circumstances—like civilian deaths—that have sufficient normative value to generate military action.[73] Entrapment occurs, but it differs from Glenn Snyder's use of the term: "being dragged into a conflict over an ally's interests that one does not share."[74] What I therefore call *manipulation* is incompatible with real friendship, is not characteristic of an alliance in the sense that formal mutual commitments are made, and may or may not be lasting in duration. Such manipulation is attractive to actors who need military muscle but do not share Western values or have strongly held goals that justify purposeful expenditure of noncombatant civilians' lives. Wealthy, liberal states are most likely to be targets because they have considerable, generally competent conventional military forces and exploitable normative vulnerabilities. This logic and recent U.S. history of manipulation by Bosnian Muslims, the KLA, and Israel, at least, differ markedly from interest-based realist expectations.

Structural Realism

These observations have very negative implications for structural realism as a theoretical perspective. Military power produced by dyadic interactions is incompatible with neorealism's concept of unitary actors' largely material and stable "capabilities" within an international system that can shift in few ways, mainly in terms of material-power defined polarity.[75] Since 1945, the nature of military power, a big part of state power, has evolved appreciably. Both material and ideational factors now produce strategically important military power at the unit level in ways that are highly contingent on actors' policies and operational context. A new ideational polarity has emerged: actors who accept restrictive, norm-based rules of warfare and those who choose to exploit them.[76] The latter, who typically are revisionist actors, have gained major systemic advantages.

Offensive and defensive realism both fare poorly. The main state goal of acquiring a preponderance of fungible power premised by offensive realism is unattainable because power is directional, dimensional, and contingent upon belligerents' interactions, not just because states may materially balance an aspiring hegemon. Similarly, states cannot simply seek security when their exploitable vulnerabilities are numerous and ways of exploiting them are limited only by prospective attackers' imagination. More reasonable are realist theories designed to capture ideational dimensions, including Stephen Walt's balance-of-threat theory, which emphasizes what I call the leadership dimension, and "neoclassical realist" theories that link domestic politics (national will and leadership dimensions) to international politics.[77]

The Vietnam, Afghanistan, and Iraq cases dramatically, and the other cases to lesser extents, demonstrate that Waltz's assertion that strong powers can afford not to learn—that they "can do the same dumb things over again"—is simply wrong.[78] The United States paid heavy prices for failing to learn. The record suggests that *skill* and Waltz's vaguely specified state *competence* are much more important than material wealth as determinants of successful foreign and military policy, and of national power generally.

Realist and constructivist claims that international socialization pressures lead states to adopt similar military practices also are wrong.[79] Waltz asserts that "[c]ompetition produces a tendency toward the sameness of the competitors" and that "[c]ontending states imitate the military innovations contrived by the country of greatest capability and ingenuity."[80] The record shows that these claims not only are untrue, but that their opposites occur. The North Vietnamese showed little interest in copying, and little ability to copy, U.S. force structures and technologies. Vietnamese military philosophy, following Mao, Sun Tzu, and their experiences in 1946–1954, told them to exploit enemy weaknesses and avoid enemy strengths. They therefore reasonably believed it was undesirable to adopt U.S. ways. More generally, materially weak actors' best way to prevail against materially stronger and technologically more "advanced" adversaries is to be different and act differently in ways that produce favorable dyadic asymmetries. Socialization that induces the weak to poorly emulate the strong, as Saddam Hussein discovered in 2003, is likely to be fatal.

The usually explicit assumption of most realist theories that military power is derived from masses of material assets is inconsistent with all wars except those in which belligerents cannot find exploitable vulnerabilities in their opponents and simply hammer each other in symmetrical, attritional warfare. While the assumption occasionally works fairly well (like the western front in 1915–1916), it does not work well even for most of the great power wars from which many materialist theorists evidently derive the lessons they embed in their theories.[81] France's Napoleon; Prussia's Frederick the Great in the Seven Years' War; Helmuth von Moltke the elder in the German unification wars of 1864, 1866, and 1870–1871; and King Gustavus Adolfus of Sweden in the Thirty Years' War all displayed generalship qualities—aspects of the leadership dimension—that enabled them repeatedly to defeat enemies with larger material endowments. And the naval orientation of Britain's "way of war" that featured small land armies but extensive use of subsidies to continental allies always generated asymmetrical British military asset relationships vis-à-vis the European continental land powers. Historian Paul Schroeder is right to question some realists' reading of history.[82]

Coercive Diplomacy

Subordination of the conduct of modern Western states' coercive diplomacy to normative considerations complicates the use of military force as a foreign policy tool.[83] Because deterrence depends on credibility as well as capability, strong military forces have diminished coercive capacities if adversaries believe norm-constrained governments will not respond, or react meekly, to situations that would lead material interest-oriented states to act vigorously.[84] The case studies illustrate that North Vietnam, Yugoslavia, Saddam Hussein's Iraq, and the Taliban at times doubted U.S. political resolve—reflected in the national will and leadership dimensions—for good reasons.

Daniel Byman and Matthew Waxman argue that coercers must consider "mechanisms" by which their actions lead to desired decisions.[85] If casualty aversion norm-driven target selection policies, ROE, force structures, and operational practices preclude effective mechanisms, coercion will fail. If the scope of acceptable violence is narrow, there also may be little room for the kind of escalatory threat Thomas Schelling postulated in his compellence theory.[86]

Foreign Policy Decision Making

These findings have implications for the literature that focuses on statesmen's misperceptions and resultant policy errors. A large literature emphasizes the importance of cognitive and emotional factors.[87] Other analysts note the power of group dynamics to push for unanimous decisions in policy deliberations, which Irving Janis calls "groupthink."[88] Bureaucracies periodically purge individuals who deliver unwelcome warnings, especially if they are accurate.[89] Organizations develop accepted ways of doing things, which Nathan Leites calls operational codes, which color organizational perceptions in systematic ways.[90] And, domestic politics sometimes push statesmen to make bad decisions.[91] All of these analyses make good sense but are incomplete when they focus on the perceptions, decisions, and actions of a single actor because international politics and warfare always feature dynamic interaction.

Vulnerability theory suggests that analysts of foreign policy decision making should focus more on three other sources of misperceptions and bad policy decisions: (1) international interactions, (2) bureaucratic institutional goals and incentives, and especially (3) ethical constraints like concerns about force protection that limit decision possibilities. Actors often do not know the specifics of interaction-related variables critical for consistently good decision making. These include how actual and potential adversaries (or friends) see mutual political-military vulnerabilities and whether they are accurate or not, adversaries' perceptions of mutual capacities to exploit vulnerabilities, and the financial and political costs of closing critical vulnerabilities. These variables make decision making much more complex, especially because adversaries often hide sensitive data and/or disseminate misleading information.[92] These analytical challenges occur independently of efforts to divine other actors' intentions—another difficult problem.

The case studies illustrate that motives like preservation of the U.S. Army in and after Vietnam, and the advancement of U.S. Air Force interests chronically, channel information collection and analysis and restrict acceptable foreign policy decisions to options favorable to organizational interests regardless of senior civilian leaders' policies or other desires. As Brian Jenkins notes, senior military officers may even be willing to lose wars to protect organizational cultures.[93] And, force protection concerns impose career-related incentives as well as military cultural constraints on tactical-level military decision making normally beyond the reach of senior civilian leaders, influencing the conduct of operations like Allied Force and indirectly affecting strategic outcomes.

The case studies suggest that claims that actor "reputation" is irrelevant to foreign policy making focus too narrowly on statesmen.[94] Popular will working through Congress repeatedly trumped presidential desires, producing a U.S. national reputation for weakness in forms like impatience and casualty sensitivity that encouraged adversaries to challenge the United States militarily, influenced their strategies and tactics, and regularly enabled their successes.[95] While the KLA manipulated Clinton personally, other adversaries ignored or accepted U.S. civilian leaders' personal reputations and attacked well-known American societal unwillingness to tolerate casualties. For example, North Vietnam regarded President Nixon's commitment to his war policies as credible and so chose successfully to defeat him by attacking weakening American national will reflected in Congress.

Regime Type and War

This analysis suggests that the diverse literature that argues that regime type and domestic politics determine wars' strategies, conduct, and outcomes is at best partly right. Numerous studies link democracies statistically to winning wars and argue that democratic accountability or cultural traits make citizens of democracies better fighters.[96] My cases are consistent with other studies showing that Americans are unlikely to support wars unless their purposes are clear and they succeed in short periods of time at small human cost.[97] They also show that American adversaries since the 1960s have been aware of these traits, seen them as weaknesses, not strengths, and repeatedly acted effectively to exploit them.

For these reasons and the broader record of military adaptation illustrated in the empirical chapters, like John Nagl I find many reasons to reject Deborah Avant's argument that regime type (presidential versus parliamentary system) accounts for military institutional flexibility—and to reject other variants of what Adam Grissom calls the civil-military model of military innovation.[98] The evidence instead indicates that military flexibility is a cultural trait largely developed within military organizations.[99] Robert Komer places much of the blame for the American defeat in Vietnam on reliance on entrenched "institutional repertoires" of the various U.S. government agencies, including the military, which fostered business-as-usual operations when appreciable learning and adaptation were needed.[100] These repertoires are mainly functions of bureaucratic interests, organizational cultures, and a national culture that excessively looks for technological solutions to political/military problems, not regime type.

Avant's argument is inconsistent with the fact that American military institutions displayed varying capacities to learn and adapt over time when the structure of the U.S. government remained effectively unchanged. For example, the operationally flexible U.S. Army that won the Philippine War was not the much more rigid one General Westmoreland commanded in Vietnam in 1964–1968. The confident army that General Franks directed into Iraq in 2003 was not the chastened, more flexible army that General Petraeus led in 2007–2008. The Philippine and Iraq cases, especially, also support Grissom's call for a bottom-up approach to assessments of military innovation.[101]

Moreover, American civilian leaders repeatedly have had trouble getting their military commanders to do their bidding. Despite micromanagement of bombing targets since 1965, senior civilians in the cases of this study did not direct the construction of military doctrine, the making of military culture, or most of the activities of military personnel. President Kennedy failed to convince the military to embrace his desire for better U.S. counterinsurgency capabilities. Presidents Kennedy and Johnson distrusted their generals, who often argued with or ignored them. In 1999, President Clinton meekly accepted military preferences for a minimalist, air-only force structure despite its severely negative implications for achievement of his humanitarian goals. In 2002 and later, Secretary Rumsfeld repeatedly clashed with his generals over Iraq war plans and policy. President Bush, however, overruled Rumsfeld and his generals in 2006 over Iraq policy. These episodes illustrate, regardless of the important but different question of whether the civilians or generals were right, that American civilian leaders rarely simply order their military forces to do anything important to the generals—including fighting wars.

The logic of vulnerability theory and the case studies offer no support to theoretical arguments that regime type determines military operational practices and outcomes, which typically offer statistically derived, general conclusions and vague causal mechanisms that are wholly unconvincing. Dan Reiter and Curtis Meek, for example, argue that democracies win with spectacularly vaguely defined "maneuver" strategies after examining alleged single annual national strategies without regard to adversary actions because, they say, knowing other sides' actions does not generate determinant predictions.[102] Exactly! Their line of reasoning has many empirical and logical problems, starting with the basic definitional fact that "maneuver" is an aspect of tactics, not strategy.[103]

Reiter and Allan Stam argue that democracies owe their alleged prowess at winning wars to:

> [T]he skeleton of democracy, those political institutions that hold democratic leaders accountable to the consent of the people, and the spirit of democracy, with its emphasis on the development of individual rights, responsibility, and initiative.[104]

Recent history suggests instead that democracies find winning wars increasingly difficult. The cases illustrate a growing variety of tactical and strategic approaches that enable nondemocracies to exploit the allegedly beneficial political/military aspects of democracies. Only in the Philippines did the U.S. Army display the "initiative" that soldiers of democracies supposedly have. U.S. troops in Vietnam and in later wars in aggregate showed little "initiative" to shift away from doctrinally conventional but ineffective operations—the prominent exception of General Petraeus notwithstanding—but they demanded "individual rights" to creature comforts, short tours, and physical safety that repeatedly damaged the institutional competence of the U.S. military as a whole. Congress took largely hands-off attitudes toward the start of the wars in Vietnam, Afghanistan, and Iraq but congressional Democrats, especially, tried to force presidents to quit when the going got tough—not the kind of "responsibility" parents typically try to teach their children. This is not the backbone of a "skeleton" or a "spirit" likely to help U.S. troops succeed in challenging situations. And, recent U.S. military operations were hobbled by norm-driven operational restrictions—aspects of the people's terms of limited "consent"—that significantly affected outcomes.

These trends and vulnerability theory refute what Michael Desch calls the "triumphalist" view of positive links between democracy and war in favor of a more pessimistic view consistent with older views like those of Alexis de Tocqueville.[105] Scott Bennett and Stam find statistical support for the hypothesis that the alleged advantages of democracy decay as conflicts age and that after 18 months of conflict autocracies are more likely to beat democracies.[106] I suspect that a reframed analysis would find that the 18-month "crossover point" has shortened in recent decades.[107]

U.S. Hegemony?

This analysis is relevant to the ongoing discussion about the nature and duration of alleged U.S. hegemony in the ostensibly unipolar post-Soviet era and the theoretical puzzle of a lack of balancing against the United States. Analysts variously argue that hard balancing has already begun or soon will begin, that soft balancing is underway, and that benign American intentions toward all but a few rogue actors preclude a need for traditional balancing.[108]

Vulnerability theory and recent history suggest another explanation for the lack of balancing: there is no hegemonic U.S. military power worth balancing. American military strengths are so narrowly focused on mid-intensity conventional combat operations—like those of March–April 2003 in Iraq—that they little affect much of the scope of political/military conflict. The broad and expanding spectrum of conflict gives competent adversaries a growing number of operational possibilities against U.S. vulnerabilities that are major, obvious to practitioners, and recurrently successfully exploited.

Because the United States neither credibly threatens nor deters a wide variety of political/military actions and because weak actors have so many ways to defy or defeat the United States, it is politically and economically costly but unnecessary to balance the functions the U.S. military performs well. Moreover, combat often does not decide the outcomes of modern limited wars, meaning weak actors reasonably can expect to prevail politically even if they fail in narrowly defined military terms—another reason why there is little reason to try to balance.

A considerable and growing body of evidence indicates that many actors actually or potentially interested in challenging American military power similarly see no American hegemony and believe they can continue to find ways to exploit America's ideational weaknesses. Indeed, they repeatedly make clear that they see an America hamstrung by weak will and impatience; many actors have called America variants of *paper tiger* since Mao used the term over half a century ago. Many U.S. adversaries, including Osama bin Laden, even bait the United States for the alleged cowardice of its force protection-motivated troops.[109]

Few are as outspoken as Chinese People's Liberation Army (PLA) theorists, who have published lucid analyses of how a materially inferior PLA can defeat the United States.[110] While disinformation is a possibility, their analyses seem plausible. Some U.S. Defense Department officials take them seriously.[111] Chinese officers urge further expansion of the scope of conflict by attacking U.S. vulnerabilities in areas that most Western states do not consider military—like international finance, the environment, drug trafficking, and others.[112] They seek to exploit technology-generated U.S. military operational vulnerabilities and America's political vulnerability to casualties.[113] Senior Colonels Qaio Liang and Wang Xiangsui argue that America suffers from a "disease" of casualty aversion that is pursued "almost to the point of morbidity"; U.S. soldiers are "like precious china bowls that people are afraid to break."[114] If one cannot defeat U.S. forces, Qaio and Wang advise enemies to kill U.S. "rank and file soldiers" for political reasons.[115] This advice is consistent with the actions of numerous U.S. enemies in recent decades.

American military history suggests that the United States has military power that is close to dominant only under highly restrictive circumstances (Figure 8.1). First, in *scope,* operations must play to the U.S. military's strengths in conventional combat operations. The United States conducts unconventional operations—including counterinsurgency, counterterrorism, and peacekeeping operations—relatively poorly. Second, the United States does well in medium *intensity* operations, especially against weak opponents. America as a whole dislikes low intensity operations and has limited capabilities to conduct them. There is no recent U.S. record of large-scale, high-intensity warfare but the supply shortages that press reports indicate began to limit operations even against Yugoslavia in 1999 suggest that the United States cannot maintain high-intensity operations for long. Third, the United States tends to do well in short *duration* conflicts. America's collective patience is modest, meaning that U.S. willingness to persevere over time is small. Such impatience repeatedly has generated U.S. political/military failures (like Vietnam and Somalia) while U.S. military forces were intact. These trends yield a roughly conical pattern of operations in which U.S. military operations are consistently successful.

The size and location of operations within this battle space are contestable. America's ability to prevail depends in significant part on its abilities, in dynamic interaction with its adversaries (and sometimes "friends"), to control the nature of its operations. The United States is advantaged if it can keep the scope of operations conventional, their intensity medium, and their duration short. Most adversaries are advantaged if they operate unconventionally at low intensity for extended periods of time, although North Korea seems to continue to deter Washington by threatening high-intensity conventional warfare as well as use of nuclear weapons.[116]

The U.S. government can improve its chances if it expands the scope of operations it can effectively conduct through institutional change; increases the flexibility, responsiveness, and cooperative spirits of government agencies; persuades the citizenry to be patient and to accept more than trivial human costs of warfare; and gets the military to abandon demands for pre-deployment exit strategies. Put differently, the United States can do better if it makes its political/military assets more fungible in all six operational dimensions and time. But the U.S. military keeps

Figure 8.1
"Battle space" in which the United States potentially has hegemonic power.

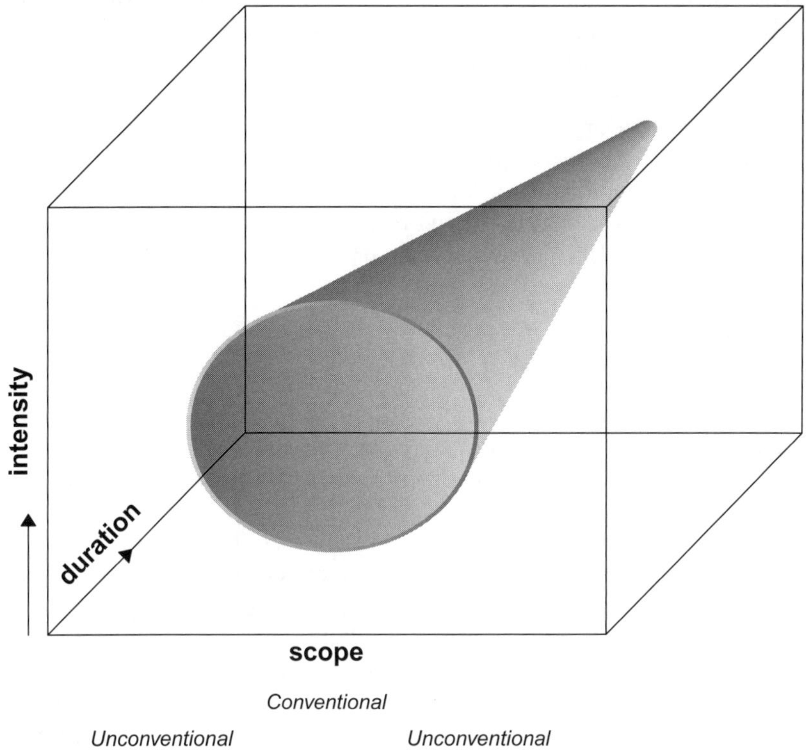

most of its traditional focus on conventional operations, American popular patience with casualties and operational failure seemingly continues to diminish, the U.S. government remains fragmented, and troops still prefer to be home.[117] In Iraq and Afghanistan, American sensitivity to casualties enabled adversaries to use fairly standard, small-unit guerrilla tactics—like roadside bombs and other ambushes that generate small but steady U.S. casualties—to influence strategically important political targets in the United States, thereby easily keeping critical operations outside the U.S. "cone of hegemony."

IMPLICATIONS FOR U.S. POLICY AND PRACTICE

Prominent exceptions like the Gulf War notwithstanding, the U.S. military has not performed very well in recent decades. The United States has made military policy and operational execution mistakes that are almost incomprehensible given American prowess in other areas.[118] And, flexible and imaginative U.S. adversaries repeatedly have made American impatience and casualty aversion centerpieces of their successful operations and are virtually certain to continue to make exploitation of these vulnerabilities central features of future operations against U.S. interests until the techniques they use no longer work and new ones they surely will invent also fail.[119]

The United States faces ongoing, and probably rising, challenges in five general areas. They reflect to some extent systemic factors but they primarily have American origins and perhaps have American solutions: (1) civilian casualty aversion, (2) force protection, (3) inapplicable force structure and doctrine, (4) the growing military welfare state, and (5) a growing lack of national will.

Civilian Casualty Aversion

Potential adversaries chronically influence American policies and military operations by threatening or producing civilian casualties despite American leaders' efforts to ameliorate vulnerabilities in this arena. Neither technology nor more material assets can solve this challenge. Only smart, politically sophisticated military and civilian leaders who can shape popular normative expectations can effectively mitigate those vulnerabilities.[120] Revised operational practices should especially include less use of air strikes, a major producer of collateral damage incidents, in unconventional operations. But tighter reins on airpower threaten two powerful bureaucratic forces: parochial air force institutional interests and the perceived force protection needs of ground troops. U.S. embrace of the R2P norm, as Clinton did implicitly in 1999 and Obama did explicitly in 2011 concerning Libya, increases U.S. vulnerability in this arena.

Force Protection

Now firmly entrenched in military and popular cultures, the force protection norm inherently inhibits mission accomplishment and generates tactical military and strategic political vulnerabilities.[121] Obvious force protection-generated vulnerabilities in 1999 and in Iraq and Afghanistan have prompted numerous actors, including bin Laden and the Chinese army, to plan to continue to exploit this vulnerability. The solution here is obvious, if hard—return to U.S. military and societal acceptance of casualties as normal and acceptable costs of war, consistent with the traditional professional military ethics the troops profess to hold.

Inapplicable Force Structure and Doctrine

The U.S. military and civilian agencies chronically have been poorly designed for the operations they conducted.[122] The dominant conventional focus of the U.S. military—in terms of force structure, training, doctrine, and a culture of "warfighters" and technophiles—has since Vietnam cost the United States dearly in unconventional operations. Some military and senior civilian officials embrace a need for military "transformation" but many of them, like former Secretary Rumsfeld, define transformation as increased reliance on "superior" technologies—not institutional change of the sort useful for improving U.S. conduct of unconventional missions.[123] Such change would require major modifications in force structure, doctrine, training, recruiting selection criteria, and culture that would take years to accomplish after senior civilian and military leaders decide that such change is needed—as yet a questionable proposition.

U.S. military operational deficiencies and challenges reflect inadequacies in the "force structure" of the U.S. government as a whole.[124] The narrow functional focus of the military means that agencies like State, Treasury, the CIA, USAID, and potentially many more should be present in complex operations like counterinsurgency wars in which military activities should be a small part of the overall U.S. effort.[125] But rival agencies with overlapping responsibilities, conflicting interests, and incompatible capabilities have chronically stymied effective policy implementation. Interagency squabbling prominently hampered U.S. military operations during World War II, in Vietnam, and in Grenada in 1983.[126] As noted in chapter 6, Secretary Rumsfeld won President Bush's approval for DoD control of Iraq in 2003–2004, but DoD performed very poorly. It willingly shared responsibilities only when they became unpleasantly burdensome, but no unified U.S. effort ever existed in Iraq or Afghanistan. The U.S. government cannot even coordinate the activities of its contractors very well.[127] The 2005 increase in State Department responsibilities in stability operations is a small step in the right direction. Significant continuing such failures of civilian and military leadership chronically, seriously degrade U.S. political/military operational performance. Given bureaucratic inertia of rest, there is no immediate prospect of more than modest improvement in this arena.

A Growing Military Welfare State

In addition to providing generous formal pay and benefits, the U.S. military since Vietnam increasingly has placed the welfare of troops and military institutions above mission accomplishment.[128] While the military services long have had parochial agendas, all of the services in Vietnam embraced short tours to limit personnel hardship and enhance career "ticket-punching" opportunities, and built increasingly comfortable bases for support troops, at least.[129] The syndrome has gotten worse. Bases in the Balkans and Iraq are opulent by earlier standards and keep troops away from local populations. The air force until recently sent its expeditionary units abroad for a grand total of three months at a time. Army troops and their families practically whined when tours in Iraq in 2007 were temporarily extended from 12 to 15 months as part of the "surge" in troop strength President Bush ordered over Pentagon objections; army chief of staff General Peter Schoomaker instead wanted to cut tours to nine months.[130] Both trends harm operational performance by preventing development of area expertise. As of this writing, tours in Afghanistan typically are 6 to 12 months; marine infantry units in 2011 go for seven-month tours.

The military's friends in Congress have tried to make the good life a goal for military personnel independent of, and arguably at the expense of, national needs. Senator James Webb and Representative John Murtha in 2007 introduced legislation that would have required the Pentagon to keep troops at home for a period as long as their last deployment before they could be sent abroad again. Not to be outdone, the air force in 2011 sought a "dwell" ratio (time in garrison to time deployed) of 4:1—which is the current ratio for army *reservists* and almost the army's 5:1 goal for reservists.[131] Possible solutions here are for the military as a whole (as the marine corps long has done) to emphasize the special nature of military service rather than try to compete with the private sector over pay, perquisites, and working conditions—or return to the draft. The Pentagon and the services want neither.

A Growing Lack of National Will

The American citizenry, and therefore Congress, seems increasingly unwilling to support virtually any military mission other than defense against overt attack on the homeland that cannot be completed quickly, successfully, and at minimal human cost, including inconvenience.[132] Americans' impatience with both casualties and slow progress places large and growing temporal and operational limits on the use of force.[133] Steven Metz argues that while Americans are willing to put up with counterinsurgency operations for only about three years, such operations typically last much longer, requiring a basic change in the way the U.S. government approaches such missions.[134] Especially for operations that do not seem to involve threats to core U.S. interests, like Balkan peacekeeping missions and Operation Allied Force in 1999, the military also opposes commitments and seeks to operate in ways, given force protection imperatives, which inhibit operational performance. This obvious lack of patience leads to a simple adversary formula for strategic success: stall, survive, impose modest costs on the Americans, and wait them out.[135]

A brief review of major U.S. wars in the last century illustrates a secular trend toward sharply reduced national acceptance of casualties. During the Philippine War and the two world wars, there were few significant, casualty-related restrictions on U.S. military operations. In 1917–1918, the United States quickly lost 116,516 men, and 292,131 U.S. troops died as a result of hostile action in World War II. But domestic reaction to U.S. combat fatalities then about 25,000 led President Johnson to decide in March 1968 against a reelection bid. And voters gave control of Congress to antiwar Democrats in November 2006 after only about 2,500 U.S. combat fatalities in Iraq. The main U.S. military goal of the very limited Operation Allied Force was no U.S. casualties at all.[136]

John Mueller makes the point in a different way.[137] He notes that at the time of the Tet Offensive of early 1968, when the percentage of Americans who thought the war in Vietnam was a mistake had risen to just over 50 percent, the number of American combat fatalities was about 20,000. When the same percentage of Americans who thought the Iraq war was a mistake was reached, U.S. combat deaths were only about 1,500.

These trends pose daunting challenges to American policy makers and military leaders. With the demise of Saddam Hussein, there are few remaining convenient incompetents the U.S. military can hope to beat quickly and easily in the medium-intensity conventional combat operations the United States clearly conducts well. This means likely dimensional movement of future conflicts further beyond the bounds of conventional operations into the specialty arenas of special operations forces, who conventional troops continue to view with some suspicion, and into areas that belong bureaucratically to agencies other than DoD. This is a formula for more bureaucratic turf battles and more military operational performance problems. History suggests there is little chance that Congress will restructure or refocus executive agencies in the absence of a major crisis—and maybe not then.

Many observers have noted U.S. government performance problems and variously recommended reforms that are far less ambitious than those implicit in the challenges noted above, with modest success.[138] Among the few successes, legislative mandates to better coordinate the activities of the services in the form of the Goldwater-Nichols Act of 1986 and to rebuild special operations forces took years of painful bureaucratic battles and remain incomplete.[139] The many failed efforts to reform the intelligence community and the massive integration travails of the Department of Homeland Security illustrate that many institutional and political obstacles to real reform remain even in the wake of the shock of the attacks of September 11, 2001.[140] Barring a catastrophic attack on the homeland that dwarfs 9/11, it is hard to envision any meaningful improvements—that is, reform that improves the actual performance of the U.S. government significantly or appreciably changes popular attitudes about military operations. Perhaps the best the United States reasonably can do is to revise foreign policy to produce fewer enemies and to get into fewer of the small wars that America fights poorly.

Notes

CHAPTER 1

1. Wright, *A Study of War*; Levy, *War in the Modern Great Power System, 1495–1975.*

2. Blainey, *The Causes of War*; Organski and Kugler, *The War Ledger*; Haas, "The Balance of Power"; Singer, "Reconstructing the Correlates of War Data Set on Material Capabilities of States, 1816–1985."

3. Morganthau, *Politics Among Nations*; Gilpin, *War and Change in World Politics*; Van Evera, *Causes of War*; Copeland, *The Origins of Major War,* 11–15; Mearsheimer, *Conventional Deterrence*; Quester, *Offense and Defense in the International System.*

4. Singer, "Reconstructing the Correlates of War Data Set on Material Capabilities of States, 1816–1985"; Cline, *World Power Assessment.*

5. Organski and Kugler, *The War Ledger.*

6. Stam, *Win, Lose, or Draw,* 73, 157–69; Bueno de Mesquito, *The War Trap*; Organski and Kugler, *The War Ledger*; Layne, *The Peace of Illusion*; Sullivan, "War Aims and War Outcomes"; Desch, *Power and Military Effectiveness.*

7. Waltz, *Theory of International Politics,* 131.

8. Ibid., 98.

9. Waltz, "Structural Realism after the Cold War," 16.

10. Grieco, "Anarchy and the Limits of Cooperation."

11. Knorr, *Military Power and Potential.*

12. For a summary of offense-defense theoretical perspectives, see Lynn-Jones, "Offense-Defense Theory and Its Critics"; Levy, "The Offense/Defense Balance of Military Technology." Other theorists include Quester, *Offense and Defense in the International System*; Van Evera, "Offense, Defense, and the Causes of War"; Glaser, "Realists as Optimists"; Adams, "Attack or Conquer?"; Posen, "Nationalism, the Mass Army, and Military Power"; Glaser and Kaufman, "What Is the Offense-Defense Balance and Can We Measure It?"

13. Glaser and Kaufman, "What Is the Offense-Defense Balance and Can We Measure It?"; Waltz, *Theory of International Politics,* 131.

14. Waltz, *Theory of International Politics*; Layne, *The Peace of Illusion*; Morgenthau, *Politics Among Nations*; Organski and Kugler, *The War Ledger*; Schweller, "Unanswered Threats."

15. Levy, "The Offense/Defense Balance of Military Technology"; Lieber, "Grasping the Technological Peace."

16. Arreguín-Toft, "How the Weak Win Small Wars," 97; Biddle, *Military Power,* 20–23; Stam, *Win, Lose, or Draw,* 73; Bennett and Stam, "The Declining Advantage of Democracy"; Sullivan, "War Aims and War Outcomes."

17. Layne, *The Peace of Illusion.*

18. Smith, *The Utility of Force,* 240–42.

19. Gentry, "Intelligence Failure Reframed."

20. Baldwin, "Power Analysis and World Politics."

21. Finel, "Taking Offense at Offense-Defense Theory," 184.

22. Beyerchen, "Clausewitz, Nonlinearity, and the Unpredictability of War."

23. Merom, *How Democracies Lose Small Wars*; Hammes, *The Sling and the Stone.*

24. National Commission on Terrorist Attacks on the United States, *The 9/11 Commission Report.*

25. Arreguín-Toft, *How the Weak Win Small Wars,* 36.

26. Van Evera, *Causes of War.*

27. Exceptions include Walt, *The Origins of Alliances.*

28. Theorists who use such terms include Arreguín-Toft, *How the Weak Win Small Wars*; Mearsheimer, *Conventional Deterrence,* 35–43; Paul, *Asymmetric Conflicts.* For a critique of concepts of strategy, see Betts, "Is Strategy an Illusion?"

29. Sullivan, "War Aims and War Outcomes."

30. Clausewitz, *On War*; Knorr, "Economic Interdependence and National Security," 16–18; Hammes, *The Sling and the Stone,* 56; Baldwin, "Power Analysis and World Politics," 163; Gamson, *Power and Discontent,* 43; Cassidy, "Why Great Powers Fight Small Wars Badly," 47–49; Cohen, "Constraints on America's Conduct of Small Wars," 155–60; Echevarria, "Deconstructing the Theory of Fourth-Generation War," 233; Betz, "The More You Know, the Less You Understand," 528; Smith, *The Utility of Force,* 242; Arreguín-Toft, *How the Weak Win Small Wars,* 34, 203.

31. A key finding of the democratic peace literature is that democracies do not fight each other but do fight autocracies. See Doyle, "Liberalism and World Politics"; Russett, *Grasping the Democratic Peace*; Owen, "How Liberalism Produces Democratic Peace." Only a few analysts disagree; see Desch, "Democracy and Victory."

32. Summers, *On Strategy,* 21–32; Hammes, *The Sling and the Stone,* 213; Howell and Pevehouse, "When Congress Stops Wars."

33. Betz, "The More You Know, the Less You Understand," 528–29.

34. Sullivan, "War Aims and War Outcomes."

35. Posen, "Nationalism, the Mass Army, and Military Power"; Glaser and Kaufman, "What Is the Offense-Defense Balance and Can We Measure It?" 57, 66–67.

36. Knorr, *Military Power and Potential.*

37. My concept of military culture is similar to the *strategic culture* of Yitzhak Klein. See Klein, "A Theory of Strategic Culture." According to U.S. Department of Defense publication JCS Pub 1–02, *Department of Defense Dictionary of Military and Associated Terms,* 189, doctrine is "fundamental principles by which the military

forces or elements thereof guide their actions in support of national military objectives. It is authoritative but requires judgment in application." This concept of doctrine differs from many theorists' use of the term to describe civilian policy statements or general military orientations like supposed offensive or defensive *doctrines*. See Brooks and Stanley, *Creating Military Power*; Posen, *Sources of Military Doctrine*; Snyder, *The Ideology of the Offensive*; Stam, *Win, Lose, or Draw*, 52, 82.

38. A rare example of prominent treatment of leadership is Shimshoni, "Technology, Military Advantage, and World War I."

39. Warden, "Success in Modern War"; Warden, *The Air Campaign*.

40. Sun-Tzu, *The Art of Warfare*.

41. Clausewitz, *On War*, 77.

42. Liddell Hart, *Strategy*, xix–xxi, 323–33; Freedman, *The Revolution in Strategic Affairs*, 14.

43. Galula, *Counterinsurgency Warfare*; Thompson, *Defeating Communist Insurgency*; Blank, *Rethinking Asymmetric Threats*, 30–31; Metz, "Strategic Asymmetry," 25–28; Barnett, *Asymmetrical Warfare*; Qaio and Wang, *Unrestricted Warfare*; Hammes, *The Sling and the Stone*; Lind et al., "The Changing Face of War."

44. Clark and Konrad, "Asymmetric Conflict."

45. Mack, "Why Big Nations Lose Small Wars"; Merom, *How Democracies Lose Small Wars*; Shimshoni, "Technology, Military Advantage, and World War I"; Biddle, *Military Power*; Biddle, "The Past as Prologue."

46. Mack, "Why Big Nations Lose Small Wars," 184.

47. Thomas, *The Ethics of Destruction*, 185.

48. Shimshoni, "Technology, Military Advantage, and World War I," 197–99.

49. Ibid., 194–97.

50. Biddle, "The Past as Prologue," 16; Biddle, "Rebuilding the Foundations of Offense-Defense Theory," 756.

51. Biddle, "The Past as Prologue," 16.

52. Biddle and Long, "Democracy and Military Effectiveness."

53. Hammes, *The Sling and the Stone*; Lind et al., "The Changing Face of War."

54. Hammes, *The Sling and the Stone*; Hoffman, "Small Wars Revisited."

55. Katzenstein, *The Culture of National Security*.

56. Kier, *Imagining War*; Rosen, *Societies and Military Power*; Pollack, *Arabs at War*; Tannenwald, "The Nuclear Taboo"; Price, "Reversing the Gun Sights."

57. Legro, "Military Culture and Inadvertent Escalation in World War II"; Legro, *Cooperation Under Fire*.

58. Farrell, "World Culture and Military Power."

59. Thomas, *The Ethics of Destruction*, 148–58.

60. Van Evera, *Causes of War*, 74–75; Van Evera, "Offense, Defense, and the Causes of War," 7, 11, 22, 43.

61. Van Evera, *Causes of War*, 73–79.

62. Posen, "The Security Dilemma and Ethnic Conflict," 29.

63. Waltz, *Theory of International Politics*, 143–46.

64. Arreguín-Toft, *How the Weak Win Small Wars*.

65. Ibid., 3, 43, 228–32.

66. Ibid., 4n6.

67. Barnett, *Asymmetrical Warfare*.

68. Arreguín-Toft, *How the Weak Win Small Wars*, 31.

69. Ibid., 31–33.

70. Ibid., 34–203.

71. Arreguín-Toft, *How the Weak Win Small Wars,* 32–33; Mack, "Why Big Nations Lose Small Wars."

72. Arreguín-Toft, *How the Weak Win Small Wars,* 18.

73. Ibid., 62–63, 128–29, 218.

74. Ibid., 30, 203.

75. Ibid., 47.

76. Ibid., 228–32.

77. T. V. Paul also uses a resource-based definition of power in examining why weak states attack asymmetrically. It is less general than Arreguín-Toft's work. See Paul, *Asymmetric Conflicts.*

78. Clausewitz, *On War;* Luttwak, *The Pentagon and the Art of War,* 270; Van Creveld, *The Transformation of War;* Beyerchen, "Clausewitz, Nonlinearity, and the Unpredictability of War"; Watts, *Clausewitzian Friction and Future War,* 86.

79. Beyerchen, "Clausewitz, Nonlinearity, and the Unpredictability of War," 86–90.

CHAPTER 2

1. The name of this war and adjectives used to describe Filipino belligerents are subjects of lively debate. Modern Filipino nationalists call Aguinaldo's people revolutionaries and nationalists, while Americans tend to call them insurgents. Modern Filipino nationalists tend to call Filipinos who allied with the United States traitors. I use terms consistent with most of the recent American scholarship on the war. See Aguinaldo, *A Second Look at America,* 113.

2. Brands, *Bound to Empire,* 21–22; Cosmas, *An Army for Empire,* 83; Linn, *The Philippine War 1899–1902,* 8.

3. Linn, *The Philippine War 1899–1902,* 8; Faust, *Campaigning in the Philippines,* 1–8.

4. The literature cites several competing political and personal interests, including concern that if Spain retained the islands rebels would be persecuted; an imperialist motive for colonies; concerns that because the Philippines were (perceived to be) too politically immature for independence civil war would break out or another imperial power, perhaps Germany or Japan, would occupy the islands if the United States did not; genuinely humanitarian motives consistent with the ethic of the "white man's burden" then extant; and arguments by a vocal if numerically small anti-imperialist group to the effect that the United States needed no colonies and that imperialism was inconsistent with America's values. McKinley finally decided to acquire the Philippines on October 26, 1898. See Joes, *America and Guerrilla Warfare,* 106; Karnow, *In Our Image,* 107–10, 125–30, 136; Welch, *Response to Imperialism,* 6–10; Gates, *Schoolbooks and Krags,* 7, 23–24; Linn, *The U.S. Army and Counterinsurgency in the Philippine War, 1899–1902,* 2, 9; Linn, *The Philippine War 1899–1902,* 3.

5. Gates, *Schoolbooks and Krags,* 18, 23–24.

6. Linn, *The Philippine War 1899–1902,* 15.

7. Gates, *Schoolbooks and Krags,* 18–23; Aguinaldo, *A Second Look at America,* 76; Karnow, *In Our Image,* 121–23.

8. Brands, *Bound to Empire,* 47; Gates, *Schoolbooks and Krags,* 20; Linn, *The Philippine War 1899–1902,* 23–25.

9. The battle of August 13 occurred because neither side knew of the agreement. Dewey earlier cut telegraph cables linking the Philippines with the rest of the world.

10. Joes, *America and Guerrilla Warfare*, 106; Aguinaldo, *A Second Look at America*, 70; Linn, *The Philippine War 1899–1902*, 21.

11. Gates, *Schoolbooks and Krags*, 33, 34, 36; Linn, *The Philippine War 1899–1902*, 30, 76; Joes, *America and Guerrilla Warfare*, 107; Karnow, *In Our Image*, 152; Faust, *Campaigning in the Philippines*, 117–18.

12. Gates, *Schoolbooks and Krags*, 16, 41; Welch, *Response to Imperialism*, 22–23.

13. Linn, *The Philippine War 1899–1902*, 53–54.

14. Gates, *Schoolbooks and Krags*, 40; Linn, *The Philippine War 1899–1902*, 36, 57–58.

15. Faust, *Campaigning in the Philippines.*

16. Silbey, *A War of Frontier and Empire*, 71.

17. Gates, *Schoolbooks and Krags*, 91; Linn, *The Philippine War 1899–1902*, 42; Joes, *America and Guerrilla Warfare*, 107.

18. Karnow, *In Our Image*, 153.

19. Linn, *The Philippine War 1899–1902*, 36, 136–38; Brands, *Bound to Empire*, 50.

20. Linn, *The Philippine War 1899–1902*, 148; Aguinaldo, *A Second Look at America*, 101.

21. Carnahan, "Lincoln, Lieber and the Laws of War," 213–31.

22. Linn, *The Philippine War 1899–1902*, 9.

23. Linn, *The U.S. Army and Counterinsurgency in the Philippine War, 1899–1902*, 20.

24. Karnow, *In Our Image*, 170.

25. Ibid., 153.

26. Joes, *America and Guerrilla Warfare*, 114.

27. Galula, *Counterinsurgency Warfare*, 88–89, 94.

28. Gates, *Schoolbooks and Krags*, 112, 128–29, 143–47.

29. Gates, *Schoolbooks and Krags*, 157; Joes, *America and Guerrilla Warfare*, 115; Karnow, *In Our Image*, 157, 177–78; Linn, *The U.S. Army and Counterinsurgency in the Philippine War, 1899–1902*, 16.

30. Gates, *Schoolbooks and Krags*, 170–71.

31. Joes, *America and Guerrilla Warfare*, 111–12.

32. Gates, *Schoolbooks and Krags*, 170–71.

33. Ibid., 174, 177–78.

34. Gates, *Schoolbooks and Krags*, 193–207; Linn, *The Philippine War 1899–1902*, 323.

35. Gates, *Schoolbooks and Krags*, 215, 225–26; Linn, *The Philippine War 1899–1902*, 215–16, 253–54; Birtle, "The U.S Army's Pacification of Marinduque, Philippine Islands, April 1900–April 1901," 255–82.

36. Joes, *America and Guerrilla Warfare*, 115–16; Linn, *The Philippine War 1899–1902*, 215.

37. Gates, *Schoolbooks and Krags*, 227–28.

38. Ibid., 228–29.

39. Birtle, "The U.S Army's Pacification of Marinduque, Philippine Islands, April 1900–April 1901."

40. Gates, *Schoolbooks and Krags*, 229–35; Linn, *The U.S. Army and Counterinsurgency in the Philippine War, 1899–1902*, 26; Silbey, *A War of Frontier and Empire*, 127.

41. Brands, *Bound to Empire*, 59.

42. Gates, *Schoolbooks and Krags*, 248–56.

43. Linn, *The U.S. Army and Counterinsurgency in the Philippine War, 1899–1902,* 26–27.

44. Linn, *The U.S. Army and Counterinsurgency in the Philippine War, 1899–1902,* 26–27; Silbey, *A War of Frontier and Empire,* 194–96.

45. Gates, *Schoolbooks and Krags,* 263.

46. Linn, *The Echo of Battle,* 122.

47. A small U.S. Marine contingent arrived late in the war, in time to help produce the Samar atrocities.

48. Linn, *The U.S. Army and Counterinsurgency in the Philippine War, 1899–1902,* 14; Linn, *The Philippine War 1899–1902,* 9.

49. Linn, *The Philippine War 1899–1902,* 132.

50. Joes, *America and Guerrilla Warfare,* 119; Linn, *The Philippine War 1899–1902,* 325.

51. Max Boot cites an even more unfavorable 1:3 force ratio for field soldiers. See Boot, *The Savage Wars of Peace,* 126. See also Galula, *Counterinsurgency Warfare,* 32; Krepinevich, *The Army and Vietnam,* 157–59; Lewy, *America in Vietnam,* 175.

52. Karnow, *In Our Image,* 119.

53. Faust, *Campaigning in the Philippines,* 201, 209.

54. Linn, *The Philippine War 1899–1902,* 323–24; Lord, "The Role of the United States in Small Wars," 91; Deady, "Lessons from a Successful Counterinsurgency."

55. Linn, *The Philippine War 1899–1902,* 325–27; Boot, *The Savage Wars of Peace,* 126–28.

56. Linn, *The U.S. Army and Counterinsurgency in the Philippine War, 1899–1902,* 68, 168–70; Silbey, *A War of Frontier and Empire,* 216; Linn, *The Echo of Battle,* 84, 85, 87.

57. Galula, *Counterinsurgency Warfare*; Thompson, *Defeating Communist Insurgency*; Boot, *The Savage Wars of Peace,* 127.

58. Linn, *The Philippine War 1899–1902,* 9, 13, 29; Linn, *The Echo of Battle,* 49–67.

59. Carnahan, "Lincoln, Lieber and the Laws of War"; Linn, *The Echo of Battle,* 75–78.

60. Boot, *The Savage Wars of Peace,* 127–28.

61. Linn, *The Philippine War 1899–1902,* 142, 167.

62. Gates, *Schoolbooks and Krags,* 258–63, 288.

63. Linn, *The Philippine War 1899–1902,* 67, 84, 326.

64. Linn, *The Philippine War 1899–1902,* 24–25, 68–69, 326; Cosmas, *An Army for Empire,* 231, 235–36, 243–44.

65. Arreguín-Toft, *How the Weak Win Small Wars,* 230.

66. Linn and Silbey similarly identify conventional and guerrilla phases of the war. See Linn, *The Philippine War 1899–1902,* and Silbey, *A War of Frontier and Empire,* xv–xvi.

67. Linn, *The Philippine War 1899–1902,* 9.

68. The *water cure*—an interrogation technique the Macabebe Scouts apparently taught American troops—involved pouring large amounts of water through a tube down a captive victim's throat. The water cure generated pain as stomachs distended and was occasionally fatal. It also often produced actionable tactical intelligence—hence its primary appeal to U.S. troops. See Karnow, *In Our Image,* 179; Boot, *The Savage Wars of Peace,* 115–16; Linn, *The Philippine War 1899–1902,* 219–24, 317–24.

69. Joes, *America and Guerrilla Warfare,* 111–12; Linn, *The Philippine War 1899–1902,* 328.

70. Linn, *The Philippine War 1899–1902*, 156, 324.

71. Arreguín-Toft, *How the Weak Win Small Wars*, 63.

72. Silbey, *A War of Frontier and Empire*, 218.

73. Linn, *The Philippine War 1899–1902*, 29–30, 59.

74. Gates, *Schoolbooks and Krags*, 27, 31; Linn, *The Philippine War 1899–1902*, 99, 122, 129, 136–38; Karnow, *In Our Image*, 180; Linn, *The U.S. Army and Counterinsurgency in the Philippine War, 1899–1902*, 133.

75. Linn, *The U.S. Army and Counterinsurgency in the Philippine War, 1899–1902*, 168.

76. Linn, *The Philippine War 1899–1902*, 59.

77. Karnow, *In Our Image*, 178.

78. Spanish priests wielded substantial political power and controlled large quantities of land, producing widespread secular grievances against the friars. See Linn, *The Philippine War 1899–1902*, 158, 221–22, 270; Boot, *The Savage Wars of Peace*, 99.

79. Arreguín-Toft, *How the Weak Win Small Wars*, 32–33.

80. Ibid., 230.

81. Alvarez, *Recalling the Revolution*; Linn, *The Philippine War 1899–1902*, 15.

82. Joes, *America and Guerrilla Warfare*, 108.

83. Ibid.

84. Karnow, *In Our Image*, 116, 131–32, 186–87.

85. Linn, *The U.S. Army and Counterinsurgency in the Philippine War, 1899–1902*, 167.

86. Galula, *Counterinsurgency Warfare*, 18–25.

87. Linn, *The Philippine War 1899–1902*, 22, 65.

88. Ibid., 33–34, 323.

89. Ibid., 19–20, 323–24.

90. Aguinaldo, *A Second Look at America*, 99; Karnow, *In Our Image*, 153.

91. Linn, *The U.S. Army and Counterinsurgency in the Philippine War, 1899–1902*, 26; Boot, *The Savage Wars of Peace*, 126–27.

92. Karnow, *In Our Image*, 176.

93. Linn, *The U.S. Army and Counterinsurgency in the Philippine War, 1899–1902*, 13–14; Linn, *The Philippine War 1899–1902*, 180.

94. Gates, *Schoolbooks and Krags*, 228, 241; Joes, *America and Guerrilla Warfare*, 113.

95. Gates, *Schoolbooks and Krags*, 95; Linn, *The Philippine War 1899–1902*, 128.

96. Linn, *The Philippine War 1899–1902*, 302–4; Silbey, *A War of Frontier and Empire*, 172–73.

97. Boot, *The Savage Wars of Peace*, 128; Linn, *The Philippine War 1899–1902*, 130–32.

98. Linn, *The Philippine War 1899–1902*, 15.

99. Joes, *America and Guerrilla Warfare*, 110; Linn, *The Philippine War 1899–1902*, 35, 130–32; Aguinaldo, *A Second Look at America*, 30.

100. Aguinaldo, *A Second Look at America*, 97.

101. Gates, *Schoolbooks and Krags*, 218; Silbey, *A War of Frontier and Empire*, 72–73.

102. Linn, *The Philippine War 1899–1902*, 35.

103. Ibid., 109.

104. Aguinaldo, *A Second Look at America*, 103–4.

105. Linn, *The Philippine War 1899–1902*, 71.

106. Silbey, *A War of Frontier and Empire*, 170.

107. Linn, *The Philippine War 1899–1902*, 136.

108. Ibid., 35.

109. Aguinaldo, *A Second Look at America,* 104.
110. Linn, *The Philippine War 1899–1902,* 23, 323–24.
111. Ibid., 324.
112. Ibid., 136.
113. Linn, *The U.S. Army and Counterinsurgency in the Philippine War, 1899–1902,* 16, 166; Linn, *The Philippine War 1899–1902,* 185–86.
114. Linn, *The Philippine War 1899–1902,* 105.
115. Ibid., 109.
116. Gates, *Schoolbooks and Krags,* 157; Linn, *The Philippine War 1899–1902,* 187, 267.
117. Linn, *The Philippine War 1899–1902,* 244; Welch, *Response to Imperialism;* Schirmer, *Republic or Empire,* 65, 230; Joes, *America and Guerrilla Warfare,* 115.
118. Linn, *The Philippine War 1899–1902,* 187, 255, 264, 272; Schirmer, *Republic or Empire,* 187; Boot, *The Savage Wars of Peace,* 113; Linn, *The U.S. Army and Counterinsurgency in the Philippine War, 1899–1902,* 23.
119. Aguinaldo, *A Second Look at America,* 115.
120. Schirmer, *Republic or Empire,* 219–20; Welch, *Response to Imperialism,* 51–56, 61, 70, 157; Brands, *Bound to Empire,* 54–55; Boot, *The Savage Wars of Peace,* 114.
121. Schirmer, *Republic or Empire,* 256.
122. Linn, *The Philippine War 1899–1902,* 156, 300, 319; Gates, *Schoolbooks and Krags,* 99.
123. Gates, *Schoolbooks and Krags,* 98–99.
124. The Spanish and British armies in Cuba and the two Boer republics, respectively, resettled rural civilians into camps in an effort to deprive guerrillas of logistical and other support, where thousands died of disease produced mainly by crowding and resultant bad sanitation—not as a result of direct mistreatment. See Slim, *Killing Civilians,* 78–80. The American practice in the Philippines of moving peasants into towns as a resource control measure bore some resemblance to contemporary Spanish and British practice. See Welch, *Response to Imperialism,* 138–39; Karnow, *In Our Image,* 179; Asprey, *War in the Shadows,* 194.
125. Gates, *Schoolbooks and Krags,* 31.
126. Joes, *America and Guerrilla Warfare,* 122.
127. Karnow, *In Our Image,* 148.
128. Gates, *Schoolbooks and Krags,* 162.
129. Linn, *The U.S. Army and Counterinsurgency in the Philippine War, 1899–1902,* 27; Karnow, *In Our Image,* 192; Silbey, *A War of Frontier and Empire,* 201–5; Miller, *"Benevolent Assimilation,"* 253–67.
130. Welch, *Response to Imperialism,* 42; Joes, *America and Guerrilla Warfare,* 120; Karnow, *In Our Image,* 140, 194.
131. Welch, *Response to Imperialism,* 42; Silbey, *A War of Frontier and Empire,* 201–5.
132. Welch, *Response to Imperialism,* 72.
133. Other insurgent groups arose periodically in the century after 1902, especially in central Luzon and among Muslims in the southern islands.

CHAPTER 3

1. Crane, *Bombs, Cities, and Civilians,* 14; Weigley, *The American Way of War,* 225.
2. In the second decade of the 20th century, the countries of Europe were experiencing rapid industrialization. Many factory workers were politically

radicalized, and many others were strongly influenced by leftist political parties that urged strikes for both economic and political reasons. Strikes intermittently afflicted defense industrial production in Britain in 1914–1918, and urban unrest contributed to Germany's decision to seek peace in November 1918. See Biddle, *Rhetoric and Reality in Air Warfare,* 29–35.

3. Douhet, *Command of the Air,* 28, 47–48, 57–58, 309; Liddell Hart, *History of the Second World War,* 589–612; Pape, *Bombing to Win,* 60–62; Biddle, "The Past as Prologue," 50; Biddle, *Rhetoric and Reality in Air Warfare.*

4. Murray, *Luftwaffe,* 298–308.

5. Biddle, *Rhetoric and Reality in Air Warfare,* 104–6.

6. Weigley, *The American Way of War,* 236.

7. Biddle, *Rhetoric and Reality in Air Warfare,* 287; Mearsheimer, *Conventional Deterrence,* 79.

8. Biddle, *Rhetoric and Reality in Air Warfare,* 37.

9. Biddle, *Rhetoric and Reality in Air Warfare,* 195; Weigley, *The American Way of War,* 355–56.

10. Biddle, *Rhetoric and Reality in Air Warfare,* 197; Pape, *Bombing to Win,* 260–62, 269–70.

11. Weigley, *The American Way of War,* 355; Murray, *Luftwaffe,* 126; Schaffer, *Wings of Judgment,* 36; Tooze, *The Wages of Destruction,* 510.

12. Crane, *Bombs, Cities, and Civilians,* 19–24; Weigley, *The American Way of War,* 354; Schaffer, *Wings of Judgment,* 30.

13. After resigning from the army in 1926, Mitchell founded the U.S. Air Force Association—a group dedicated to fostering strategic bombardment and parochial air force interests. See Perret, *Winged Victory,* 26; Weigley, *The American Way of War,* 233–37; Biddle, *Rhetoric and Reality in Air Warfare,* 130–31, 139; Crane, *Bombs, Cities, and Civilians,* 19, 23–24; Murray, *Luftwaffe,* 308–13; Horowitz and Reiter, "When Does Bombing Work? " 147–73.

14. Schaffer, *Wings of Judgment,* 31.

15. This economic vulnerability rationale for strategic bombing remains part of the U.S. Air Force rationale for strategic bombing and played a major role in targeting decisions in Vietnam (chapter 4), the Gulf War of 1991, and in NATO's U.S.-led war against Yugoslavia in 1999 (chapter 5). See Crane, *Bombs, Cities, and Civilians,* 19–27; Neillands, *The Bomber War,* 295; Weigley, *The American Way of War,* 224.

16. The USAAF became the independent U.S. Air Force per the National Security Act of 1947. See Weigley, *The American Way of War,* 240–41; Schaffer, *Wings of Judgment.*

17. Watts, *The Foundations of U.S. Air Doctrine,* 19.

18. Crane, *Bombs, Cities, and Civilians,* 26, 103; Weigley, *The American Way of War,* 337; Schaffer, *Wings of Judgment,* 32–34.

19. Strategic air forces—the focus of this chapter—used heavy, long-range bombers. Medium and light bombers and fighters were used primarily for tactical purposes—including interdiction, close air support, and aerial combat.

20. Neillands, *The Bomber War,* 27–28; Murray and Knox, "Conclusion," 184.

21. In 1942, the RAF retired Hampden, Whitley, Manchester, and Wellington bombers. See Harris, *Bomber Offensive,* 101.

22. Some *Luftwaffe* officers wanted to be able to do just that—but war with the USSR and industrial constraints got in the way. See Murray, *Luftwaffe,* 10–15.

23. Murray, *Luftwaffe,* 129; Murray and Knox, "Conclusion," 184–85.

24. Weigley, *The American Way of War,* 334–35; Perret, *Winged Victory,* 28.

25. Weigley, *The American Way of War,* 241.

26. Perret, *Winged Victory,* 28, 127–31.

27. Weigley, *The American Way of War,* 335.

28. Murray, "British Military Effectiveness in the Second World War," 95, 114–15.

29. Williamson Murray and MacGregor Knox argue that the RAF failed to learn two obvious lessons of World War I: the importance of air superiority for strategic bombing and the problem of navigating to reach targets. See Murray and Knox, "Conclusion," 182–83; Biddle, *Rhetoric and Reality in Air Warfare,* 88–94.

30. Harris, *Bomber Offensive,* 95.

31. Murray, "British Military Effectiveness in the Second World War," 123.

32. Neillands, *The Bomber War,* 27–30, 365.

33. Neillands, *The Bomber War,* 269; Weigley, *The American Way of War,* 336.

34. Weigley, *The American Way of War,* 336; Schaffer, *Wings of Judgment,* 36; Beck, *Under the Bombs,* 2.

35. Neillands, *The Bomber War,* 395.

36. Mierzejewski, *The Collapse of the German War Economy, 1944–1945,* 64.

37. Biddle, *Rhetoric and Reality in Air Warfare,* 215; Perret, *Winged Victory,* 244; Hastings, *Bomber Command,* 184–88, 273; Watts, *The Foundations of U.S. Air Doctrine,* 134–50.

38. Schaffer, *Wings of Judgment,* 37.

39. Neillands, *The Bomber War,* 369–70; Biddle, *Rhetoric and Reality in Air Warfare.*

40. Crane, *Bombs, Cities, and Civilians,* 43; Perret, *Winged Victory,* 298.

41. Weigley, *The American Way of War,* 339.

42. Biddle, *Rhetoric and Reality in Air Warfare,* 206; Pape, *Bombing to Win,* 62–63, 258–59.

43. Watts, *The Foundations of U.S. Air Doctrine,* 107–8; Perret, *Winged Victory,* 262–82.

44. Davis, *Carl A. Spaatz and the Air War in Europe.*

45. In World War II, German aircraft shot down some 4,300 American planes and German AAA downed nearly 5,300 American aircraft. See Perret, *Winged Victory,* 330.

46. Perret, *Winged Victory,* 265–69.

47. Crane, *Bombs, Cities, and Civilians,* 51.

48. Watts, *The Foundations of U.S. Air Doctrine,* 71, 85.

49. Biddle, *Rhetoric and Reality in Air Warfare,* 213; Perret, *Winged Victory,* 281–84; Tooze, *The Wages of Destruction,* 625.

50. Watts, *The Foundations of U.S. Air Doctrine,* 85; Millett, "The United States Armed Forces in the Second World War," 45–89, 69; Murray, "British Military Effectiveness in the Second World War," 123.

51. In August 1942, Eaker also thought American bombers could put 40 percent of their bombs within 500 yards of targets—a degree of accuracy American bombers never achieved in World War II. See Hastings, *Bomber Command,* 182; Murray, *Luftwaffe,* 78.

52. Murray, *Luftwaffe,* 215.

53. Perret, *Winged Victory,* 253, 267, 276.

54. Perret, *Winged Victory,* 119–20; Weigley, *The American Way of War,* 341–42; Murray, *Luftwaffe,* 214.

55. Murray, *Luftwaffe,* 213–14.

56. Weigley, *The American Way of War*, 342; Watts, *The Foundations of U.S. Air Doctrine*, 76; Perret, *Winged Victory*, 284.

57. Neillands, *The Bomber War*, 388.

58. Mierzejewski, *The Collapse of the German War Economy, 1944–1945*, 68.

59. Perret, *Winged Victory*, 361.

60. Perret, *Winged Victory*, 288; Tooze, *The Wages of Destruction*, 626.

61. Tooze, *The Wages of Destruction*, 629.

62. Pape, *Bombing to Win*, 317.

63. Galland, *The First and the Last*, 199; Tooze, *The Wages of Destruction*, 627; Weigley, *The American Way of War*, 342–43.

64. Watts, *The Foundations of U.S. Air Doctrine*, 81.

65. Galland, *The First and the Last*, 195; Murray, *Luftwaffe*, 198, 229–32, 241–42, 262, 286, 291.

66. Perret, *Winged Victory*, 300.

67. Ibid., 295.

68. Tooze, *The Wages of Destruction*, 649.

69. Mierzejewski, *The Collapse of the German War Economy, 1944–1945*, 67.

70. The USAAF also raided Japan with incendiaries designed to raze cities, especially after March 1945, and used atomic weapons on two Japanese cities in August 1945. See Schaffer, *Wings of Judgment*, 54–58.

71. Crane, *Bombs, Cities, and Civilians*, 103–8, 118; Schaffer, *Wings of Judgment*, 67, 98–99.

72. Casualty estimates for the Berlin raid range from 3,000 to 25,000. See Schaffer, *Wings of Judgment*, 97; Crane, *Bombs, Cities, and Civilians*, 105–8; Perret, *Winged Victory*, 369.

73. Crane, *Bombs, Cities, and Civilians*, 105–8, 118, 161.

74. Glaser and Kaufman, "What Is the Offense-Defense Balance and Can We Measure It?"

75. Adams, "Attack or Conquer?" 60.

76. Mierzejewski, *The Collapse of the German War Economy, 1944–1945*, 80, 84, 101–2, 122, 163.

77. Gudmundsson, *Stormtroop Tactics*, 47–52, 77–81, 86; Strachan, *The First World War*, 187.

78. Pape, *Bombing to Win*, 61–62; Biddle, *Rhetoric and Reality in Air Warfare*, 198, 202, 257–58, 260–61, 287.

79. Arreguín-Toft, *How the Weak Win Small Wars*, 228–32.

80. Pape, *Bombing to Win*, 258.

81. Overy, *Why the Allies Won*, 199; Mierzejewski, *The Collapse of the German War Economy, 1944–1945*, 176.

82. Beck, *Under the Bombs*, 29, 55, 101; Tooze, *The Wages of Destruction*, 603.

83. Pape, *Bombing to Win*, 273.

84. Schaffer, *Wings of Judgment*, 57–58; Horowitz and Reiter, "When Does Bombing Work?"; Neillands, *The Bomber War*; Pape, *Bombing to Win*, 269–73, 316.

85. Biddle, *Rhetoric and Reality in Air Warfare*, 29–35.

86. Biddle, *Rhetoric and Reality in Air Warfare*, 190–91; Murray, *Luftwaffe*, 60.

87. Weigley, *The American Way of War*, 237–38.

88. Biddle, *Rhetoric and Reality in Air Warfare*, 193–94; Murray, *Luftwaffe*, 60, 126.

89. Tooze, *The Wages of Destruction*, 603; Beck, *Under the Bombs*, 18, 38, 51–52, 95, 102, 142, 145, 166–67.

90. Beck, *Under the Bombs*, 142–43; Tooze, *The Wages of Destruction*, 628.

91. Overy, *Why the Allies Won*, 205.

92. Beck, *Under the Bombs*, 137, 176, 196, 199–200; Pape, *Bombing to Win*, 255–56; Slim, *Killing Civilians*, 142, 234–35.

93. The political irrelevance of physical vulnerability points to a weakness in Robert Pape's mildly quantitative analysis of the efficacy of strategic bombing. He explicitly codes *civilian vulnerability* in physical, not political terms. His analysis is thereby strikingly apolitical given that he is trying to assess the use of bombing as a tool of coercion, and he can reach the same conclusion that punishment bombing is ineffective via psychological reasoning, as he recognizes. But quantifying political vulnerability is difficult at best, and he may have opted for a simpler but less persuasive line of reasoning as a matter of necessity. See Weigley, *The American Way of War*, 358; Pape, *Bombing to Win*; Watts, "Ignoring Reality," 115–71.

94. Biddle, *Rhetoric and Reality in Air Warfare*; Mierzejewski, *The Collapse of the German War Economy, 1944–1945*, 162, 170–71, 181–82; Watts, *The Foundations of U.S. Air Doctrine*.

95. Harris, *Bomber Offensive*; Murray, "British Military Effectiveness in the Second World War," 110; Biddle, *Rhetoric and Reality in Air Warfare*, 199, 260; Crane, *Bombs, Cities, and Civilians*, 103–18; Neillands, *The Bomber War*, 346–50.

96. Schaffer, *Wings of Judgment*, 35, 69–70, 80, 86–104.

97. Mierzejewski, *The Collapse of the German War Economy, 1944–1945*, 61, 76–80, 98.

98. Mierzejewski, *The Collapse of the German War Economy, 1944–1945*; Murray, *Luftwaffe*, 166, 183; Harris, *Bomber Offensive*, 220; Crane, *Bombs, Cities, and Civilians*, 41.

99. Neillands, *The Bomber War*, 315, 320, 349.

100. Mierzejewski, *The Collapse of the German War Economy, 1944–1945*, 84, 121–22; Davis, *Carl A. Spaatz and the Air War in Europe*, 384–400.

101. Speer, *Inside the Third Reich*, 417; Perret, *Winged Victory*, 367; Biddle, *Rhetoric and Reality in Air Warfare*, 245–53; Weigley, *The American Way of War*, 356.

102. Neillands, *The Bomber War*, 326.

103. Biddle, *Rhetoric and Reality in Air Warfare*, 280.

104. Perret, *Winged Victory*, 297, 367–69; Speer, *Inside the Third Reich*, 414; Tooze, *The Wages of Destruction*, 650–51; Neillands, *The Bomber War*, 313, 344; Weigley, *The American Way of War*, 357–58; Mierzejewski, *The Collapse of the German War Economy, 1944–1945*, 81–83, 180.

105. Biddle, *Rhetoric and Reality in Air Warfare*, 243.

106. *Circular error probable* is a measure of bomb and missile accuracy—the radius from a target within which half of ordnance expended can be expected to fall. See Neillands, *The Bomber War*, 389.

107. Perret, *Winged Victory*, 367.

108. Beck, *Under the Bombs*, 131.

109. Overy, *Why the Allies Won*, 204.

110. Tooze, *The Wages of Destruction*, 646.

111. Ibid., 600.

112. Biddle, *Rhetoric and Reality in Air Warfare*, 274; Watts, "Ignoring Reality," 162.

113. Biddle, *Rhetoric and Reality in Air Warfare*, 274.

114. Synthetic fuel plants converted Germany's plentiful low-grade brown coal into liquid fuels usable in military vehicles. See Murray, *Luftwaffe*, 258; Perret, *Winged Victory*, 367.

115. Biddle, *Rhetoric and Reality in Air Warfare*, 276.

116. Weigley, *The American Way of War*, 357; Speer, *Inside the Third Reich*, 280; Murray, *Luftwaffe*, 134–35, 162.

117. Galland, *The First and the Last*, 178, 208; Murray, *Luftwaffe*, 257–58.

118. Tooze, *The Wages of Destruction*, 573–74.

119. Pape, *Bombing to Win*, 263, 273, 275.

120. Ibid., 317.

121. Murray, *Luftwaffe*, 182; Watts, "Ignoring Reality," 162.

122. Meilinger, "The Origins of Effects-Based Operations," 119; Millett, "The United States Armed Forces in the Second World War," 76; Perret, *Winged Victory*, 297.

123. Biddle, *Rhetoric and Reality in Air Warfare*, 276–77.

124. Biddle, *Rhetoric and Reality in Air Warfare*, 234–37, 284; Pape, *Bombing to Win*, 10, 21–29.

125. Taylor, *Dresden*, 369.

126. Schaffer, *Wings of Judgment*, 55–59.

127. Taylor, *Dresden*, 369.

128. Crane, *Bombs, Cities, and Civilians*, 114.

129. Crane, *Bombs, Cities, and Civilians*, 116–17; Neillands, *The Bomber War*, 368.

130. The USAAF was much less reluctant to target Japanese civilians than German civilians. Surveys showed that after Pearl Harbor Americans favored attacks on Japanese cities. U.S. raids on dozens of cities killed some 800,000 Japanese civilians in 1945. See Crane, *Bombs, Cities, and Civilians*, 29, 32–37, 75–76, 102, 131, 137, 161–62; Pape, *Bombing to Win*, 262, 270, 271; Schaffer, *Wings of Judgment*, 154–55.

131. Crane, *Bombs, Cities, and Civilians*, 158–59.

132. Ibid., 161–62.

133. Schaffer, *Wings of Judgment*, 152, 154, 166, 169, 171–72, 175–76.

134. Neillands, *The Bomber War*, 329.

135. Meilinger, "The Origins of Effects-Based Operations."

136. Of 55,573 RAF fatalities, 38,462 were Britons; the rest were airmen from Canada, Australia, New Zealand, and other Allied countries. See Hastings, *Bomber Command*, 11; Neillands, *The Bomber War*, 379.

137. Neillands, *The Bomber War*, 379.

138. Millett, "The United States Armed Forces in the Second World War," 61.

139. Weigley, *The American Way of War*, 336; Murray, *Luftwaffe*, 17, 38–39.

140. Biddle, *Military Power*.

141. Weigley, *The American Way of War*, 356; Hastings, *Bomber Command*, 273; Murray, *Luftwaffe*, 168.

142. Murray and Knox, "Conclusion," 184–85.

143. Murray, "British Military Effectiveness in the Second World War," 114.

144. Schaffer, *Wings of Judgment*, 32–38, 86–104.

145. Gray, "National Styles in Strategy," 32–33.

146. Neillands, *The Bomber War*, 30–31; Horowitz and Reiter, "When Does Bombing Work?"

147. Shimshoni, "Technology, Military Advantage, and World War I"; Biddle, *Military Power*.

148. Merom, *How Democracies Lose Small Wars*.

CHAPTER 4

1. Bui, *Following the Ho Chi Minh Trail*, 23.

2. The Geneva accords, including the provision for elections, mainly were nonbinding, unsigned agreements. GVN and U.S. delegates at Geneva rejected the election provisions. See Pike, *Viet Cong*, 52; Herring, *America's Longest War*, 67; Sharp, *Strategy for Defeat*, 13–14.

3. The NLF was led by communists and directed by Hanoi, but it included a variety of non-communist individuals and groups with various political, religious, and economic orientations but shared opposition to the GVN.

4. Lewy, *America in Vietnam*, 17; Pike, *PAVN*, 44.

5. Pike, *Viet Cong*, 74–78.

6. Hamilton, *The Art of Insurgency*, 118; Herring, *America's Longest War*, 67.

7. Pike, *PAVN*, 213–53.

8. Pike, *Viet Cong*, 76, 107; Bui, *Following the Ho Chi Minh Trail*, 42–46.

9. Pike, *Viet Cong*, 36, 135.

10. Galula, *Counterinsurgency Warfare*, 104–6.

11. A Vietnamese village usually consists of several hamlets spread over an appreciable area. There were 2,561 villages in South Vietnam, organized into 247 districts, 44 provinces, and four administrative regions equivalent to the army corps boundaries. There also were 11 autonomous cities. See Pike, *Viet Cong*, 110.

12. Pike, *War, Peace, and the Viet Cong*, 116–17; Karnow, *Vietnam*, 238.

13. For statistics on Viet Cong assassinations, see Lewy, *America in Vietnam*, 272–73; Hamilton, *The Art of Insurgency*, 117; Pike, *Viet Cong*, 102; Herring, *America's Longest War*, 67; Sorley, *A Better War*, 221.

14. Truong, *A Vietcong Memoir*, 153; Moyar, *Phoenix and the Birds of Prey*, 305–6.

15. Herring, *America's Longest War*, 88; Krepinevich, *The Army and Vietnam*, 78–89; Hamilton, *The Art of Insurgency*, 139–40, 148; Nagl, *Learning to Eat Soup with a Knife*, 133–34.

16. Herring, *America's Longest War*, 128; Lewy, *America in Vietnam*, 39–40.

17. Johnson, *The Vantage Point*, 137.

18. McMaster, *Dereliction of Duty*, 120–36.

19. Sharp, *Strategy for Defeat*, 44.

20. Lewy, *America in Vietnam*, 33–34.

21. Herring, *America's Longest War*, 79, 126, 129; Krepinevich, *The Army and Vietnam*, 99.

22. Lewy, *America in Vietnam*, 375; Gelb and Betts, *The Irony of Vietnam*, 105.

23. Krepinevich, *The Army and Vietnam*, 180.

24. McNamara, *In Retrospect*, 244.

25. Krepinevich, *The Army and Vietnam*, 157–59; Lewy, *America in Vietnam*, 175; Rose, *How Wars End*, 159–95.

26. Lewy, *America in Vietnam*, 162–89.

27. Ibid., 205–15.

28. Vien, *The Final Collapse*.

29. Gelb and Betts, *The Irony of Vietnam*, 71–72, 105–6; Record, *The Wrong War*, x; McNamara, *In Retrospect*, 102.

30. Gelb and Betts, *The Irony of Vietnam*, 348–49.

31. Record, *The Wrong War*, 41.

32. Gelb and Betts, *The Irony of Vietnam*, 249–50; Record, *The Wrong War*, 4.

33. Kattenburg, *The Vietnam Trauma in American Foreign Policy, 1945–75*, 246.

34. McNamara, *In Retrospect*, 236. For similar thoughts from McGeorge Bundy, see Sharp, *Strategy for Defeat*, 167–68.

35. Khong, *Analogies at War*; Tilford, *Setup*, 73–74; McNamara, *In Retrospect*, 109.

36. Krepinevich, *The Army and Vietnam*, 137.

37. Gelb and Betts, *The Irony of Vietnam*, 268, 271, 295, 307, 319–20, 330.

38. Johnson, *The Vantage Point*, 245.

39. Tilford, *Setup*, 103–6; Drew, "U.S. Airpower Theory and the Insurgent Challenge," 818–19.

40. Many senior U.S. Air Force officers of the early 1960s flew bombers against Germany in 1943–1945. For example, Chief of Staff Curtis LeMay served with distinction in Eighth Air Force and then designed and commanded city raids on Japan in 1945. See Sharp, *Strategy for Defeat*, 52, 69, 125–26; Tilford, *Setup*, 92–93; Gacek, *The Logic of Force*, 204, 207.

41. Krepinevich, *The Army and Vietnam*, 94.

42. Sharp, *Strategy for Defeat*; McMaster, *Dereliction of Duty*, 93.

43. Rostow worked on strategic bombing targeting of Germany during World War II. See Schaffer, *Wings of Judgment*, 42.

44. Osgood, *Limited War*; Schelling, *The Strategy of Conflict*; Lewy, *America in Vietnam*, 30; Moyar, *Triumph Forsaken*, 306–7.

45. Krepinevich, *The Army and Vietnam*, 96, 98, 136; Gacek, *The Logic of Force*, 184–85.

46. Gacek, *The Logic of Force*, 189.

47. Tilford, *Setup*, 105.

48. Ibid., 106.

49. Krepinevich, *The Army and Vietnam*, 146.

50. Sharp, *Strategy for Defeat*, 99.

51. Ibid., 95, 99–104.

52. Tilford, *Setup*, 92–93; Gacek, *The Logic of Force*, 182–84; Clodfelter, *The Limits of Air Power*, 127.

53. Tilford, *Setup*, 71.

54. Lewy, *America in Vietnam*, 375.

55. Tilford, *Setup*, 121; Clodfelter, *The Limits of Air Power*, 127.

56. Tilford, *Setup*, 142.

57. Sharp, *Strategy for Defeat*, 86.

58. Ibid., 116.

59. Tilford, *Setup*, 116–20.

60. Sharp, *Strategy for Defeat*, 160–61.

61. Gelb and Betts, *The Irony of Vietnam*, 147–48, 167–68; Komer, *Bureaucracy at War*, 53–55; Tilford, *Setup*, 115–16, 133, 173; Sharp, *Strategy for Defeat*, 207.

62. George Ball worked on the U.S. Strategic Bombing Survey's assessment of the effectiveness of Allied bombing of Germany in World War II, presumably accounting for some of his skepticism about airmen's claims for strategic bombing. See Krepinevich, *The Army and Vietnam*, 133.

63. Tilford, *Setup*, 98.

64. Gelb and Betts, *The Irony of Vietnam*, 122; Gacek, *The Logic of Force*, 205–6.

65. Pape, *Bombing to Win*, 187.

66. Lewy, *America in Vietnam*, 396; Tilford, *Setup*, 101–3: Gelb and Betts, *The Irony of Vietnam*, 168.

67. Tilford, *Setup*, 153.

68. Bui, *Following the Ho Chi Minh Trail*, 49–50; Truong, *A Vietcong Memoir*, 232, 240–42.

69. Tilford, *Setup*, 101–3; Gelb and Betts, *The Irony of Vietnam*, 168.

70. Krepinevich, *The Army and Vietnam*, 149; Gacek, *The Logic of Force*, 207.

71. Krepinevich, *The Army and Vietnam*, 154–55, 166.

72. Krepinevich, *The Army and Vietnam*, 166; Linn, *The Echo of Battle*, 152.

73. Komer, *Bureaucracy at War*, 50.

74. Galula, *Counterinsurgency Warfare*; Thompson, *Defeating Communist Insurgency*.

75. Krepinevich, *The Army and Vietnam*, 179; McNamara, *In Retrospect*, 238.

76. Westmoreland, *A Soldier Reports*, 153.

77. McNamara, *In Retrospect*, 236.

78. Krepinevich, *The Army and Vietnam*, 169.

79. Jenkins, *The Unchangeable War*, 3; Krepinevich, *The Army and Vietnam*; Lewy, *America in Vietnam*, 67–68; Boot, *The Savage Wars of Peace*, 316.

80. Military History Institute of Vietnam, *Victory in Vietnam*, 67; Boot, *The Savage Wars of Peace*, 303.

81. In addition, some 137,000 ARVN troops were killed in action. The United States lost 47,406 troops to hostile causes and 58,193 deaths to all causes in the theater. Allies also experienced appreciable combat losses: the Republic of Korea lost 4,907 troops; Australia and New Zealand together lost 409; and, Thailand lost about 300. See Macdonald, *Giap*, 291, 332; Boot, *The Savage Wars of Peace*, 303; Mueller, "The Search for the 'Breaking Point' in Vietnam," 507–9; Military History Institute of Vietnam, *Victory in Vietnam*, 299.

82. MACV statistics in early 1967 showed that the enemy initiated 88 percent of fights with U.S. armored units. The Pentagon calculated similar figures for all engagements involving company-sized American units. See Krepinevich, *The Army and Vietnam*, 177–78, 188, 190; Jenkins, *The Unchangeable War*, 4; Lewy, *America in Vietnam*, 82–83; Komer, *Bureaucracy at War*, 56–60.

83. McNamara recognized in 1966 that North Vietnam easily replaced annual NVA losses then estimated at 60,000 troops. See Krepinevich, *The Army and Vietnam*, 184; Herring, *America's Longest War*, 154; Lewy, *America in Vietnam*, 82–84.

84. Krepinevich, *The Army and Vietnam*, 159–62, 255–56.

85. McNamara, *In Retrospect*, 236, 238.

86. Bui, *Following the Ho Chi Minh Trail*, 61–62.

87. Lewy, *America in Vietnam*, 67; Krepinevich, *The Army and Vietnam*, 237–39.

88. Krepinevich, *The Army and Vietnam*, 249; Clodfelter, *The Limits of Air Power*, 139.

89. Herring, *America's Longest War*, 183–89; Clodfelter, *The Limits of Air Power*, 139; Truong, *A Vietcong Memoir*, 154; Bui, *Following the Ho Chi Minh Trail*, 63.

90. Lewy, *America in Vietnam*, 75–76.

91. Boot, *The Savage Wars of Peace*, 309.

92. Military History Institute of Vietnam, *Victory in Vietnam*, 223–24.

93. Truong, *A Vietcong Memoir*, 284–86.

94. Rose, *How Wars End*, 181–85.

95. Clodfelter, *The Limits of Air Power*, 168; Lewy, *America in Vietnam*, 196–98.

96. Lewy, *America in Vietnam*, 196; Truong, *A Vietcong Memoir*, 204.

97. Clodfelter, *The Limits of Air Power*, 169.

98. Ibid., 159, 164.

99. Sharp, *Strategy for Defeat,* 118–19, 189–90, 253–54; Clodfelter, *The Limits of Air Power,* 168. Linebacker also reflected fruits of Kissinger's diplomacy. Nixon did not fear that heavy attacks on the DRV would trigger strong Soviet or Chinese reactions, so he ordered bombing nearer to China and the mining of Haiphong harbor. See Tilford, *Setup,* 246–47, 261.

100. Lewy, *America in Vietnam,* 202–5.

101. Lewy, *America in Vietnam,* 412; Clodfelter, *The Limits of Air Power,* 180.

102. Sharp, *Strategy for Defeat,* 259–61.

103. Truong, *A Vietcong Memoir,* 209–18, 220–22; Moyar, *Phoenix and the Birds of Prey,* 275.

104. Sharp, *Strategy for Defeat,* 96; Lewy, *America in Vietnam,* 416–17; Momyer, *Air Power in Three Wars,* 318; Rose, *How Wars End,* 174.

105. Pape, *Bombing to Win,* 195–205.

106. Westmoreland, *A Soldier Reports,* 214–15.

107. Moyar, *Phoenix and the Birds of Prey,* 48–49.

108. During his tour (1968–1972) as commander, Abrams encouraged U.S. troops to directly support pacification efforts as the U.S. ground combat role declined after 1969. As army vice chief of staff, Abrams had participated in, and approved the findings of, a study sponsored by Chief of Staff General Harold Johnson called "A Program for the Pacification and Long-Term Development of Vietnam," or PROVN, which argued for priority to the pacification effort. The Pentagon quashed the study. See Sorley, *A Better War,* 6, 7, 10, 18, 20, 60–61, 123–25, 189, 192, 219; Spector, *After Tet,* 280; Lewy, *America in Vietnam,* 133, 146.

109. Blaufarb, *The Counterinsurgency Era,* 269–73.

110. Herring, *America's Longest War,* 227.

111. Vien, *The Final Collapse.*

112. Truong, *A Vietcong Memoir,* 220, 231.

113. Lewy, *America in Vietnam,* 204–5.

114. Linn, *The Echo of Battle,* 152, 174, 182.

115. Nagl, *Learning to Eat Soup with a Knife,* 175.

116. Tilford, *Setup,* 33.

117. Ibid., 35.

118. Clodfelter, *The Limits of Air Power,* 134.

119. Tilford, *Setup,* 130.

120. Pentagon Papers, *The Defense Department History of United States Decisionmaking on Vietnam,* vol. IV, 136.

121. Krepinevich, *The Army and Vietnam,* 197–98.

122. Boot, *The Savage Wars of Peace,* 300–301.

123. Tilford, *Setup,* 97; Watts, *The Foundations of U.S. Air Doctrine,* 106.

124. Wendt, "Using a Sledgehammer to Kill a Gnat."

125. These gun ships fired mini-guns (multi-barrel machine guns), 20-mm and 40-mm cannon, and 105-mm howitzers from the sides of aircraft. Used mainly at night, they were effective against a variety of targets.

126. Blaufarb, *The Counterinsurgency Era,* 119; Tilford, *Setup,* 72.

127. Momyer, *Air Power in Three Wars.*

128. Ibid.

129. Linn, *The Echo of Battle,* 182, 191, 195.

130. Warner, "'US Intelligence in Vietnam.'"

131. Nagl, *Learning to Eat Soup with a Knife,* 141.

132. McNamara, *In Retrospect*, 252.

133. Nagl, *Learning to Eat Soup with a Knife.*

134. As quoted in *Newsweek*, June 6, 1983, and cited in Cohen, "Constraints on America's Conduct of Small Wars," 181.

135. Hamilton, *The Art of Insurgency*, 155–61.

136. Krepinevich, *The Army and Vietnam*, 197.

137. Tilford, *Setup*, 113; Clodfelter, *The Limits of Air Power*, 165.

138. Wendt, "Using a Sledgehammer to Kill a Gnat."

139. Corum and Johnson, *Airpower in Small Wars*, 262; Wendt, "Using a Sledgehammer to Kill a Gnat."

140. Tilford, *Setup*, 286–87.

141. Ibid., 216.

142. Herring, *America's Longest War*, 131.

143. Krepinevich, *The Army and Vietnam*, 206.

144. Author discussion with Special Forces troops who conducted the training.

145. Summers, *On Strategy*; Momyer, *Air Power in Three Wars*; Sharp, *Strategy for Defeat.*

146. Krepinevich, *The Army and Vietnam.*

147. Palmer, *The 25-Year War*, 201–3; McMaster, *Dereliction of Duty*; McNamara, *In Retrospect*, 212; Nagl, *Learning to Eat Soup with a Knife.*

148. Jenkins, *The Unchangeable War*, 3.

149. Luttwak, *The Pentagon and the Art of War*, 41–42.

150. Wirtz, "Intelligence to Please?" 257–58, 263.

151. Momyer, *Air Power in Three Wars*, 103–8.

152. Krepinevich, *The Army and Vietnam*, 70–73.

153. Linn, *The Echo of Battle*, 193.

154. Herring, "America's Strategy in Vietnam: The Post-War Debate," 57–63.

155. Krepinevich, *The Army and Vietnam*, 275; Luttwak, *The Pentagon and the Art of War*; Cohen, "Constraints on America's Conduct of Small Wars," 171; Nagl, *Learning to Eat Soup with a Knife*, 9, 177–81, 205–8; Record, "Why the Strong Lose Wars," 29; Lord, "The Role of the United States in Small Wars," 94; Drew, "U.S. Airpower Theory and the Insurgent Challenge."

156. This experience differs markedly from that of the German army of the 1920s. Its commander, General Hans von Seekt, appointed at least 57 committees to assess lessons of the Great War and instilled in the army as a whole the cultural outlook of the general staff. The resulting reforms set the stage for Germany's impressive victories in 1939–1941. See Murray, "Does Military Culture Matter?"; Cassidy, "Back to Street Without Joy," 73–75.

157. Nagl, *Learning to Eat Soup with a Knife*, 206; Ricks, *Fiasco*, 133; Gacek, *The Logic of Force*, 224; Linn, *The Echo of Battle*, 213.

158. Arreguín-Toft, *How the Weak Win Small Wars*, 144–68.

159. Ibid., 152–54.

160. Pape, *Bombing to Win*; Horowitz and Reiter, "When Does Bombing Work?"

161. Tilford, *Setup*, 109.

162. Ibid., 74–75, 106.

163. Pape, "Coercive Air Power in the Vietnam War," 105, 115; Pape, *Bombing to Win*, 175, 180–81; Clodfelter, *The Limits of Air Power*, 127.

164. Clodfelter, *The Limits of Air Power*, 136.

165. McNamara, *In Retrospect*, 269.

166. Sorley, *A Better War*, 95.

167. Ibid., 151–52.

168. Sorley, *A Better War*, 45–46.

169. Clodfelter, *The Limits of Air Power*, 136.

170. Truong, *A Vietcong Memoir*, 243.

171. Karnow, *Vietnam*, 458–59; Bui, *Following the Ho Chi Minh Trail*, 63–65; Clodfelter, *The Limits of Air Power*, 138.

172. Clodfelter, *The Limits of Air Power*, 138; Hoeffding, *Bombing North Vietnam*, 17; Lewy, *America in Vietnam*, 396.

173. Arreguín-Toft, *How the Weak Win Small Wars*, 153.

174. McNamara, *In Retrospect*, 311–15; Karnow, *Vietnam*, 510; Sharp, *Strategy for Defeat*, 198.

175. Boot, *The Savage Wars of Peace*, 291.

176. Clodfelter, *The Limits of Air Power*, 144–46; Sharp, *Strategy for Defeat*; Momyer, *Air Power in Three Wars*.

177. Arreguín-Toft, *How the Weak Win Small Wars*, 154–55.

178. According to Lewy, the casualties were 56 Americans and 505 North Vietnamese killed in action. See Krepinevich, *The Army and Vietnam*, 256–57; Lewy, *America in Vietnam*, 144–45.

179. Lewy, *America in Vietnam*, 144–46.

180. Lewy, *America in Vietnam*, 146; Jenkins, *The Unchangeable War*, 8; Nagl, *Learning to Eat Soup with a Knife*, 172–73.

181. McNamara, *In Retrospect*, 262, 277; Lewy, *America in Vietnam*, 67.

182. Summers, *On Strategy*, 1.

183. Arreguín-Toft, *How the Weak Win Small Wars*, 155–56.

184. Ibid., 156–57.

185. Herring, *America's Longest War*, 85; Moyar, *Triumph Forsaken*, 155–59.

186. Pike, *Viet Cong*, 64–65; Blaufarb, *The Counterinsurgency Era*, 110, 112; Tran, *Our Endless War*, 81–83; Thompson, *Defeating Communist Insurgency*, 125–26.

187. Colby, *Lost Victory*, 100–101; Asprey, *War in the Shadows*, 983.

188. Pike, *Viet Cong*, 347n; Blaufarb, *The Counterinsurgency Era*, 115; Sharp, *Strategy for Defeat*, 20; Komer, *Bureaucracy at War*, 64.

189. Thompson, *Defeating Communist Insurgency*, 121–40.

190. Hamilton, *The Art of Insurgency*, 141–44; Nagl, *Learning to Eat Soup with a Knife*, 130–31.

191. FitzGerald, *Fire in the Lake*, 155–57, 339.

192. Pike, *Viet Cong*, 67.

193. Herring, *America's Longest War*, 88–90; Pike, *Viet Cong*, 66.

194. Blaufarb, *The Counterinsurgency Era*, 115.

195. Thompson, *Defeating Communist Insurgency*, 137; Blaufarb, *The Counterinsurgency Era*, 120.

196. Blaufarb, *The Counterinsurgency Era*, 120.

197. Thompson, *Defeating Communist Insurgency*, 133–37; Moyar, *Triumph Forsaken*, 109, 207–8.

198. Lewy, *America in Vietnam*, 25.

199. Moyar, *Triumph Forsaken*, 182–84, 283.

200. Thompson, *Defeating Communist Insurgency*, 139; Moyar, *Triumph Forsaken*, 281–84.

201. Moyar, *Triumph Forsaken*, 258, 284.

202. Moyar, *Triumph Forsaken,* 284; Military History Institute of Vietnam, *Victory in Vietnam,* 109–10, 113, 117, 121–22, 134–35.

203. Asprey, *War in the Shadows,* 1007–10, 1034, 1037; Shafer, *Deadly Paradigms,* 266–68.

204. Karnow, *Vietnam,* 255–58, 323–24.

205. Hamlets in the delta consisted of strings of houses along waterways, not concentrations of houses. There could be no meaningful fortification without moving some people to new homes. A VC agent working as a senior GVN official pushed the program in ways he designed to alienate villagers; see Berman, *Perfect Spy,* 148–50. Contrary to Arreguín-Toft (p. 156), the agroville program was not French; the French left Vietnam years earlier. See Sheehan, *A Bright Shining Lie,* 309–11; Moyar, *Triumph Forsaken,* 82–83, 158.

206. Kolko, *Anatomy of a War,* 133.

207. Krepinevich, *The Army and Vietnam,* 82–83.

208. Krepinevich, *The Army and Vietnam,* 73; Nagl, *Learning to Eat Soup with a Knife,* 128–29.

209. For example, the 1971 joint pacification plan included 25 named activities in three broad functional areas—security, self-government, and development; Phoenix was one of the 25. In 1970, CORDS had about 7,600 staffers. Colby says Phoenix eventually received 500 military officers to staff Phoenix offices at provincial and district levels. John Prados says U.S. military involvement in Phoenix peaked in 1970 at 704 personnel and fell thereafter due to General Abrams's opposition to it; participation by CORDS civilians other than CIA people peaked at 20 in 1969 and included a single individual in 1972, while CIA strength is reported in rough terms of about 100 people. While the statistics are not directly comparable, they suggest that Phoenix was a small part of CORDS—and small in absolute terms. See Colby, *Lost Victory,* 254–76, 280–82, 296–97; Blaufarb, *The Counterinsurgency Era,* 269; Prados, *The Hidden History of the Vietnam War,* 217.

210. Herrington, *Silence Was a Weapon,* 4, 13.

211. Among the Vietnamese-manned police, pacification, and military security agencies at various times were the National Police, Police Field Force, the Special Branch police, Provincial Reconnaissance Units, Counter-Terror Teams, Provincial Interrogation Centers, Kit Carson Scouts, the Census Grievance program, Revolutionary Development Cadre, Armed Propaganda Teams, and the Military Security Service. Conventional South Vietnamese armed forces organizations included the ARVN, Regional Forces (province-level militia), Popular Forces (district-level militia), and People's Self-Defense Forces (village-level militia). U.S. organizations included the conventional military, military special operations forces (Army Special Forces and Navy SEALs), CIA, and CORDS. See Valentine, *The Phoenix Program,* 128–31; Moyar, *Phoenix and the Birds of Prey,* 51–52.

212. Herrington, *Silence Was a Weapon.*

213. Thompson, *Peace Is Not at Hand,* 50.

214. Valentine, *The Phoenix Program,* 118–19.

215. Ibid., 148.

216. Illegal VC were South Vietnamese living away from home in an unregistered status, were known VC on the run, or were infiltrated North Vietnamese without GVN papers. Legal VC were registered citizens of the RVN who were covert NLF cadre. See Moyar, *Phoenix and the Birds of Prey,* 17, 59, 66, 226–27; Herrington, *Silence Was a Weapon,* 13.

217. Valentine, *The Phoenix Program*, 137; Moyar, *Phoenix and the Birds of Prey*, 52–53.

218. Valentine, *The Phoenix Program*, 122, 148; Colby, *Lost Victory*, 240.

219. Blaufarb, *The Counterinsurgency Era*, 246.

220. Valentine, *The Phoenix Program*, 142–43.

221. Moyar, *Phoenix and the Birds of Prey*, 127.

222. Colby, *Lost Victory*, 244–46.

223. Colby, *Lost Victory*, 240; Sorley, *A Better War*, 64; Truong, *A Vietcong Memoir*, 146–47.

224. Colby, *Lost Victory*, 247.

225. Colby, *Lost Victory*, 247; Valentine, *The Phoenix Program*, 154.

226. Valentine, *The Phoenix Program*, 159–73; Spector, *After Tet*, 287–88.

227. Herrington, *Silence Was a Weapon*, 13.

228. Ibid., 236

229. Thompson, *Peace Is Not at Hand*, 51.

230. Prados, *The Hidden History of the Vietnam War*, 214, 219.

231. Moyar, *Phoenix and the Birds of Prey*, 224–30.

232. Ibid.

233. Herrington, *Silence Was a Weapon*, 12, 59, 193–97; Race, *War Comes to Long An*, 237–42.

234. There are legitimate questions about the extent to which GVN agencies and programs with which Phoenix interacted killed civilians, inappropriately arrested people, tortured prisoners, took bribes, and so on. Mark Moyar (*Phoenix and the Birds of Prey*) and Douglas Valentine (*The Phoenix Program*) discuss these at length. Because these issues have little direct bearing on Arreguín-Toft's issue, I little discuss them here.

235. Blaufarb, *The Counterinsurgency Era*, 274; Lewy, *America in Vietnam*, 284–85; Komer, "Impact of Pacification on Insurgency in South Vietnam," 53; Komer, *Bureaucracy at War*, 61–62.

236. Blaufarb, *The Counterinsurgency Era*, 246; Herrington, *Silence Was a Weapon*; Moyar, *Phoenix and the Birds of Prey*, 130.

237. Lewy, *America in Vietnam*, 282, 291.

238. Blaufarb, *The Counterinsurgency Era*, 247; Lewy, *America in Vietnam*, 283–84; Moyar, *Phoenix and the Birds of Prey*, 181–88.

239. Blaufarb, *The Counterinsurgency Era*, 290; Maclear, *The Ten Thousand Day War*, 303–15; Colby, *Lost Victory*, 280; Lewy, *America in Vietnam*, 283–84; Valentine, *The Phoenix Program*, 126, 174–78; Prados, *The Hidden History of the Vietnam War*, 215; Moyar, *Phoenix and the Birds of Prey*, 81–82, 118, 194–97.

240. Lewy, *America in Vietnam*, 283; Race, *War Comes to Long An*, 237–42; Herrington, *Silence Was a Weapon*, 193–97; Prados, *The Hidden History of the Vietnam War*, 215.

241. Valentine, *The Phoenix Program*, 145–46; Herrington, *Silence Was a Weapon*, 193.

242. Colby, *Lost Victory*, 280; Karnow, *Vietnam*, 602; Moyar, *Phoenix and the Birds of Prey*, 116; Lewy, *America in Vietnam*, 184–86, 289–91; Herrington, *Silence Was a Weapon*, 181, 197; Sheehan, *A Bright Shining Lie*, 742–43.

243. Colby, *Lost Victory*, 280; Valentine, *The Phoenix Program*, 154.

244. Moyar, *Phoenix and the Birds of Prey*, 206; Tran, *Our Endless War*, 158–59; Herrington, *Silence Was a Weapon*, 197.

245. Colby, *Lost Victory*, 259–60.

246. Moyar, *Phoenix and the Birds of Prey*, 55; Herrington, *Silence Was a Weapon*, 177, 180–81.

247. Moyar, *Phoenix and the Birds of Prey*, 244–54; Boot, *The Savage Wars of Peace*, 310.

248. Karnow, *Vietnam*, 602; Kolko, *Anatomy of a War*, 397.

249. Moyar, *Phoenix and the Birds of Prey*, 245–46.

250. Bui, *Following the Ho Chi Minh Trail*, 63.

251. For example, the North Vietnamese claimed the GVN held 200,000 political prisoners in 1973; the U.S. Embassy in Saigon found only about 35,000 prisoners incarcerated for all causes. See Moyar, *Phoenix and the Birds of Prey*, 210; McCoy, "Torture in the Crucible of Counterinsurgency," 230–62, 233–45; FitzGerald, *Fire in the Lake*, 515–18; Blaufarb, *The Counterinsurgency Era*, 275; Sheehan, *A Bright Shining Lie*, 732.

252. Moyar, *Phoenix and the Birds of Prey*, 210.

253. Weiner, *Legacy of Ashes*, 340; Arreguín-Toft, *How the Weak Win Small Wars*, 157; Herring and Rangwala, *Iraq in Fragments*, 178–79.

254. Blaufarb, *The Counterinsurgency Era*, 275; Karnow, *Vietnam*, 602; Lewy, *America in Vietnam*, 279–80, 287–88, 318; Colby, *Lost Victory*, 286–88; Moyar, *Phoenix and the Birds of Prey*, 182–83.

255. Moyar, *Phoenix and the Birds of Prey*, 213–20.

256. Colby, *Lost Victory*, 246.

257. Truong, *A Vietcong Memoir*, 201–2.

258. Lewy, *America in Vietnam*, 272–73.

259. Moyar, *Phoenix and the Birds of Prey*, 305–6; Truong, *A Vietcong Memoir*, 153; Pike, *War, Peace, and the Viet Cong*.

260. Prados, *The Hidden History of the Vietnam War*, 209–20.

261. Truong, *A Vietcong Memoir*, 206, 209, 210; Krepinevich, *The Army and Vietnam*; Jenkins, *The Unchangeable War*; Nagl, *Learning to Eat Soup with a Knife*.

262. Johnson, *The Vantage Point*, 132–34.

263. Jenkins, *The Unchangeable War*; Pike, *Viet Cong*.

264. Gelb and Betts, *The Irony of Vietnam*.

265. Sharp, *Strategy for Defeat*, 33; Tilford, *Setup*, 60, 101–3.

266. McNamara, *In Retrospect*, 203.

267. Clodfelter, *The Limits of Air Power*, 135.

268. Tilford, *Setup*, 90.

269. Ibid., 74–75.

270. Krepinevich, *The Army and Vietnam*, 168; Tilford, *Setup*, 88; McNamara, *In Retrospect*, 220.

271. Karnow, *Vietnam*, 597; Record, "Why the Strong Lose Wars," 18; Clodfelter, *The Limits of Air Power*, 140.

272. Tilford, *Setup*, 195; Clodfelter, *The Limits of Air Power*, 140.

273. Gelb and Betts, *The Irony of Vietnam*, 139; Mueller, "The Search for the 'Breaking Point' in Vietnam."

274. The United States attacked dikes in North Korea during the Korean War of 1950–1953.

275. Sharp, *Strategy for Defeat*, 117–18; Moyar, *Phoenix and the Birds of Prey*, 360–61; Clodfelter, *The Limits of Air Power*, 126, 140, 205.

276. Krepinevich, *The Army and Vietnam*, 37.

277. Palmer, *The 25-Year War*, 175–76; Krepinevich, *The Army and Vietnam*.

278. Neustadt and May, *Thinking in Time*, 84; Palmer, *The 25-Year War*, 175–76; Mueller, "The Search for the 'Breaking Point' in Vietnam," 497; Kissinger, *Diplomacy*, 678; McNamara, *In Retrospect*, 321–22; Record "Why the Strong Lose Wars," 18.

279. Gacek, *The Logic of Force*, 213–19.

280. Record, *The Wrong War*, 53; Boot, *The Savage Wars of Peace*, 316.

281. Wirtz, "The Balance of Power Paradox," 136.

282. Military History Institute of Vietnam, *Victory in Vietnam*, 206, 232, 231–33.

283. Gelb and Betts, *The Irony of Vietnam*, 13, 335; Boot, *The Savage Wars of Peace*, 316.

284. Truong, *A Vietcong Memoir*, 86–87.

285. Ibid., 150–55.

286. Sorley, *A Better War*, 93.

287. Ibid., 93–94.

288. Truong, *A Vietcong Memoir*, 212.

289. Pike, *War, Peace, and the Viet Cong*, 116.

290. Lewy, *America in Vietnam*, 335.

291. Bui, *Following the Ho Chi Minh Trail*, 83–84; Berman, *Perfect Spy*, 30, 146–47, 156–62, 177, 199–200.

292. Karnow, *Vietnam*, 598.

293. Lewy, *America in Vietnam*, 223–25.

294. Karnow, *Vietnam*, 489–90; Sharp, *Strategy for Defeat*, 56; Lewy, *America in Vietnam*, 334.

295. Moyar, *Phoenix and the Birds of Prey*, 210.

296. Lewy, *America in Vietnam*, 294–99, 332–42.

297. Clodfelter, *The Limits of Air Power*, 136.

298. Pike, *War, Peace, and the Viet Cong*, 115–16.

299. Truong, *A Vietcong Memoir*, 146–47, 150–55.

300. Bui, *Following the Ho Chi Minh Trail*, 91.

301. Truong, *A Vietcong Memoir*, 194, 210; Lewy, *America in Vietnam*, 387; Bui, *Following the Ho Chi Minh Trail*, 89–91.

302. Johnson, *The Vantage Point*, 252, 254; McNamara, *In Retrospect*, 251; Truong, *A Vietcong Memoir*, 194, 209–18; Sharp, *Strategy for Defeat*, 145, 147–48, 228.

303. Truong, *A Vietcong Memoir*, 194.

304. Ibid., 204.

305. McNamara, *In Retrospect*, 262, 277.

306. Krepinevich, *The Army and Vietnam*, 76; Tilford, *Setup*, 75; Lewy, *America in Vietnam*, 52–53, 59.

307. Lewy, *America in Vietnam*, 72.

308. Nagl, *Learning to Eat Soup with a Knife*, 200.

309. Ibid., 155.

310. Lewy, *America in Vietnam*, 59; Sharp, *Strategy for Defeat*, 116.

311. Cassidy, *Peacekeeping in the Abyss*, 17.

312. Gelb and Betts, *The Irony of Vietnam*, 13, 216; Karnow, *Vietnam*, 414; Turner, *Lyndon Johnson's Dual War*.

313. Johnson, *The Vantage Point*, 151, 236.

314. Karnow, *Vietnam*, 482.

315. Sharp, *Strategy for Defeat*, 109.

316. Gelb and Betts, *The Irony of Vietnam*, 114.

317. Johnson, *The Vantage Point*, 153; Neustadt, and May, *Thinking in Time*, 77; Lewy, *America in Vietnam*, 36.

318. Neustadt, and May, *Thinking in Time*, 78–79; Turner, *Lyndon Johnson's Dual War*, 111–33.

319. Palmer, *The 25-Year War*, 177.

320. Gelb and Betts, *The Irony of Vietnam*, 140–42; Palmer, *The 25-Year War*, 177; Sharp, *Strategy for Defeat*, 126.

321. Gelb and Betts, *The Irony of Vietnam*, 122.

322. McNamara, *In Retrospect*, 262.

323. Johnson, *The Vantage Point*, 132–34.

324. Ibid., 132, 240–41.

325. Johnson, *The Vantage Point*, 132–34; Sharp, *Strategy for Defeat*, 75–76.

326. Lord, "The Role of the United States in Small Wars," 91.

327. Sharp, *Strategy for Defeat*, 106–7.

328. Sharp, *Strategy for Defeat*, 123, 128, 129, 227; Olson, "The Effect of Civilian Casualties on USAF Bombing Policy in Vietnam."

329. Bui, *Following the Ho Chi Minh Trail*, 70; Truong, *A Vietcong Memoir*, 212, 253.

330. Military History Institute of Vietnam, *Victory in Vietnam*, 299.

331. Military History Institute of Vietnam, *Victory in Vietnam*, 431; Pike, *PAVN*.

332. Lewy, *America in Vietnam*, 173–74.

333. Turner, *Burn Before Reading*, 103–21.

334. Warner, "'US Intelligence in Vietnam.'"

335. Spector, *After Tet*, 313.

CHAPTER 5

1. A note on terminology: In 1999, the Federal Republic of Yugoslavia (FRY) consisted of two republics—Montenegro and Serbia. Kosovo was a region of Serbia. I designate police and other security forces in Kosovo as Serbian and FRY military and civilian leaders as Yugoslav.

2. Other norms also played roles—especially the norm of noninterference in the internal affairs of established states—which led many to call NATO's war illegal. See Martel, *Victory in War*, 202; Finnemore and Sikkink, "International Norm Dynamics and Political Change."

3. Daalder and O'Hanlon, *Winning Ugly*, 101–8; Tilford, "Operation Allied Force and the Role of Air Power."

4. Byman and Waxman, *The Dynamics of Coercion*.

5. Daalder and O'Hanlon, *Winning Ugly*, 147–50; Strickland, "USAF Aerospace-Power Doctrine," 13–25, 16–20.

6. Thomas, *The Ethics of Destruction*, 148–58; Olson, "The Effect of Civilian Casualties on USAF Bombing Policy in Vietnam."

7. Petraeus, "Lessons of History and Lessons of Vietnam," 45; Record, "Force-Protection Fetishism."

8. Cohen, "The Unequal Dialogue," 456.

9. For example, as discussed in chapter 3, the RAF during World War II shifted to night bombing, and the USAAF for a time in late 1943 reduced operations over Germany not to protect individual aviators' lives, but because effective German air defenses threatened to render both bomber forces combat ineffective as institutions.

10. Dallaire, *Shake Hands with the Devil*, 143–44, 239–40, 255–56, 293–95, 318.

11. Smith, *The Utility of Force*, 354.

12. Drew, *On the Edge,* 325–30; Lord, "The Role of the United States in Small Wars," 96–97.

13. Clinton, *My Life,* 552, 554.

14. Riehm, "The USS *Harlan County* Affair"; Halberstam, *War in a Time of Peace,* 271; Soderberg, *The Superpower Myth,* 44.

15. Myers, "U.S. Wields Defter Weapon Against Iraq: Concrete Bomb."

16. White House Office of the Press Secretary, "Address by the President to the Nation on Iraq Air Strike," 3; Daalder, *Getting to Dayton,* 184; Knights, *Cradle of Conflict,* 200–210.

17. Holbrooke, *To End a War,* 65–67. For a different view, see Daalder, *Getting to Dayton,* 55–59, 169.

18. Bosnian Croats and Muslim engaged in the same sorts of activities—if less frequently and in less well-publicized incidents. See Halberstam, *War in a Time of Peace,* 376, 388; Albright, *Madame Secretary,* 177–78.

19. Ricks, *Fiasco,* 70.

20. Summers, *On Strategy*; Cohen, "Constraints on America's Conduct of Small Wars," 167–69; Petraeus, "Lessons of History and Lessons of Vietnam," 46–47; Gordon and Trainor, *Cobra II*; Gacek, *The Logic of Force,* 225; Hoffman, *Decisive Force,* 19–37, 77–78, 82–84; Handel, *Masters of War,* 185–203; Linn, *The Echo of Battle,* 193, 195.

21. Cohen, "Constraints on America's Conduct of Small Wars," 173; Gacek, *The Logic of Force,* 247; Kitfield, *Prodigal Soldiers,* 149–51.

22. Weinberger's tests were the following: vital national interests must be present, overwhelming force must be used to ensure victory, objectives must be clear, force structures and dispositions must be adjusted as necessary, there must be a reasonable chance of popular and congressional support, and force must be a last resort. See Weinberger, "The Uses of Military Power." See also Kohn, "Out of Control," 12.

23. Powell, *My American Journey,* 303; Woodward, *The Commanders,* 117–18.

24. Powell, "U.S. Forces: Challenges Ahead."

25. Gordon and Trainor, *The Generals' War,* viii; Cassidy, *Peacekeeping in the Abyss,* 218; Hoffman, *Decisive Force,* xii, 99–101; Linn, *The Echo of Battle,* 198.

26. Petraeus, "Lessons of History and Lessons of Vietnam."

27. Gordon and Trainor, *The Generals' War,* ix, 52, 130–31, 133; Schwarzkopf, *It Doesn't Take a Hero,* 421.

28. Gordon and Trainor, *The Generals' War,* 123–58.

29. Ibid., 250.

30. Keaney and Cohen, *Gulf War Air Power Survey (GWAPS) Summary Report,* 249; Kitfield, *Prodigal Soldiers,* 149.

31. Schwarzkopf also personally reviewed Baghdad targets. See Keaney and Cohen, *Gulf War Air Power Survey (GWAPS) Summary Report,* 59; Gordon and Trainor, *The Generals' War,* 325–27; Cooper, "The Politics of Air Strikes," 78; Freedman and Karsh, *The Gulf Conflict 1990–1991,* 324–29.

32. Cooper, "The Politics of Air Strikes," 77; Cohen, "The Mystique of Air Power," 121.

33. Gordon and Trainor, *The Generals' War,* 414–16, 476–77; Schwarzkopf, *It Doesn't Take a Hero,* 468–72.

34. Kohn, "The Early Retirement of Gen Ronald R. Fogelman, Chief of Staff, United States Air Force"; Feaver, *Armed Servants,* 216–18.

35. Kohn, "The Early Retirement of Gen Ronald R. Fogelman, Chief of Staff, United States Air Force," 6, 16–17; Gentry, "Military Force in an Age of National Cowardice."

36. Warden, "Employing Air Power in the Twenty-First Century," 57, 61–62.

37. U.S. Army, *Operations*, 1–4, 1–5, 2–1.

38. U.S. Department of Defense, *Annual Report to the President and to the Congress*, 7.

39. Foreman, "The Casualty Myth," 40.

40. Caires, "Force Protection Requires Cultural Change, General Says."

41. Matthews, "Shali's Warning."

42. Fitzsimmons, "The Coming Military Revolution."

43. Conversino, "Sawdust Superpower," 21.

44. Dunlap, "Organizational Change and the New Technologies of War," as quoted in Hyde, "Casualty Aversion," 25. See also Eikenberry, "Take No Casualties."

45. Moore, "Inside Looking Out."

46. Kretchik, "Force Protection Disparities."

47. Moore, "Inside Looking Out"; Gentry, "Military Force in an Age of National Cowardice."

48. Galula, *Counterinsurgency Warfare*, 113; Smith, "A G.I.'s Home Is His Fortress"; author's discussions with British Army officers in Sarajevo, 1996.

49. Cohen, "Distant Battles," 159.

50. Foreman, "The Casualty Myth," 40.

51. General Mohsen Rezai, quoted in *Resalat*, September 25, 1995, as cited in deSutter, *Denial and Jeopardy*, 65–66.

52. Gellman, "U.S. and China Nearly Came to Blows in 1996."

53. Ibid.

54. Daalder and O'Hanlon, *Winning Ugly*, 94.

55. Thurow, "Serbs Bet that the West Won't Risk the Thing They Fear: Ground Troops."

56. Gordon and Trainor, *The Generals' War*, 107.

57. Clark, *Waging Modern War*, 144; Daalder and O'Hanlon, *Winning Ugly*, 160; Halberstam, *War in a Time of Peace*, 451.

58. Sell, *Slobodan Milosevic and the Destruction of Yugoslavia*, 299–304; Daalder and O'Hanlon, *Winning Ugly*, 19–20, 91–96, 208–10; Djukić, *Milošević and Marković*, 120–30; Doder and Branson, *Milosevic*, 253–54; Kay, "NATO, the Kosovo War and Neoliberal Theory," 266.

59. Posen, "The War for Kosovo."

60. For an overview of this period, see Judah, *Kosovo*.

61. Human Rights Watch, *A Week of Terror in Drenica*.

62. Judah, *Kosovo*, 178.

63. Human Rights Watch, *Yugoslav Government War Crimes in Racak*. Serbs claimed that Kosovar Albanians killed the civilians in order to blame them, a common charge at the time. See Talentino, *Military Intervention after the Cold War*, 252.

64. Albright, *Madame Secretary*, 386–96; Soderberg, *The Superpower Myth*, 89–90.

65. Independent International Commission on Kosovo, *Kosovo Report*, 83.

66. Kuperman, "The Moral Hazard of Humanitarian Intervention," 56.

67. Sell, *Slobodan Milosevic and the Destruction of Yugoslavia*, 296; Daalder and O'Hanlon, *Winning Ugly*, 77–89; Independent International Commission on Kosovo, *Kosovo Report*, 157–58.

68. Martel, *Victory in War,* 212–13; Talentino, *Military Intervention after the Cold War,* 241, 244–49.

69. Hosmer, *Why Milosevic Decided to Settle When He Did.*

70. Clinton, *My Life,* 850–51; Albright, *Madame Secretary,* 394.

71. Bono, *NATO's 'Peace Enforcement' Tasks and 'Policy Communities': 1990–1999,* 119.

72. Statement by William J. Clinton, March 24, 1999, quoted in Nardulli et al., *Disjointed War,* 22–23. The Clinton administration had help in their decision to run an air-only operation; congressmen of both parties made clear they opposed a ground war. See Lyon, "Operation Allied Force," 58–59.

73. Statement of Assistant Secretary of Defense Kenneth H. Bacon, March 27, 1999, quoted in Nardulli, *Disjointed War,* 23.

74. Lyon, "Operation Allied Force," 59.

75. Clinton, *My Life,* 851.

76. Clark, *Waging Modern War,* 288–89, 305.

77. Soderberg, *The Superpower Myth,* 91.

78. Posen, "The War for Kosovo," 44n10; Byman and Waxman, "Kosovo and the Great Air Power Debate," 15–16.

79. Independent International Commission on Kosovo, *Kosovo Report,* 86; Albright, *Madame Secretary,* 406–7.

80. Clinton, *My Life,* 851; Albright, *Madame Secretary,* 394; Soderberg, *The Superpower Myth,* 209–11.

81. Lyon, "Operation Allied Force," 61.

82. Nardulli, *Disjointed War,* 27.

83. Clark, *Waging Modern War,* 426–30; Reveron, "U.S. European Command," 108–12; Strickland, "USAF Aerospace-Power Doctrine," 20.

84. Cordesman, *The Lessons and Non-Lessons of the Air and Missile Campaign in Kosovo,* 74–75.

85. Priest, "France Balked at NATO Targets"; Clark, *Waging Modern War,* 304–17.

86. Clark, *Waging Modern War,* 201–2, 224, 225, 228, 275, 279, 339, 364.

87. Strickland, "USAF Aerospace-Power Doctrine," 17.

88. Independent International Commission on Kosovo, *Kosovo Report,* 92–93; Clark, *Waging Modern War,* 203.

89. Lyon, "Operation Allied Force," 64–67.

90. Daalder and O'Hanlon, *Winning Ugly,* 122.

91. Sell, *Slobodan Milosevic and the Destruction of Yugoslavia,* 304–6; Talentino, *Military Intervention after the Cold War,* 254.

92. Clark, *Waging Modern War,* 171, 208.

93. Tilford, "Operation Allied Force and the Role of Air Power."

94. Daalder and O'Hanlon, *Winning Ugly,* 210–12.

95. Martel, *Victory in War,* 214.

96. VIP Daily News Report (Belgrade), June 11, 1999, as reported in Judah, *Kosovo,* 285.

97. Cordesman, *The Lessons and Non-Lessons of the Air and Missile Campaign in Kosovo,* 103.

98. Priest, "A Decisive Battle That Never Was"; Cordesman, *The Lessons and Non-Lessons of the Air and Missile Campaign in Kosovo,* 95; Byman and Waxman, "Kosovo and the Great Air Power Debate," 24; Daalder and O'Hanlon, *Winning Ugly,* 153–55.

99. Lyon, "Operation Allied Force," 60.

100. Ibid., 74.

101. See Short's October 21, 1999, testimony titled "Lessons Learned from Military Operations and Relief Efforts in Kosovo." See Strickland, "USAF Aerospace-Power Doctrine," 17–21; Kay, "NATO, the Kosovo War and Neoliberal Theory," 273–74. For other discussion of this commonly held airmen's view, see Pape, *Bombing to Win*; Biddle, *Rhetoric and Reality in Air Warfare*; Byman and Waxman, "Kosovo and the Great Air Power Debate"; Byman and Waxman, *The Dynamics of Coercion.*

102. Clark, *Waging Modern War,* 243–45.

103. Thomas, *The Ethics of Destruction,* 165.

104. Matlary, *Values and Weapons,* 115.

105. Cooper, "The Politics of Air Strikes," 74; Clark, *Waging Modern War,* 254.

106. Lyon, "Operation Allied Force," 61; Thomas, *The Ethics of Destruction,* 163–64.

107. Cordesman, *The Lessons and Non-Lessons of the Air and Missile Campaign in Kosovo,* 123; Clark, *Waging Modern War,* 296.

108. Clark, *Waging Modern War,* 298.

109. Ibid., 334.

110. Daalder and O'Hanlon, *Winning Ugly,* 141, 156–60; Kay, "NATO, the Kosovo War and Neoliberal Theory," 270; Soderberg, *The Superpower Myth,* 92; Harris and Graham, "Clinton Is Reassessing Sufficiency of Air War."

111. Martin and Brawley, *Alliance Politics, Kosovo, and NATO's War.*

112. Daalder and O'Hanlon, *Winning Ugly,* 101–3, 192–93.

113. Stigler, "A Clear Victory for Air Power."

114. Byman and Waxman, "Kosovo and the Great Air Power Debate," 5–7, 19–20, 25–28; Cordesman, *The Lessons and Non-Lessons of the Air and Missile Campaign in Kosovo,* 32–34, 169, 246–47; Lambeth, *NATO's Air War for Kosovo,* 76; Clark, *Waging Modern War,* 425; Daalder and O'Hanlon, *Winning Ugly,* 155–61, 184, 198–206; Hosmer, *Why Milosevic Decided to Settle When He Did*; Halberstam, *War in a Time of Peace,* 475; Talentino, *Military Intervention after the Cold War,* 255.

115. Drew, "U.S. Airpower Theory and the Insurgent Challenge, 810; Kissinger, "The Vietnam Negotiations," 214.

116. Lambeth, *NATO's Air War for Kosovo*; Clark, *Waging Modern War,* 459–60; Cordesman, *The Lessons and Non-Lessons of the Air and Missile Campaign in Kosovo,* 18; Daalder and O'Hanlon, *Winning Ugly.*

117. Strickland, "USAF Aerospace-Power Doctrine," 21–23.

118. Lyon, "Operation Allied Force," 57, 63, 67; Kay, "NATO, the Kosovo War and Neoliberal Theory," 273–74.

119. Strickland, "USAF Aerospace-Power Doctrine."

120. Ignatieff, *Virtual War,* 62.

121. Thomas, *The Ethics of Destruction,* 167–68.

122. Judah, *Kosovo,* 250–54; Clark, *Waging Modern War,* 443–44; Hosmer, *Why Milosevic Decided to Settle When He Did,* 37–42; Daalder and O'Hanlon, *Winning Ugly,* 19–20, 58–59; Independent International Commission on Kosovo, *Kosovo Report,* 89.

123. Talentino, *Military Intervention after the Cold War,* 265–66.

124. Clark, *Waging Modern War,* 183–84, 346; Clark, "The Strength of an Alliance," 253; Byman and Waxman, "Kosovo and the Great Air Power Debate," 15–16; Daalder and O'Hanlon, *Winning Ugly,* Appendix C; Independent International Commission on Kosovo, *Kosovo Report,* 85.

125. Freedman and Karsh, *The Gulf Conflict 1990–1991*, 329.

126. Layne, "Collateral Damage in Yugoslavia," 54; Daalder and O'Hanlon, *Winning Ugly*, 240–42; Independent International Commission on Kosovo, *Kosovo Report*, 94.

127. Cordesman, *The Lessons and Non-Lessons of the Air and Missile Campaign in Kosovo*, 120.

128. Sell, *Slobodan Milosevic and the Destruction of Yugoslavia*, 306n95; Judah, *Kosovo*, 310; Daalder and O'Hanlon, *Winning Ugly*, 110; Independent International Commission on Kosovo, *Kosovo Report*, 91, Annex 1.

129. Independent International Commission on Kosovo, *Kosovo Report*, 90; Cordesman, *The Lessons and Non-Lessons of the Air and Missile Campaign in Kosovo*, 90–93.

130. Thomas, *The Ethics of Destruction*, 164.

131. Halberstam, *War in a Time of Peace*, 451; Lambeth, *NATO's Air War for Kosovo*, 236; Lyon, "Operation Allied Force," 62, 70.

132. Thomas, *The Ethics of Destruction*, 171–73, 175.

133. Stephen Biddle's force employment theory is an exception to this generalization. See Biddle, *Military Power*.

134. Strickland, "USAF Aerospace-Power Doctrine."

135. Kay, "NATO, the Kosovo War and Neoliberal Theory," 267.

136. U.S. General Accounting Office, *Kosovo Air Operations*; Strickland, "USAF Aerospace-Power Doctrine"; Lyon, "Operation Allied Force."

137. Sciolino and Myers, "Bush Says 'Time Is Running Out;' " Kay, "NATO, the Kosovo War and Neoliberal Theory," 271–72, 274–76.

138. Arreguín-Toft, *How the Weak Win Small Wars*, 232.

139. Martel, *Victory in War*, 213.

140. Ibid., 213–14.

141. Ibid.

142. Daalder and O'Hanlon, *Winning Ugly*, 55.

143. Daalder and O'Hanlon, *Winning Ugly*, 97, 159; Lambeth, *NATO's Air War for Kosovo*, 12, 236.

144. Kay, "NATO, the Kosovo War and Neoliberal Theory."

145. U.S. General Accounting Office, *Kosovo Air Operations*.

146. Thomas, *The Ethics of Destruction*, 167–68.

147. Talentino, *Military Intervention after the Cold War*, 256.

148. Cordesman, *The Lessons and Non-Lessons of the Air and Missile Campaign in Kosovo*, 121–22, 239; Goff, ed., *The Kosovo News and Propaganda War*, 13–14.

149. Clark, *Waging Modern War*, 298.

150. Gocić, "Symbolic Warfare," 90.

151. Clark, *Waging Modern War*, 314, 316, 339, 444; Strickland, "USAF Aerospace-Power Doctrine"; 22–23.

152. Judah, *Kosovo*, 138; Albright, *Madame Secretary*, 391; Talentino, *Military Intervention after the Cold War*, 242, 251–52, 259–60.

153. Gellman, "How the U.S. and Allies Went to War"; Halberstam, *War in a Time of Peace*, 397.

154. Holbrooke on *Newsnight*, August 20, 1999, as quoted in Judah, *Kosovo*, 191.

155. Kuperman, "Suicidal Rebellions and the Moral Hazard of Humanitarian Intervention," 11–12; Buckley, *Kosovo*, 287; Kuperman, "The Moral Hazard of Humanitarian Intervention," 64–71.

156. Kuperman, "Suicidal Rebellions and the Moral Hazard of Humanitarian Intervention," 10–11; Kuperman, "The Moral Hazard of Humanitarian Intervention," 56–64.

157. Rose, *Fighting for Peace*, 43–44, 48–49, 162, 197, 198; Wiebes, *Intelligence and the War in Bosnia, 1992–1995*, 68.

158. Author's conversations with NATO officers in Sarajevo, 1996. See Albright, *Madame Secretary*, 190; Wiebes, *Intelligence and the War in Bosnia*, 68; Sremac, *War of Words*, 183. In contrast, Ivo Daalder claims NATO determined that Serbs fired the round. See Daalder, *Getting to Dayton*, 130.

159. Wiebes, *Intelligence and the War in Bosnia*, 68–69.

160. Sell, *Slobodan Milosevic and the Destruction of Yugoslavia*, 308–17; Byman and Waxman, "Defeating US Coercion," 111; Cordesman, *The Lessons and Non-Lessons of the Air and Missile Campaign in Kosovo*, 239; Sapolsky and Shapiro, "Casualties, Technology, and America's Future Wars," 124; Bowden, *Black Hawk Down*, 43, 46.

161. Kuperman, "The Moral Hazard of Humanitarian Intervention."

162. Posen, "Military Responses to Refugee Disasters," 87.

163. Metz, "Strategic Asymmetry."

CHAPTER 6

1. A large literature describes why the Bush administration chose to go to war and assesses the many controversies surrounding the decision to invade and the occupation administration. See Bush, *Decision Points*, 223–71, 355–94; Bodansky, *The Secret History of the Iraq War*; Bremmer, *My Year in Iraq*; Diamond, *Squandered Victory*; Feith, *War and Decision*; Gordon and Trainor, *Cobra II*; Isikoff and Corn, *Hubris*; Ricks, *Fiasco*; Suskind, *The One Percent Doctrine*; Rose, *How Wars End*, 236–76; Rumsfeld, *Known and Unknown*; 411–541, 659–60, 671–80; Woodward, *Bush at War*; Woodward, *Plan of Attack*; Woodward, *State of Denial*.

2. Bush, *Decision Points*, 268; Bodansky, *The Secret History of the Iraq War*, 178; Hendrickson and Tucker, *Revisions in Need of Revising*; Bensahel, "Mission Not Accomplished," 456–58; Ricks, *Fiasco*, 58–59.

3. Ricks, *Fiasco*, 256.

4. The U.S. military chronically has a considerable number of crude war plans available for use as planning aids if an actual mission arises. These contingency plans typically require significant refinement before they are operationally usable. In this case, OPLAN 1003 underwent many refinements before Rumsfeld was satisfied with it.

5. Rumsfeld, "Transforming the Military"; Kitfield, *War and Destiny*, 143; Ricks, *Fiasco*, 75, 128–29; Ucko, *The New Counterinsurgency Era*, 51–53; Rumsfeld, *Known and Unknown*, 425.

6. Hersch, *Chain of Command*, 250–51; Fallows, *Blind Into Baghdad*, 74–77; Ricks, *Fiasco*, 70–71.

7. Kitfield, *War and Destiny*, 276.

8. Scarborough, *Rumsfeld's War*, 45–46; Ricks, *Fiasco*, 128–29.

9. Scarborough, *Rumsfeld's War*, 45–46.

10. Franks, *American Soldier*, 428–31.

11. Cordesman, *The Iraq War*, 36–37; Franks, *American Soldier*, 433, 474.

12. Franks, *American Soldier*, 415.

13. Gordon and Trainor, *Cobra II*, 110.

14. Record, *Dark Victory*, 76.

15. Bensahel, "Mission Not Accomplished," 466–68; Sanchez, *Wiser in Battle*; Fallows, *Blind Into Baghdad*; Ricks, *Fiasco*, 128–29.

16. Gelpi, Feaver, and Reifler, *Paying the Human Costs of War*.

17. Bush, *Decision Points*, 389–91.

18. Rose, *How Wars End*, 276.

19. Cordesman, *The Iraq War*, 177; Franks, *American Soldier*, 382–431; Keegan, *The Iraq War*, 236; Bodansky, *The Secret History of the Iraq War*, 169.

20. Boyne, *Operation Iraqi Freedom*, 38–39; Franks, *American Soldier*.

21. Cordesman, *The Iraq War*, 40, 45–46.

22. Ibid., 20, 22.

23. Keegan, *The Iraq War*, 129–30; Cordesman, *The Iraq War*, 47; Franks, *American Soldier*, 451; Bodansky, *The Secret History of the Iraq War*, 76.

24. Allawi, *The Occupation of Iraq*, 89; Hashim, *Insurgency and Counter-Insurgency in Iraq*, 12–13.

25. Cordesman, *The Iraq War*, 40–41.

26. Ibid., 22–33.

27. Ibid., 22–24.

28. Ibid., 20.

29. Ibid., 48–52.

30. Donnelly, *Operation Iraqi Freedom*, 38.

31. Atkinson, *In the Company of Soldiers*, 289.

32. Franks, *American Soldier*, 415.

33. Ullman and Wade, *Shock and Awe*.

34. Kitfield, *War and Destiny*, 140.

35. Atkinson, *In the Company of Soldiers*, 185–86; Rumsfeld, *Known and Unknown*, 467.

36. Franks, *American Soldier*, 391, 417; Keegan, *The Iraq War*, 187; Atkinson, *In the Company of Soldiers*, 185–86.

37. Cordesman, *The Iraq War*, 31.

38. Franks, *American Soldier*, 476; Hashim, *Insurgency and Counter-Insurgency in Iraq*, 12.

39. Boyne, *Operation Iraqi Freedom*, 55–56; Record, *Dark Victory*, 99; Meilinger, "The Origins of Effects-Based Operations."

40. Franks, *American Soldier*, 476–77; Kahl, "How We Fight," 91.

41. Boyne, *Operation Iraqi Freedom*, 55–56; Record, *Dark Victory*, 99.

42. Kitfield, *War and Destiny*, 205.

43. Ibid., 139.

44. Franks, *American Soldier*, 465–66; Kitfield, *War and Destiny*, 172–73, 187, 189.

45. Kitfield, *War and Destiny*, 187.

46. Donnelly, *Operation Iraqi Freedom*, 41.

47. Cordesman, *The Iraq War*, 475–86; Murray and Scales, *The Iraq War*; Bodansky, *The Secret History of the Iraq War*.

48. Keegan, *The Iraq War*, 186.

49. Boyne, *Operation Iraqi Freedom*, 163.

50. Franks, *American Soldier*, 486, 508.

51. Bodansky, *The Secret History of the Iraq War*, 76, 97–98; Tyson, "FBI Agent: Saddam Didn't Expect Invasion"; Rumsfeld, *Known and Unknown*, 674–75.

52. Franks, *American Soldier,* 519; Bodansky, *The Secret History of the Iraq War,* 205; Hashim, *Insurgency and Counter-Insurgency in Iraq,* 15–17.

53. Gordon and Trainor, *Cobra II,* 311.

54. Franks, *American Soldier,* 490.

55. Ricks, *Fiasco,* 126.

56. Franks, *American Soldier,* 524.

57. Record, *Dark Victory;* Ricks, *Fiasco,* 58–59, 97, 106.

58. Record, *Dark Victory.*

59. Kitfield, *War and Destiny,* 281; Krepinevich, *The Army and Vietnam.*

60. Knights, *Cradle of Conflict,* 331; Allawi, *The Occupation of Iraq,* 173–75; Cordesman, *The Iraq War,* 238; Herring and Rangwala, *Iraq in Fragments,* 167–68; Bodansky, *The Secret History of the Iraq War;* Packer, *The Assassin's Gate,* 299.

61. Allawi, *The Occupation of Iraq,* 135–36; Metz, *Rethinking Insurgency,* 30–32.

62. Ricks, *Fiasco,* 337–38.

63. Packer, *The Assassin's Gate,* 299; Hashim, *Insurgency and Counter-Insurgency in Iraq,* 15–17, 33.

64. Packer, *The Assassin's Gate,* 307; Ricks, *Fiasco,* 338; Suskind, *The One Percent Doctrine,* 52.

65. Atkinson, "Left of Boom," Part 2.

66. Atkinson, "Left of Boom," Part 1.

67. Herring and Rangwala, *Iraq in Fragments,* 201–6.

68. Hashim, *Insurgency and Counter-Insurgency in Iraq,* 77.

69. Hashim, *Insurgency and Counter-Insurgency in Iraq,* 99; Allawi, *The Occupation of Iraq,* 242–43.

70. Benjamin and Simon, *The Age of Sacred Terror,* 286–87; Knights, *Cradle of Conflict,* 332–37.

71. Murray and Scales, *The Iraq War,* 236.

72. Packer, *The Assassin's Gate,* 310; Hashim, *Insurgency and Counter-Insurgency in Iraq,* 9, 32–33, 178–80; Herring and Rangwala, *Iraq in Fragments,* 170; Allawi, *The Occupation of Iraq,* 175–76; Hashim, *Insurgency and Counter-Insurgency in Iraq,* 197–99.

73. Allawi, *The Occupation of Iraq,* 175–76.

74. Herring and Rangwala, *Iraq in Fragments,* 67, 70; Foulk, *The Battle for Fallujah,* 172–74.

75. Foulk, *The Battle for Fallujah,* 27, 35–36, 191–92, 224; West, *No True Glory,* 113, 117.

76. Donnelly, *Operation Iraqi Freedom,* 97; Kitfield, *War and Destiny,* 263, 265; Foulk, *The Battle for Fallujah,* 142.

77. Allawi, *The Occupation of Iraq,* 116–17; Feith, *War and Decision,* 435.

78. Atkinson, "Left of Boom," Part 1.

79. Kitfield, *War and Destiny,* 310; Atkinson, "Left of Boom," Parts 1 and 2.

80. West, *No True Glory,* 94.

81. Bodansky, *The Secret History of the Iraq War,* 432; Suskind, *The One Percent Doctrine,* 235; Kepel and Milelli, *Al Qaeda in Its Own Words,* 52, 55, 59, 60–65.

82. Hoffman, "Scarier Than Bin Laden"; Kepel and Milelli, *Al Qaeda in Its Own Words,* 203.

83. Gunaratna, *Inside Al Qaeda,* 299; Bodansky, *The Secret History of the Iraq War,* 305–6; Hashim, *Insurgency and Counter-Insurgency in Iraq,* 212.

84. Kepel and Milelli, *Al Qaeda in Its Own Words,* 204; Slim, *Killing Civilians,* 157; Hashim, *Insurgency and Counter-Insurgency in Iraq,* 180.

85. Pape, *Dying to Win,* 122–24, 130, 190–97, 253–63; Benjamin and Simon, *The Age of Sacred Terror,* 119.

86. Record, *Dark Victory,* 50.

87. Hashim, *Insurgency and Counter-Insurgency in Iraq,* 196–97; Suskind, *The One Percent Doctrine,* 302–4; Cohen, "Distant Battles," 147; Packer, *The Assassin's Gate,* 310.

88. Phillips, *Losing Iraq,* 178–68.

89. Donnelly, *Operation Iraqi Freedom,* 96, 100.

90. Kitfield, *War and Destiny,* 302, 330–31; Byman, *The Five Front War,* 178.

91. Kitfield, *War and Destiny,* 314, 321, 331; Phillips, *Losing Iraq,* 152.

92. Hoffman, *Al Qaeda, Trends in Terrorism and Future Potentialities,* 11; Slim, *Killing Civilians,* 148.

93. Gunaratna, *Inside Al Qaeda,* 300; Kepel and Milelli, *Al Qaeda in Its Own Words,* 193.

94. Benjamin and Simon, *The Age of Sacred Terror,* 120; Gunaratna, *Inside Al Qaeda,* 297–302.

95. Hashim, *Insurgency and Counter-Insurgency in Iraq,* 195–96.

96. Knickmeyer, "Surge in Deaths Reveals More Lethal Foe."

97. Packer, *The Assassin's Gate,* 311; Ricks, *Fiasco,* 141; Foulk, *The Battle for Fallujah,* 190.

98. Allawi, *The Occupation of Iraq,* 170–71, 181; Donnelly, *Operation Iraqi Freedom,* 96, 100.

99. Keegan, *The Iraq War,* 207; Phillips, *Losing Iraq,* 9, 163, 182; Soderberg, *The Superpower Myth,* 225–28.

100. Donnelly, *Operation Iraqi Freedom,* 99; Phillips, *Losing Iraq,* 167–68.

101. Kitfield, *War and Destiny,* 352; Phillips, *Losing Iraq,* 202.

102. Phillips, *Losing Iraq,* 202.

103. Jones, *Terrorist Beheadings,* 5–7.

104. Foulk, *The Battle for Fallujah,* 50.

105. Record, *Dark Victory,* 115.

106. Hashim, *Insurgency and Counter-Insurgency in Iraq,* 47, 181; Ricks, *Fiasco,* 436; Allawi, *The Occupation of Iraq,* 234, 238–40; Byman, *The Five Front War,* 225–27.

107. Ricks, *Fiasco,* 436.

108. Kepel and Milelli, *Al Qaeda in Its Own Words,* 7, 204, 237–67.

109. Glasser and Coll, "The Web as Weapon; Gunaratna, *Inside Al Qaeda,* 308; Jones, *Terrorist Beheadings*; Packer, *The Assassin's Gate,* 311; Byman, *The Five Front War,* 175–80.

110. Foulk, *The Battle for Fallujah,* 56–61.

111. Glasser and Coll, "The Web as Weapon."

112. Ricks, *Fiasco,* 414.

113. Knickmeyer, "Surge in Deaths Reveals More Lethal Foe."

114. Kitfield, *War and Destiny,* 351.

115. Knickmeyer, "Surge in Deaths Reveals More Lethal Foe."

116. Michaels, "19,000 insurgents killed in Iraq since '03."

117. Kitfield, *War and Destiny,* 305; Herring and Rangwala, *Iraq in Fragments,* 75.

118. Bush, *Decision Points,* 363–84; Woodward, *The War Within*; Ricks, *The Gamble.*

119. U.S. Army, *Counterinsurgency,* Field Manual 3–24; Ricks, *The Gamble*; Kilcullen, *The Accidental Guerrilla,* 133–35; Ucko, *The New Counterinsurgency Era,* 65, 109–12.

120. Bush, *Decision Points,* 381.

121. Kilcullen, *The Accidental Guerrilla,* 159–60, 171–73.

122. Rumsfeld, *Known and Unknown,* 701.

123. Al-Jabouri and Jensen, "The Iraqi and AQI Roles in the Sunni Awakening."

124. Fallows, *Blind Into Baghdad,* 163; Feith, *War and Decision,* 498–500.

125. Al-Jabouri and Jensen, "The Iraqi and AQI Roles in the Sunni Awakening"; Hashim, *Insurgency and Counter-Insurgency in Iraq,* 275–88.

126. Atkinson, *In the Company of Soldiers,* 301; Record, *Dark Victory;* Scarborough, *Rumsfeld's War,* 46; Ricks, *Fiasco,* 33, 58–59, 109–11.

127. Franks, *American Soldier,* 331; Ricks, *Fiasco,* 116.

128. Donnelly, *Operation Iraqi Freedom,* 46, 87–88, 91, 92; Kitfield, *War and Destiny,* 248; Ricks, *Fiasco,* 109–11; Scarborough, *Rumsfeld's War,* 178, 180–82; Packer, *The Assassin's Gate,* 119–20; Ucko, *The New Counterinsurgency Era,* 60.

129. Per standard U.S. military practice, a subordinate commander, like the 3rd Infantry Division's commander, could have told his staff to draft a Phase IV plan even if the CENTCOM or corps orders under which he operated did not have one. See Bush, *Decision Points,* 268; Kitfield, *War and Destiny,* 226; Phillips, *Losing Iraq,* 133–34, 136.

130. Donnelly, *Operation Iraqi Freedom,* 46, 60, 87–88, 91; Kitfield, *War and Destiny,* 224; O'Hanlon, *Defense Strategies for the Post-Saddam Era,* 37; Fallows, *Blind Into Baghdad,* 101.

131. Phillips, *Losing Iraq,* 135, 192.

132. Hashim, *Insurgency and Counter-Insurgency in Iraq,* 163; Knights, *Cradle of Conflict,* 330; Ricks, *Fiasco,* 145–46.

133. Allawi, *The Occupation of Iraq,* 140; Atkinson, "Left of Boom," Part 1.

134. Packer, *The Assassin's Gate,* 310; Hashim, *Insurgency and Counter-Insurgency in Iraq,* 340; Allawi, *The Occupation of Iraq,* 184.

135. Galula, *Counterinsurgency Warfare,* 36.

136. Ricks, *Fiasco,* 232–33, 237–38, 279–80; Kahl, "How We Fight," 93; Foulk, *The Battle for Fallujah,* 11, 13–14, 17, 192–93; Miles, *Al-Jazeera,* 281.

137. Allawi, *The Occupation of Iraq,* 169; Ricks, *Fiasco,* 367.

138. Ricks, *Fiasco,* 140–44, 234, 250.

139. Alwyn-Foster, "Changing the Army for Counterinsurgency Operations"; Hashim, *Insurgency and Counter-Insurgency in Iraq,* 79; Ricks, *Fiasco,* 266; Allawi, *The Occupation of Iraq,* 186; West, *No True Glory,* 92.

140. Hashim, *Insurgency and Counter-Insurgency in Iraq,* 275–88; Ricks, *Fiasco,* 238–39; Paley and Partlow, "Evolution of a U.S. General in Iraq."

141. Herring and Rangwala, *Iraq in Fragments,* 186–92; Packer, *The Assassin's Gate,* 325–27.

142. Ricks, *Fiasco,* 139, 233, 329; Foulk, *The Battle for Fallujah,* 17.

143. Ricks, *Fiasco,* 256–58; Hashim, *Insurgency and Counter-Insurgency in Iraq,* 107–8, 325, 346–47; Herring and Rangwala, *Iraq in Fragments,* 182; Foulk, *The Battle for Fallujah,* 17, 92, 96–98; Slim, *Killing Civilians,* 166; Kahl, "In the Crossfire or the Crosshairs."

144. Yet, U.S. military officers happy with force protection policies and standard operating procedures that produce significant numbers of civilian deaths also regularly criticize security contractors who kill far fewer Iraqi civilians than do U.S. military personnel. See Cordesman, *The Iraq War,* 502; Hendrickson and Tucker, *Revisions in Need of Revising,* 28; Byman, *The Five Front War,* 234.

145. White, "U.S. Boosts Its Use of Airstrikes in Iraq."

146. Foulk, *The Battle for Fallujah,* 42, 159, 198; Kahl, "How We Fight," 93.

147. Allawi, *The Occupation of Iraq,* 244.

148. Ricks, *Fiasco,* 80–81, 101–7, 154–55; Feith, *War and Decision,* 349–50.

149. Some blame Bremer for these decisions, but they appear to have been made at least in part in the Pentagon at the Rumsfeld level and delivered to him for announcement and implementation in Iraq. See Bremmer, *My Year in Iraq,* 39, 40, 54–57; Hashim, *Insurgency and Counter-Insurgency in Iraq,* 76, 93; Feith, *War and Decision;* Al-Jabouri and Jensen, "The Iraqi and AQI Roles in the Sunni Awakening," Phillips, *Losing Iraq,* 9, 140–41, 144–45, 152, 158, 170; Fallows, *Blind Into Baghdad,* 102–3; Rose, *How Wars End,* 249–51.

150. Quoted in Allawi, *The Occupation of Iraq,* 105, from the *Guardian,* March 14, 2006.

151. Chandrasekaren, *Imperial Life in the Emerald City;* Diamond, *Squandered Victory;* Packer, *The Assassin's Gate,* 325.

152. Packer, *The Assassin's Gate,* 306–7; Fallows, *Blind Into Baghdad,* 147–86.

153. Donnelly, *Operation Iraqi Freedom,* 110; Phillips, *Losing Iraq,* 192; Fallows, *Blind Into Baghdad,* 61; Bensahel, "Mission Not Accomplished," 462–66; Feith, *War and Decision;* Sanchez, *Wiser in Battle.*

154. Bensahel, "Mission Not Accomplished," 458–62.

155. President Bush gave the Defense Department authority for postwar relief, reconstruction, and administration in National Security Presidential Directive 24 and established ORHA under the supervision of Under Secretary of Defense Douglas Feith on January 21, 2003. See Phillips, *Losing Iraq,* 126–31; Cordesman, *The Iraq War,* 498; Ricks, *Fiasco,* 102–3.

156. Atkinson, *In the Company of Soldiers,* 299; Phillips, *Losing Iraq,* 8, 126–31, 138, 140–41; Scarborough, *Rumsfeld's War,* 46, 51, 53, 59, 178, 180–82.

157. George W. Bush, "Management of Interagency Efforts Concerning Reconstruction and Stabilization"; and U.S. Department of Defense, Gordon England, Deputy Secretary, "Military Support for Stability, Security, Transition and Reconstruction (SSTR) Operations." See Ucko, *The New Counterinsurgency Era,* 96–100.

158. Bodansky, *The Secret History of the Iraq War,* 168–77.

159. Phillips, *Losing Iraq,* 73–75; Ricks, *Fiasco,* 272; Rumsfeld, *Known and Unknown.*

160. Kitfield, *War and Destiny,* 318–19; Scarborough, *Rumsfeld's War,* 46–47, 58, 59.

161. Bush, *Decision Points,* 268; Record, *Dark Victory,* 118; Feith, *War and Decision;* author discussion with military officers.

162. Atkinson, *In the Company of Soldiers,* 160, 162, 175–76; Bodansky, *The Secret History of the Iraq War,* 507–8; Donnelly, *Operation Iraqi Freedom,* 87–88; O'Hanlon, *Defense Strategies for the Post-Saddam Era,* 37–38; Murray and Scales, *The Iraq War,* 182.

163. Phillips, *Losing Iraq,* 73–75, 155, 160; Gentry, "Doomed to Fail"; Packer, *The Assassin's Gate,* 303–4; Galula, *Counterinsurgency Warfare,* 72.

164. Kitfield, *War and Destiny,* 248, 295.

165. The U.S. military had an appreciable nation-building cadre, at least by international military standards, in its special operations forces, especially Special Forces and civil affairs units. Civil affairs units did not have much influence on policy or military operational decisions, however. A measure of the priority given this work was then placement of almost all of the army's civil affairs personnel in the Army Reserve.

166. Record, *Dark Victory,* 119–20; Ricks, *Fiasco,* 132–33; Fallows, *Blind Into Baghdad,* 89, 182–83.

167. Record, *Dark Victory,* 118.

168. Murray and Scales, *The Iraq War,* 183.

169. Fallows, *Blind Into Baghdad.*

170. Newbold later went public with his criticisms of war policy, becoming part of what became known as the revolt of the generals. See Ricks, *Fiasco,* 40, 55, 67; Rumsfeld, *Known and Unknown,* 485.

171. Gordon and Trainor, *Cobra II,* 97–98.

172. Ibid., 116–17.

173. Ricks, *Fiasco,* 35–36; Gordon and Trainor, *Cobra II,* 92.

174. Ricks, *Fiasco,* 34, 96–100; Fallows, *Blind Into Baghdad,* 78, 80, 97–99.

175. Fallows, *Blind Into Baghdad,* 130; Feith, *War and Decision*; Woodward, *The War Within.*

176. Cordesman, *The Iraq War,* 238.

177. I prefer this measure of casualties rather than overall U.S. military deaths in the region. The DoD counts as casualties the deaths of military personnel for any reason, including road accidents, heart attacks, suicides, and air crashes that had nothing to do with the war that occurred in Iraq and several surrounding countries and bodies of water, including Oman, the Arabian Sea, and the Red Sea. Many such deaths regularly occur in peacetime, garrison situations in the United States. As a result, DoD exaggerates the actual number of war casualties. Given the significant political sensitivity of U.S. military casualties and journalists' habit of reporting the larger figure, it is odd that DoD overreports them as a matter of accounting policy. For running tallies and a description of DoD casualty accounting practices, see U.S. Department of Defense, "Operation Enduring Freedom U.S. Casualty Status." See also Phillips, *Losing Iraq,* 167–68; Gentry, "Casualty Management."

178. "Iraq Casualties," *Washington Post,* October 13, A13.

179. Ibid.

180. See http://www.defenselink.mil/news/casualty.pdf (accessed April 11, 2011).

181. Hammes, *Private Contractors in Conflict Zones,* 3.

182. Schooner, "Why Contractor Fatalities Matter," 81–82, 88–89.

183. *Washington Post,* "Iraq Casualties," October 13, 2007, A13.

184. U.S. Central Command figures, as reported in Beckett, *Insurgency in Iraq,* 8.

185. Kitfield, *War and Destiny,* 356.

186. Atkinson, "Left of Boom," Parts 1 and 2.

187. "Iraq Casualties," *Washington Post,* October 13, 2007, A13.

188. The United States refuses to regularly report enemy combatant or civilian deaths as a direct result of its bad experience in Vietnam, where body count was a dubious measure of operational success, had the negative affect of encouraging some commanders to kill civilians to boost the apparent operational effectiveness of their units, and drew much public scorn. For other casualty estimates, see Herring and Rangwala, *Iraq in Fragments,* 183–85; and Michaels, "19,000 insurgents killed in Iraq since '03."

189. Herring and Rangwala, *Iraq in Fragments,* 184; Kahl, "How We Fight," 86–88.

190. Herring and Rangwala, *Iraq in Fragments,* 184.

191. Ibid.

192. Running counts and selective analyses are available at Iraq Body Count's website, http://www.iraqbodycount.org. See also Herring and Rangwala, *Iraq in Fragments,* 184.

193. "Iraq Casualties," *Washington Post*, October 13, 2007, A13.

194. Iraq Body Count, http://www.iraqbodycount.org/ (accessed February 1, 2011).

195. Sebti, "What I Risked as an Iraqi Journalist."

196. Murray and Scales, *The Iraq War*, 74–75; Ullman and Wade, *Shock and Awe*.

197. Donnelly, *Operation Iraqi Freedom*, 43.

198. Murray and Scales, *The Iraq War*, 162.

199. Cordesman, *The Iraq War*, 52–53; Kitfield, *War and Destiny*, 141; Glaser and Kaufman, "What Is the Offense-Defense Balance and Can We Measure It?" 75n1; Mearsheimer, *Conventional Deterrence*, 46, 47, 51, 222n48; Biddle, *Military Power*.

200. Atkinson, "Left of Boom," Part 1.

201. Manhattan Project refers to the massive U.S. effort during World War II to develop the atomic bomb. See Atkinson, "Left of Boom," Part 3.

202. Atkinson, "Left of Boom," Part 3.

203. Ricks, *The Gamble*, 192–93.

204. Miles, *Al-Jazeera*, 130.

205. Adams, "Attack or Conquer?"; Lieber, *War and the Engineers*.

206. Glaser and Kaufman, "What Is the Offense-Defense Balance and Can We Measure It?" 62.

207. Atkinson, "Left of Boom," Part 1; Foulk, *The Battle for Fallujah*, 56–61; Hoffman, "Neoclassical Counterinsurgency?"; Miles, *Al-Jazeera*, 130; Betz, "The More You Know, the Less You Understand," 525–28; Kepel and Milelli, *Al Qaeda in Its Own Words*, 7.

208. Glasser and Coll, "The Web as Weapon"; Hoffman, "Scarier Than Bin Laden."

209. Bensahel, "Mission Not Accomplished"; Woodward, *The War Within*.

210. Krepinevich, "How to Win in Iraq," 89.

211. Sanchez, *Wiser in Battle*; Packer, *The Assassin's Gate*, 325; Ricks, *Fiasco*, 225–28; West, *No True Glory*, 90; Linn, *The Echo of Battle*, 239.

212. Ricks, *Fiasco*, 352.

213. Packer, *The Assassin's Gate*, 304–7; Ricks, *Fiasco*, 367–73; Paley and Partlow, "Evolution of a U.S. General in Iraq."

214. Ricks, *Fiasco*, 397, 414; Ucko, *The New Counterinsurgency Era*, 76.

215. Ricks, *Fiasco*, 398.

216. The army and marine corps jointly produced *Counterinsurgency* as U.S. Army Field Manual 3–24 and as Marine Corps Warfighting Publication 3–33.5, dated December 15, 2006. See Hoffman, "Neoclassical Counterinsurgency?"; Ucko, *The New Counterinsurgency Era*, 109–12.

217. Nagl, *Learning to Eat Soup with a Knife*.

218. Galula, *Counterinsurgency Warfare*; Thompson, *Defeating Communist Insurgency*.

219. Kahl, "How We Fight."

220. Ricks, *Fiasco*, 417.

221. Fallows, *Blind Into Baghdad*, 170–71, 182–84; Kilcullen, "Countering Global Insurgency."

222. Cordesman, *The Iraq War*, 502; Herring and Rangwala, *Iraq in Fragments*, 182, 207; Alwyn-Foster, "Changing the Army for Counterinsurgency Operations," 6, 10, 15; Byman, *The Five Front War*, 234.

223. Foulk, *The Battle for Fallujah*, 34.

224. Kitfield, *War and Destiny*, 60.

225. Hoffman, *Al Qaeda, Trends in Terrorism and Future Potentialities*, 12.

226. Hoffman, "Scarier Than Bin Laden"; Byman, *The Five Front War*, 17.

227. Al-Jabouri and Jensen, "The Iraqi and AQI Roles in the Sunni Awakening."

228. Arreguín-Toft, *How the Weak Win Small Wars*, 20, 232.

229. Metz, "Insurgency and Counterinsurgency in Iraq."

230. Feith, *War and Decision*, 366–68.

231. Allawi, *The Occupation of Iraq*, 241–42.

232. The other imperative was saving the oil fields, which U.S. planners expected to be the major source of reconstruction funding. It was therefore a vulnerability amelioration measure in the *resource mobilization* dimension. See Murray and Scales, *The Iraq War*, 88–89.

233. Gordon and Trainor, *Cobra II*, 110.

234. Atkinson, *In the Company of Soldiers*, 185–86; Murray and Scales, *The Iraq War*, 56.

235. Bodansky, *The Secret History of the Iraq War.*

236. Gordon and Trainor, *Cobra II*, 322–24; Gentry, "Doomed to Fail," 91.

237. Murray and Scales, *The Iraq War*, 103; Feith, *War and Decision*, 185, 331, 339.

238. Boyne, *Operation Iraqi Freedom*, 56; Franks, *American Soldier*, 489–90; Keegan, *The Iraq War*, 159; Kitfield, *War and Destiny*, 81, 148–49, 190; Murray and Scales, *The Iraq War*, 96, 103, 168, 178–79, 209; Freedman and Karsh, *The Gulf Conflict 1990–1991*, 156, 305, 306, 325.

239. Franks, *American Soldier*, 509–10.

240. Lord, "The Role of the United States in Small Wars," 96.

241. Boyne, *Operation Iraqi Freedom*, 179; Kitfield, *War and Destiny*, 141, 148.

242. Donnelly, *Operation Iraqi Freedom*, 76; Murray and Scales, *The Iraq War*, 147–48.

243. Levy, "Learning and Foreign Policy," 284; Murray and Scales, *The Iraq War*, 37, 236; Record, *Dark Victory*, 13; Wirtz, "The Balance of Power Paradox," 141, 145; Gordon and Trainor, *Cobra II*, 65–66; Feith, *War and Decision*, 185, 331, 339; Rumsfeld, *Known and Unknown*, 414, 467.

244. Galula, *Counterinsurgency Warfare.*

245. Hoffman, "Neoclassical Counterinsurgency?"

246. Ricks, *The Gamble.*

247. Ibid., 192–93.

248. Allawi, *The Occupation of Iraq*, 241–42.

249. Woodward, *The War Within*, 370.

250. Hoffman, *Al Qaeda, Trends in Terrorism and Future Potentialities*, 6, 12.

251. Atkinson, "Left of Boom," Part 3.

252. Ibid.

253. Suskind, *The One Percent Doctrine*, 302–4.

254. Gunaratna, *Inside Al Qaeda*, 300–301.

255. National Commission on Terrorist Attacks on the United States, *The 9/11 Commission Report*, 169–71.

256. Kitfield, *War and Destiny*, 320.

257. Foulk, *The Battle for Fallujah*, 176.

258. Franks, *American Soldier*, 555.

259. Ignatius, "Next Challenge in Iraq."

260. Hashim, *Insurgency and Counter-Insurgency in Iraq*, 99–120.

261. Ibid., 121–24.

262. Suskind, *The One Percent Doctrine.*

263. Feith, *War and Decision,* 294.

264. Woodward, *The War Within,* 379–81; Ricks, *The Gamble.*

265. Waltz, *Theory of International Politics,* 16.

266. Glaser, "Realists as Optimists," 58, 60–64.

267. Hammes, "War Evolving into the Fourth Generation"; Echevarria, "Deconstructing the Theory of Fourth-Generation War," 233–41; Hammes, *The Sling and the Stone*; Freedman, *The Revolution in Strategic Affairs,* 9.

CHAPTER 7

1. Because the unconventional phase of this war continues as I write, this chapter's discussion of the insurgency phase of this war is more tentative than previous case study chapters.

2. Johnson and Mason, "No Sign until the Burst of Fire"; Rashid, *Taliban,* 75, 77, 88–89, 219.

3. Graham, *By His Own Rules,* 289–310; Rumsfeld, *Known and Unknown,* 369–70.

4. Tenet, *At the Center of the Storm,* 174–89; Graham, *By His Own Rules,* 286–91, 294–96, 300–310.

5. Jones, *Counterinsurgency in Afghanistan,* 89–93.

6. Jones, *In the Graveyard of Empires*; Ucko, *The New Counterinsurgency Era,* 56–59.

7. Kilcullen, *The Accidental Guerrilla,* 43; Nagl, "A Better War in Afghanistan," 39; Graham, *By His Own Rules,* 501, 644; Jalali, "Winning in Afghanistan"; Woodward, *Obama's Wars,* 34, 82, 117–18; Rashid, *Taliban,* 176; Rumsfeld, *Known and Unknown,* 683.

8. Bush, *Decision Points,* 218.

9. Woodward, *Obama's Wars,* 185, 268–69.

10. Ibid., 187.

11. Ibid., 184, 290.

12. Ibid., 253, 260, 271, 300, 301, 303, 351.

13. Biddle, "Is There a Middle Way?" 28–31; Nagl, "A Better War in Afghanistan," 32–33; Iqbal, "An Appraisal of the Afghanistan-Pakistan Strategy to Counter Terrorism," 21.

14. Woodward, *Obama's Wars,* 176; Isby, *Afghanistan,* 13; O'Hanlon and Sherjan, *Toughing It Out in Afghanistan.*

15. Rothstein, *Afghanistan & the Troubled Future of Unconventional Warfare,* 99–100.

16. Schroen, *First In*; Crumpton, "Intelligence and War."

17. Tanner, *Afghanistan,* 289–320.

18. Rashid, *Taliban,* 100, 220; O'Hanlon and Sherjan, *Toughing It Out in Afghanistan,* 22; Rumsfeld, *Known and Unknown,* 371.

19. Tanner, *Afghanistan,* 289–320.

20. Rashid, *Taliban,* 219–20.

21. Rashid, *Taliban,* 221; Semple, *Reconciliation in Afghanistan.*

22. Oliker, *Aid During Conflict,* 78.

23. Isby, *Afghanistan,* 136–49; Jones, *Counterinsurgency in Afghanistan,* 38; O'Hanlon and Sherjan, *Toughing It Out in Afghanistan,* 31–32.

24. Kilcullen, *The Accidental Guerrilla*, 52.
25. Jones, *Counterinsurgency in Afghanistan*, 38–41.
26. Ibid., 41–42.
27. Isby, *Afghanistan*, 141.
28. Jones, *Counterinsurgency in Afghanistan*, 43–46.
29. Gul, *The Most Dangerous Place*, 191–92.
30. Ibid., 127.
31. Isby, *Afghanistan*, 141; Rashid, *Taliban*, 224; Gul, *The Most Dangerous Place*, 11–12, 89.
32. Rashid, *Taliban*, 224, 227, 237–42.
33. Johnson and Mason, "No Sign until the Burst of Fire."
34. Crews, "Introduction," 24–28; Rashid, *Taliban*, 21; Jones, "The Rise of Afghanistan's Insurgency."
35. Isby, *Afghanistan*, 13.
36. Ibid., 149–58.
37. Kilcullen, *The Accidental Guerrilla*, 49, 52, 58; Isby, *Afghanistan*, 153; Jalali, "Winning in Afghanistan," 13; Karzai, "Strengthening Security in Contemporary Afghanistan," 61–64.
38. Rashid, *Taliban*, 224.
39. Karzai, "Strengthening Security in Contemporary Afghanistan," 63–64.
40. Isby, *Afghanistan*, 16; Oliker, *Aid During Conflict*.
41. Neumann, *The Other War*, 58; Biddle, "Is There a Middle Way?" 30; Semple, *Reconciliation in Afghanistan*; Woodward, *Obama's Wars*, 128, 162.
42. Kilcullen, *The Accidental Guerrilla*, 83–87.
43. O'Hanlon and Sherjan, *Toughing It Out in Afghanistan*, 9, 152, 154.
44. Karzai, "Strengthening Security in Contemporary Afghanistan," 74.
45. Jones, *Counterinsurgency in Afghanistan*, 107; Jalali, "Winning in Afghanistan"; Younossi, *The Long March*.
46. Oliker, *Aid During Conflict*, 79.
47. Jalali, "The Legacy of War and the Challenge of Peace Building," 43.
48. Rothstein, *Afghanistan & the Troubled Future of Unconventional Warfare*, 110.
49. Neumann, *The Other War*, 15.
50. Woodward, *Obama's Wars*, 51–53; Meckler and Entous, "U.S. Forces Kill Osama Bin Laden."
51. Younossi, *The Long March*.
52. Caldwell and Finney, "Building Police Capacity in Afghanistan," 121; O'Hanlon and Sherjan, *Toughing It Out in Afghanistan*, 13, 27–28.
53. Neumann, *The Other War*, 167.
54. Ibid., 119–20.
55. Rumsfeld, *Known and Unknown*, 685–86.
56. Rashid, *Taliban*, 229.
57. Jalali, "The Legacy of War and the Challenge of Peace Building," 33.
58. Fazey, "Responding to the Opium Dilemma."
59. Rashid, *Taliban*, 226.
60. Arsala, "Revitalizing Afghanistan's Economy," 140.
61. Rashid, *Taliban*, 120–21.
62. Cited in Isby, *Afghanistan*, 180.
63. Rashid, *Taliban*, 226–27.

64. Jalali, "The Legacy of War and the Challenge of Peace Building," 38–39; O'Hanlon and Sherjan, *Toughing It Out in Afghanistan,* 54–57, 104.

65. Isby, *Afghanistan,* 218; Rashid, *Taliban,* 13.

66. Neumann, *The Other War,* 44, 80–83.

67. Neumann, *The Other War,* 44–45, 49–50; Karzai, "Strengthening Security in Contemporary Afghanistan," 68.

68. Oliker, *Aid During Conflict,* 82–88.

69. Chandrasekaren, "Marine unit gone rogue, or leading the way?"

70. Barno, "Fighting 'The Other War.'"

71. Woodward, *Obama's Wars,* 262.

72. Jalali, "The Legacy of War and the Challenge of Peace Building," 43.

73. Kilcullen, *The Accidental Guerrilla,* 70–109.

74. Wright and Tohid, "Drones Push Taliban from a Pakistani Haven."

75. Isby, *Afghanistan,* 202–8.

76. Peters, *Seeds of Terror*; Fazey, "Responding to the Opium Dilemma."

77. Jones, *Counterinsurgency in Afghanistan,* 68–72; Isby, *Afghanistan,* 318–23; Younossi, *The Long March.*

78. Isby, *Afghanistan,* 158.

79. Kilcullen, *The Accidental Guerrilla*; Jones, *Counterinsurgency in Afghanistan.*

80. Isby, *Afghanistan,* 301; Woodward, *Obama's Wars,* 133.

81. Jones, *Counterinsurgency in Afghanistan,* 39; Isby, *Afghanistan,* 148; Woodward, *Obama's Wars,* 133; O'Hanlon and Sherjan, *Toughing It Out in Afghanistan,* 26.

82. Kilcullen, *The Accidental Guerrilla,* 48–49.

83. Jones, *Counterinsurgency in Afghanistan,* 48–53.

84. Ibid.

85. Jones, *In the Graveyard of Empires,* 293–95; Gul, *The Most Dangerous Place,* 132–38.

86. Rashid, *Taliban,* 229.

87. Karzai, "Strengthening Security in Contemporary Afghanistan," 64–65.

88. Rashid, *Taliban,* 229.

89. Partlow, "Oft-targeted Kandahar police chief is killed."

90. Isby, *Afghanistan,* 160.

91. Jones, *Counterinsurgency in Afghanistan,* 51–53.

92. Rashid, *Taliban,* 234.

93. Isby, *Afghanistan,* 164–67; Karzai, "Strengthening Security in Contemporary Afghanistan," 65.

94. Karzai, "Strengthening Security in Contemporary Afghanistan," 67.

95. Karzai, "Strengthening Security in Contemporary Afghanistan," 65; O'Hanlon and Sherjan, *Toughing It Out in Afghanistan,* 130–31.

96. Rashid, *Taliban,* 229; Isby, *Afghanistan,* 255.

97. Rashid, *Taliban,* 23.

98. Jones, *In the Graveyard of Empires,* 106, 148–49, 207, 230–31, 283; Isby, *Afghanistan,* 136, 153.

99. "Dutch troops end Afghanistan deployment."

100. Austen, "Canada Poised for 2011 Withdrawal from Afghanistan."

101. "In the line of duty: Canada's casualties."

102. Wintour and Norton-Taylor, "British troops to withdraw from Sangin in Afghanistan."

103. UK Ministry of Defence, "Operations in Afghanistan: British Fatalities." See also O'Hanlon and Sherjan, *Toughing It Out in Afghanistan*, 132–33.

104. Woodward, *Obama's Wars*.

105. U.S. Department of Defense, "Operation Enduring Freedom U.S. Casualty Status."

106. Count as reported on the web site of iCasualties.org at http://icasualties.org/oef/, accessed September 24, 2011.

107. Isby, *Afghanistan*, 329.

108. Rogers and Sedghi, "Afghanistan civilian casualties: year by year, month by month."

109. UN Assistance Mission in Afghanistan, "Shifting Tactics Drive Record High Afghan Civilian Death Toll in First Half of 2011."

110. "'Troops in Contact.'"

111. Rashid, *Taliban*, 227.

112. Isby, *Afghanistan*, 160; Woodward, *Obama's Wars*, 68–69.

113. Christia and Semple, "Flipping the Taliban"; Isby, *Afghanistan*, 331; Semple, *Reconciliation in Afghanistan*; Kilcullen, *The Accidental Guerrilla*, 46; Biddle, "Is There a Middle Way?"

114. Rashid, *Taliban*, 35–36.

115. Isby, *Afghanistan*, 34; Rashid, *Taliban*, 79–80, 95–104.

116. Nagl, "A Better War in Afghanistan," 36.

117. Arreguín-Toft, *How the Weak Win Small Wars*, 232.

118. Jalali, "The Legacy of War and the Challenge of Peace Building," 44.

119. Partlow, "Karzai calls on U.S. to lighten troop presence."

120. Rashid, *Taliban*, 223; Semple, *Reconciliation in Afghanistan*.

121. Gul, *The Most Dangerous Place*, 175–76.

122. Whitlock, "U.S. terrorist hunter had long prepared for call."

123. Constable, "Pakistan reckons with fallout from CIA drone strikes."

124. Whitlock, "U.S. terrorist hunter had long prepared for call."

125. Woodward, *Obama's Wars*, 150.

126. O'Hanlon and Sherjan, *Toughing It Out in Afghanistan*, 27, 44.

127. Isby, *Afghanistan*, 163–64; Gul, *The Most Dangerous Place*, 191–93; Rashid, *Taliban*, 59.

128. Isby, *Afghanistan*, 167–69.

129. Jones, *Counterinsurgency in Afghanistan*.

130. Rothstein, *Afghanistan & the Troubled Future of Unconventional Warfare*; O'Hanlon and Sherjan, *Toughing It Out in Afghanistan*.

131. Isby, *Afghanistan*, 85, 111–13.

132. Biddle, "Is There a Middle Way?"; Woodward, *Obama's Wars*, 346, 348; O'Hanlon and Sherjan, *Toughing It Out in Afghanistan*, 32–34, 68–69, 89; Rashid, *Taliban*, 176.

133. Isby, *Afghanistan*, 299.

134. Jones, *Counterinsurgency in Afghanistan*, 89–91.

135. Isby, *Afghanistan*, 146.

136. Kilcullen, *The Accidental Guerrilla*, 44.

137. Nagl, "A Better War in Afghanistan," 33.

138. Woodward, *Obama's Wars*, 308–9.

139. Ibid., 336.

140. Nagl, "A Better War in Afghanistan," 37.

141. Isby, *Afghanistan,* 306–7, 328; Rothstein, *Afghanistan & the Troubled Future of Unconventional Warfare,* 98, 111, 114.

142. Flynn, Pottinger, and Batchelor, *Fixing Intel*; Woodward, *Obama's Wars,* 150; Blas, "Intelligence in Afghanistan."

143. Nagl, "A Better War in Afghanistan," 34.

144. Isby, *Afghanistan,* 106–7, 166–67; Rothstein, *Afghanistan & the Troubled Future of Unconventional Warfare,* 164; Woodward, *Obama's Wars,* 106; O'Hanlon and Sherjan, *Toughing It Out in Afghanistan,* 19.

145. Bennett, "Pressure builds to end the Afghan war."

146. O'Hanlon and Sherjan, *Toughing It Out in Afghanistan.*

CHAPTER 8

1. Slater, "Ideology vs. the National Interest."
2. Mearsheimer and Walt, *The Israel Lobby and U.S. Foreign Policy,* 231–43.
3. Lord, "The Role of the United States in Small Wars," 89–100.
4. Pape, *Bombing to Win*; Hoffman, "Neoclassical Counterinsurgency?" 82.
5. Kilcullen, "Countering Global Insurgency," 609.
6. Pape, *Dying to Win*; Hoffman, "Neoclassical Counterinsurgency?"
7. Freedman, *The Revolution in Strategic Affairs,* 66.
8. Gunaratna, *Inside Al Qaeda,* 318; Hoffman, "Small Wars Revisited."
9. Miles, *Al-Jazeera*; Hoffman, "Neoclassical Counterinsurgency?"
10. Echevarria, "Deconstructing the Theory of Fourth-Generation War," 233.
11. Gentry, "Complex Civil-Military Operations."
12. Hammes, *Private Contractors in Conflict Zones.*
13. Metz, "Three Years and You're Out."
14. Gunaratna, *Inside Al Qaeda,* 300.
15. Slim, *Killing Civilians,* 173–74, 284.
16. Cohen, "Distant Battles."
17. Kaldor, *The Baroque Arsenal,* 5; Gentry, "Doomed to Fail"; Betz, "The More You Know, the Less You Understand," 511–12.
18. Murray and Scales, *The Iraq War,* 183.
19. Adams, "Attack or Conquer?"
20. Hoffman, "Small Wars Revisited"; Hoffman, "Neoclassical Counterinsurgency?"; Kilcullen, "Countering Global Insurgency," 601–2; Betz, "The More You Know, the Less You Understand," 510.
21. Freedman, *The Revolution in Strategic Affairs,* 76.
22. Cohen, "Distant Battles."
23. Freedman, *The Revolution in Strategic Affairs,* 29.
24. Ucko, *The New Counterinsurgency Era.*
25. Gentry, "Intelligence Learning and Adaptation."
26. Linn, *The Echo of Battle,* 115, 129, 156–57, 162–63, 191.
27. Cohen, *Supreme Command.*
28. Gansler, "Protecting Cyberspace"; Gentry, "Doomed to Fail"; Klein, "The Complex Crux of Wireless Warfare"; Freedman, *The Revolution in Strategic Affairs,* 23, 50, 52–64, 69.
29. Snyder, *Myths of Empire,* 30; Linn, *The Echo of Battle,* 87–88, 129, 157, 162–63, 191.
30. Arreguín-Toft, *How the Weak Win Small Wars,* 33–34, 41n, 223–24.

31. Kilcullen, "Countering Global Insurgency," 606–8, 612, 615; Hammes, *The Sling and the Stone*; Metz, *Rethinking Insurgency*; Hoffman, "Neoclassical Counterinsurgency?"

32. Galula, *Counterinsurgency Warfare*, 58–62; Metz, *Rethinking Insurgency*, 48.

33. Arreguín-Toft, *How the Weak Win Small Wars*, 35.

34. Ibid., 225.

35. Hamilton, *The Art of Insurgency*, 20, 22.

36. Asprey, *War in the Shadows*.

37. Arreguín-Toft, *How the Weak Win Small Wars*, 18, 36–37.

38. Biddle, *Military Power*, 20–23; Arreguín-Toft, "How the Weak Win Small Wars," 97.

39. Arreguín-Toft, *How the Weak Win Small Wars*, 218.

40. Ibid., 6, 223.

41. Gentry, "Norms as Weapons of War."

42. Katzenstein, *The Culture of National Security*, 5.

43. Finnemore and Sikkink, "International Norm Dynamics and Political Change," 891.

44. Hoffman, "Neoclassical Counterinsurgency?"

45. The constructivist literature on broadly military issues is discussed in chapter 1.

46. Legro, *Cooperation Under Fire*, 238–39.

47. Finnemore and Sikkink, "International Norm Dynamics and Political Change."

48. Keck and Sikkink, *Activists Beyond Borders*, 13.

49. Farrell, "World Culture and Military Power"; Katzenstein, *The Culture of National Security*; Ikenberry and Kupchan, "Socialization and Hegemonic Power."

50. Gentry, "Norms as Weapons of War."

51. Farrell, "World Culture and Military Power," 486.

52. Yugoslavia recognized the threshold concept in 1999. Before the war, a Yugoslav diplomat noted the importance of gradualism in ethnic cleansing by quipping, "a village a day keeps NATO away," as cited in Geyer, "Kosovo Countdown."

53. Lyon, "Operation Allied Force."

54. Handel, *Masters of War*, 99–117.

55. Baldwin, "Power Analysis and World Politics," 192.

56. Smith, *The Utility of Force*, 242.

57. Blainey, *The Causes of War*, 114; Wohlforth, "The Stability of a Unipolar World," 24.

58. Baldwin, "Power Analysis and World Politics," 166, 168n.

59. Knorr, *The Power of Nations*, 45; Baldwin, "Power Analysis and World Politics"; Art, "American Foreign Policy and the Fungibility of Force"; Arreguín-Toft, *How the Weak Win Small Wars*, 37.

60. Waltz, "Structural Realism after the Cold War," 16; Desch, *Power and Military Effectiveness*, 171.

61. Baldwin, "Power Analysis and World Politics," 163–70.

62. Barnett, *Asymmetrical Warfare*; Hammes, *The Sling and the Stone*; Blank, *Rethinking Asymmetric Threats*; Metz, "Three Years and You're Out."

63. Williamson, *The Economic Institutions of Capitalism*.

64. Jenkins, *The Unchangeable War*; Krepinevich, *The Army and Vietnam*; McMaster, *Dereliction of Duty*; Cassidy, "Why Great Powers Fight Small Wars Badly"; Cassidy, *Peacekeeping in the Abyss*; Nagl, *Learning to Eat Soup with a Knife*; Alwyn-Foster,

"Changing the Army for Counterinsurgency Operations"; Linn, *The Echo of Battle*; Hoffman, "Small Wars Revisited"; Ucko, *The New Counterinsurgency Era*; Allison and Zelikow, *Essence of Decision.*

65. Komer, *Bureaucracy at War.*

66. For example, Alexander Downes argues, using data from 1816–2003, that targeting civilians in war is common and that democracies often do so. His finding that U.S. actions in Iraq in 1991 and 2003 are inconsistent with his generalization appears only as an anomaly in the general pattern—not a reflection of the fundamental change in the conduct of war since 1945. See Stam, *Win, Lose, or Draw,* 77; Waltz, *Theory of International Politics,* 76–77, 127; Downes, *Targeting Civilians in War.*

67. Cohen, "Toward Better Net Assessment," 86–87; Friedberg, "The Assessment of Military Power," 194.

68. Levy, "The Offense/Defense Balance of Military Technology"; Smith, *The Utility of Force,* 228–29.

69. Qaio and Wang, *Unrestricted Warfare.*

70. Galula, *Counterinsurgency Warfare,* 32.

71. Arreguín-Toft, *How the Weak Win Small Wars,* 217.

72. Jervis, "Cooperation Under the Security Dilemma." For descriptions of the difficulties several states had in determining net national capabilities in the 1930s, see Murray and Millet, *Calculations.*

73. Christensen and Snyder, "Chain Gangs and Passed Bucks."

74. Snyder, "The Security Dilemma in Alliance Politics," 467.

75. Waltz, *Theory of International Politics,* 68.

76. Qaio and Wang, *Unrestricted Warfare,* 35, 154.

77. Walt, *The Origins of Alliances*; Schweller, *Unanswered Threats*; Snyder, *Myths of Empire*; Rose, "Neoclassical Realism and Theories of Foreign Policy."

78. Waltz, *Theory of International Politics,* 195; Posen, "Nationalism, the Mass Army, and Military Power," 82.

79. Farrell, "World Culture and Military Power"; Katzenstein, *The Culture of National Security.*

80. Waltz, *Theory of International Politics,* 127.

81. During World War I, slogging attritional warfare was characteristic of the fighting on the western front in 1914–1917 but not of the western front in 1917–1918 or the eastern front for most of the war. See Biddle, *Military Power*; Gudmundsson, *Stormtroop Tactics.*

82. Schroeder, "Historical Reality vs. Neo-realist Theory."

83. Byman and Waxman, "Defeating US Coercion"; Mack, "Why Big Nations Lose Small Wars."

84. Lambeth, *NATO's Air War for Kosovo,* 231–32.

85. Byman and Waxman, *The Dynamics of Coercion,* 48–86, 134–37.

86. Schelling, *Arms and Influence*; Merom, *How Democracies Lose Small Wars.*

87. Jervis, *Perception and Misperception in International Politics*; Lebow, *Between Peace and War,* 101–47; Axelrod, *Framework for a General Theory of Cognition and Choice.*

88. Janis, *Victims of Groupthink.*

89. Lebow, *Between Peace and War,* 153–69.

90. Leites, *A Study of Bolshevism.*

91. Lebow, *Between Peace and War,* 169–91.

92. Friedberg, "The Assessment of Military Power."

93. Jenkins, *The Unchangeable War,* 3.

94. Press, *Calculating Credibility.*

95. Ibid.

96. Doyle, "Liberalism and World Politics"; Russett, *Grasping the Democratic Peace*; Owen, "How Liberalism Produces Democratic Peace"; Bennett and Stam, "The Declining Advantage of Democracy"; Reiter and Stam, *Democracies at War*; Feaver and Gelpi, *Choosing Your Battles*.

97. Bennett and Stam, "The Declining Advantage of Democracy"; Gelpi, Feaver, and Reifler, *Paying the Human Costs of War*.

98. Nagl, *Learning to Eat Soup with a Knife*, 214–15; Avant, *Political Institutions and Military Change*, 21–36, 49–75; Grissom, "The Future of Innovation Studies," 908–10.

99. Jenkins, *The Unchangeable War*; Komer, *Bureaucracy at War*, 69–78; Cassidy, *Peacekeeping in the Abyss*; Alwyn-Foster, "Changing the Army for Counterinsurgency Operations"; Hoffman, "Small Wars Revisited"; Ucko, *The New Counterinsurgency Era*, 169–82.

100. Komer, *Bureaucracy at War*, 48–52.

101. Grissom, "The Future of Innovation Studies," 920–30.

102. Reiter and Meek, "Determinants of Military Strategy, 1903–1994."

103. Desch, "Democracy and Victory."

104. Reiter and Stam, *Democracies at War*, 193.

105. Desch, *Power and Military Effectiveness*; de Tocqueville, *Democracy in America*.

106. Bennett and Stam, "The Declining Advantage of Democracy."

107. Metz, "Three Years and You're Out."

108. For a summary of the academic debate on the issue, see four articles in *International Security* 30, no. 1 (Summer 2005) by Robert Pape, T. V. Paul, Stephen Brooks and William Wohlforth, and Keir Lieber and Gerard Alexander. See also Layne, "The Unipolar Illusion"; Wohlforth, "The Stability of a Unipolar World"; Layne, "Impotent Power."

109. Kepel and Milelli, *Al Qaeda in Its Own Words*, 60, 204, 254.

110. Shambaugh, *Modernizing China's Military*; Godwin, "The PLA Faces the Twenty-First Century."

111. Pillsbury, "Chinese Views of Future Warfare."

112. Qaio and Wang, *Unrestricted Warfare*, xii, xiii, 123, 177.

113. Gentry, "Doomed to Fail"; Klein, "The Complex Crux of Wireless Warfare."

114. Qaio and Wang, *Unrestricted Warfare*, 74–76.

115. Ibid., 76.

116. Soderberg, *The Superpower Myth*, 251–53, 260–61; O'Hanlon and Mochizuki, *Crisis on the Korean Peninsula*.

117. Rogers, "I Love It."

118. Rose, *How Wars End*, 284–87.

119. Larson and Savych, *Misfortunes of War*, 22–26, 71–74, 161–62; Bowden, *Black Hawk Down*, 43, 46; Gentry, "Norms and Military Power."

120. Gentry, "Casualty Management."

121. Linn, *The Echo of Battle*, 222; Betz, "The More You Know, the Less You Understand," 513.

122. Hoffman, *Conflict in the 21st Century*.

123. Rumsfeld, "Transforming the Military."

124. Bensahel, "Mission Not Accomplished"; Feith, *War and Decision*, 86–87, 139–40, 149, 158–59, 519–20.

125. Galula, *Counterinsurgency Warfare*, 89.

126. Komer, *Bureaucracy at War*; Luttwak, *The Pentagon and the Art of War*, 271; Feith, *War and Decision*, 349–50.

127. Hammes, *Private Contractors in Conflict Zones.*

128. Luttwak, "From Vietnam to *Desert Fox*."

129. Spector, *After Tet*, 260–78; Linn, *The Echo of Battle*, 199–200; Rumsfeld, *Known and Unknown*, 714.

130. Rogers, "I Love It"; Woodward, *The War Within*, 288; Rumsfeld, *Known and Unknown*, 710, 716.

131. Maze, "Amid global unrest, lawmakers ask: Will troops get enough dwell time?"

132. Freedman, *The Revolution in Strategic Affairs*, 34–35, 77–78; Luttwak, "A Post-Heroic Military Policy"; Bennett and Stam, "The Declining Advantage of Democracy"; Luttwak, "From Vietnam to *Desert Fox*:"

133. For a bibliography and analysis, see Feaver and Gelpi, *Choosing Your Battles*, 102.

134. Metz, "Three Years and You're Out."

135. Betz, "The More You Know, the Less You Understand," 527–29.

136. Halberstam, *War in a Time of Peace*, 451; Lambeth, *NATO's Air War for Kosovo*, 236; Lyon, "Operation Allied Force," 62, 70.

137. Mueller, "The Iraq Syndrome."

138. Cohen, "Constraints on America's Conduct of Small Wars"; Krepinevich, *The Army and Vietnam*; Lord, "The Role of the United States in Small Wars"; Ricks, *Fiasco*; Luttwak, *The Pentagon and the Art of War*; Shafer, *Deadly Paradigms*; Nagl, *Learning to Eat Soup with a Knife.*

139. Locher, *Victory on the Potomac*; Marquis, *Unconventional Warfare.*

140. Ervin, *Open Target.*

Bibliography

Adams, Karen Ruth. 2003/04. "Attack or Conquer? International Anarchy and the Offense-Defense-Deterrence Balance." *International Security* 28, no. 3 (Winter): 45–83.

Aguinaldo, Emilio. 1957. *A Second Look at America.* With Vicente Albano Pacis. New York: Robert Speller & Sons.

Albright, Madeleine. 2003. *Madame Secretary.* New York: Miramar.

Allawi, Ali A. 2007. *The Occupation of Iraq: Winning the War: Losing the Peace.* New Haven, CT: Yale University Press.

Al-Jabouri, Najim Abed, and Sterling Jensen. 2010. "The Iraqi and AQI Roles in the Sunni Awakening." *Prism* 2, no. 1 (November): 3–18.

Allison, Graham, and Philip Zelikow. 1999. *Essence of Decision: Explaining the Cuban Missile Crisis.* 2nd ed. New York: Longman.

Alvarez, Santiago V. 1992. *Recalling the Revolution: Memoirs of Filipino General,* translated by Paula Carolina S. Malay. Madison: University of Wisconsin Center for Southeast Asia Studies.

Alwyn-Foster, Nigel. 2005. "Changing the Army for Counterinsurgency Operations." *Military Review* 85, no. 6 (November–December): 2–15.

Arreguín-Toft, Ivan. 2001. "How the Weak Win Small Wars: A Theory of Asymmetric Conflict." *International Security* 26, no. 1 (Summer): 93–128.

Arreguín-Toft, Ivan. 2005. *How the Weak Win Small Wars: A Theory of Asymmetric Conflict.* Cambridge: Cambridge University Press.

Arsala, Hedayat Amin. 2007. "Revitalizing Afghanistan's Economy: The Government's Plan." In *Building a New Afghanistan,* ed. Robert I. Rotberg, 134–54. Cambridge, MA: World Peace Foundation.

Art, Robert J. 1996. "American Foreign Policy and the Fungibility of Force." *Security Studies* 5, no. 4 (Summer): 7–42.

Asprey, Robert B. 1975. *War in the Shadows: The Guerrilla in History.* Garden City, NJ: Doubleday.

Atkinson, Rick. 2004. *In the Company of Soldiers*. New York: Henry Holt.
Atkinson, Rick. 2007. "Left of Boom: The Struggle to Defeat Roadside Bombs," Part 1. *Washington Post* (September 30): A1.
Atkinson, Rick. 2007. "Left of Boom: The Struggle to Defeat Roadside Bombs," Part 2. *Washington Post* (October 1): A1.
Atkinson, Rick. 2007. "Left of Boom: The Struggle to Defeat Roadside Bombs," Part 3. *Washington Post* (October 2): A1.
Atkinson, Rick. 2007. "Left of Boom: The Struggle to Defeat Roadside Bombs," Part 4. *Washington Post* (October 3): A1.
Austen, Ian. 2008. "Canada Poised for 2011 Withdrawal from Afghanistan." *New York Times* (October 18), http://atwar.blogs.nytimes.com/2010/10/18/canada-poised-for-2011-withdrawal-from-afghanistan. Accessed August 14, 2011.
Avant, Deborah D. 1994. *Political Institutions and Military Change: Lessons from Peripheral Wars*. Ithaca, NY: Cornell University Press.
Axelrod, Robert. 1972. *Framework for a General Theory of Cognition and Choice*. Berkeley, CA: Institute of International Studies.
Baldwin, David. A. 1979. "Power Analysis and World Politics: New Trends versus Old Tendencies." *World Politics* 31, no. 2 (January): 161–94.
Barnett, Roger W. 2003. *Asymmetrical Warfare: Today's Challenge to U.S. Military Power*. Washington, DC: Brassey's.
Barno, David W. 2006. "Challenges in Fighting a Global Insurgency." *Parameters* 36, no. 2 (Summer): 15–29.
Barno, David W. 2007. "Fighting 'The Other War': Counterinsurgency Strategy in Afghanistan, 2003–2005," *Military Review* 87, no. 5 (September–October): 32–44.
Barno, David W. 2009. "Military Adaptation in Complex Operations." *prism* 1, no. 1 (December): 27–36.
Beck, Earl R. 1986. *Under the Bombs: The German Home Front in 1942–1945*. Lexington: University Press of Kentucky.
Beckett, Ian F. 2005. *Insurgency in Iraq: An Historical Perspective*. Carlisle, PA: U.S. Army War College Strategic Studies Institute.
Benjamin, Daniel, and Steven Simon. 2003. *The Age of Sacred Terror: Radical Islam's War Against America*. New York: Random House.
Bennett, D. Scott, and Allan C. Stam. 1998. "The Declining Advantage of Democracy: A Combined Model of War Outcomes and Duration." *Journal of Conflict Resolution* 42, no. 3 (June): 344–66.
Bennett, John T. 2011. "Pressure Builds to End the Afghan War." *The Hill*, 4 May, 1.
Bensahel, Nora. 2006. "Mission Not Accomplished: What Went Wrong with Iraqi Reconstruction." *Journal of Strategic Studies* 29, no. 3 (June): 453–73.
Berman, Larry. 2007. *Perfect Spy: The Incredible Double Life of Pham Xuan An*. New York: HarperCollins.
Betts, Richard K. 2000. "Is Strategy an Illusion?" *International Security* 25, no. 2 (Fall): 5–50.
Betz, David J. 2006. "The More You Know, the Less You Understand: The Problem with Information Warfare." *Journal of Strategic Studies* 29, no. 3 (June): 505–33.
Beyerchen, Alan. 1992–1993. "Clausewitz, Nonlinearity, and the Unpredictability of War." *International Security* 17, no. 3 (Winter): 59–90.
Biddle, Stephen. 1998. "The Past as Prologue: Assessing Theories of Future Warfare." *Security Studies* 8, no. 1 (Autumn): 1–74.

Biddle, Stephen. 2004. *Military Power: Explaining Victory and Defeat in Modern Battle.* Princeton, NJ: Princeton University Press.

Biddle, Stephen. 2009. "Is There a Middle Way?" *New Republic* (November 4): 28–31.

Biddle, Stephen, Fotini Christia, and J. Alexander Their. 2010. "Defining Success in Afghanistan: What Can the United States Accept?" *Foreign Affairs* 89, no. 4 (July–August): 48–60.

Biddle, Stephen, and Stephen Long. 2004. "Democracy and Military Effectiveness." *Journal of Conflict Resolution* 48, no. 3 (June): 525–45.

Biddle, Stephen D. 2001. "Rebuilding the Foundations of Offense-Defense Theory." *Journal of Politics* 63, no. 3 (August): 741–74.

Biddle, Tami Davis. 2002. *Rhetoric and Reality in Air Warfare.* Princeton, NJ: Princeton University Press.

Birtle, Andrew J. 1997. "The U.S Army's Pacification of Marinduque, Philippine Islands, April 1900–April 1901." *Journal of Military History* 61, no. 2 (April): 255–82.

Blainey, Geoffrey. 1973. *The Causes of War.* New York: Free Press.

Blank, Stephen J. 2003. *Rethinking Asymmetric Threats.* Carlisle, PA: U.S. Army War College.

Blas, Anne K. 2009. "Intelligence in Afghanistan: Why It Is Hard." *American Intelligence Journal* 27, no. 1 (Fall): 49–54.

Blaufarb, Douglas S. 1977. *The Counterinsurgency Era: U.S. Doctrine and Performance 1950 to the Present.* New York: Free Press.

Bodansky, Yossef. 2004. *The Secret History of the Iraq War.* New York: ReganBooks.

Bono, Giovanna. 2003. *NATO's 'Peace Enforcement' Tasks and 'Policy Communities': 1990–1999.* Aldershot, UK: Ashgate.

Boot, Max. 2002. *The Savage Wars of Peace.* New York: Basic.

Bowden, Mark. 1999. *Black Hawk Down.* New York: Atlantic Monthly Press.

Boyne, Walter J. 2003. *Operation Iraqi Freedom: What Went Right, What Went Wrong, and Why.* New York: Forge.

Brands, W. W. 1992. *Bound to Empire: The United States and the Philippines.* New York: Oxford University Press.

Bremer, L. Paul, III. 2006. *My Year in Iraq: The Struggle to Build a Future of Hope.* New York: Simon & Schuster.

Brooks, Risa A., and Elizabeth A. Stanley. 2007. *Creating Military Power: The Sources of Military Effectiveness.* Stanford, CA: Stanford University Press.

Brooks, Stephen G., and William C. Wohlforth. 2005. "Hard Times for Soft Balancing." *International Security* 30, no. 1 (Summer): 72–108.

Buckley, William Joseph, ed. 2000. *Kosovo: Contemporary Voices on Balkan Interventions.* Grand Rapids, MI: Eerdmans.

Bueno de Mesquito, Bruce. 1981. *The War Trap.* New Haven, CT: Yale University Press.

Bui Tin. 1995. *Following the Ho Chi Minh Trail: Memoirs of a North Vietnamese Colonel.* Honolulu: University of Hawaii Press.

Bui Tin. 2002. *From Enemy to Friend.* Annapolis, MD: Naval Institute Press.

Burchett, Wilfred G. 1966. *Vietnam North.* New York: International Press.

Burk, James. 1999. "Public Support for Peacekeeping in Lebanon and Somalia: Assessing the Casualties Hypothesis." *Political Science Quarterly* 114, no. 1 (Spring): 53–78.

Bush, George W. 2005. "Management of Interagency Efforts Concerning Reconstruction and Stabilization." National Security Presidential Directive 44 (NSPD-44), December 7. Washington DC: The White House.

Bush, George W. 2010. *Decision Points*. New York: Crown.

Byman, Daniel. 2008. *The Five Front War: The Better Way to Fight Global Jihad*. Hoboken, NJ: John Wiley & Sons.

Byman, Daniel, and Matthew Waxman. 1999. "Defeating US Coercion." *Survival* 41, no. 2 (Summer): 107–20.

Byman, Daniel, and Matthew Waxman. 2000. "Kosovo and the Great Air Power Debate." *International Security* 24, no. 4 (Spring): 5–38.

Byman, Daniel, and Matthew Waxman. 2002. *The Dynamics of Coercion: American Foreign Policy and the Limits of Military Power*. Cambridge: Cambridge University Press.

Caires, Greg. 1997. "Force Protection Requires Cultural Change, General Says." *Defense Daily* (August 25) http://findarticles.com/p/articles/mi_6712/is_n40_v196/ai_n28689231/, accessed September 19, 2011.

Caldwell, William B. IV, and Nathan K. Finney. 2010. "Building Police Capacity in Afghanistan: The Challenges of a Multilateral Approach." *Prism* 2, no. 1 (November): 121–30.

Carnahan, Burrus M. 1998. "Lincoln, Lieber and the Laws of War: The Origins and Limits of the Principle of Military Necessity." *American Journal of International Law* 92, no. 2 (April): 213–31.

Cassidy, Robert M. 2002. "Why Great Powers Fight Small Wars Badly." *Military Review* 82, no. 5 (September–October): 41–53.

Cassidy, Robert M. 2004. "Back to Street Without Joy: Counterinsurgency Lessons from Vietnam and Other Small Wars." *Parameters* 34, no. 2 (Summer): 73–83.

Cassidy, Robert M. 2004. *Peacekeeping in the Abyss: British and American Peacekeeping Doctrine and Practice After the Cold War*. Westport, CT: Praeger.

Chandrasekaren, Rajiv. 2006. *Imperial Life in the Emerald City: Inside Iraq's Green Zone*. New York: Alfred A. Knopf.

Chandrasekaren, Rajiv. 2010. "Marine unit gone rogue, or leading the way? *Washington Post* (March 14): A1.

Christensen, Thomas J., and Jack Snyder. 1990. "Chain Gangs and Passed Bucks: Predicting Alliance Patterns in Multipolarity." *International Organization* 44, no. 2 (Spring): 137–68.

Christia, Fotini, and Michael Semple. 2009. "Flipping the Taliban: How to Win in Afghanistan." *Foreign Affairs* 88, no. 4 (July/August): 34–45.

Clark, Derek J., and Kai A. Konrad. 2007. "Asymmetric Conflict: Weakest Link against Best Shot." *Journal of Conflict Resolution* 51, no. 3 (June): 457–69.

Clark, Wesley K. 2000. "The Strength of an Alliance." In *Kosovo: Contending Voices on Balkan Intervention*, ed. William Joseph Buckley. Grand Rapids, MI: Eerdmans: 253–55.

Clark, Wesley K. 2001. *Waging Modern War*. New York: PublicAffairs.

Clausewitz, Carl von. 1976. *On War*, edited and translated by Michael Howard and Peter Paret. Princeton, NJ: Princeton University Press.

Cline, Ray. 1975. *World Power Assessment*. Washington, DC: Center for Strategic and International Studies.

Clinton, William J. 2004. *My Life*. New York: Alfred A. Knopf.

Clodfelter, Mark. 2006. *The Limits of Air Power: The American Bombing of North Vietnam*. Lincoln: University of Nebraska Press.

Cohen, Eliot A. 1984. "Constraints on America's Conduct of Small Wars." *International Security* 9, no. 2 (Autumn): 151–81.

Cohen, Eliot A. 1986. "Distant Battles: Modern War in the Third World." *International Security* 10, no. 4 (Spring): 143–77.

Cohen, Eliot A. 1988. "Toward Better Net Assessment: Rethinking the European Conventional Balance." *International Security* 13, no. 1 (Summer): 50–89.

Cohen, Eliot A. 1994. "The Mystique of Air Power," *Foreign Affairs* 73, no. 1 (January/February): 109–24.

Cohen, Eliot A. 2001. "The Unequal Dialogue: The Theory and Reality of Civil-Military Relations and the Use of Force." In *Soldiers and Civilians: The Civil-Military Gap and American National Security*, ed. Peter D. Feaver and Richard H. Kohn, 429–57. Cambridge, MA: MIT Press.

Cohen, Eliot A. 2002. *Supreme Command: Soldiers, Statesmen, and Leadership in Wartime*. New York: Free Press.

Colby, William. 1989. *Lost Victory*. Chicago: Contemporary.

Constable, Pamela. 2011. "Pakistan reckons with fallout from CIA drone strikes." *Washington Post* (April 16): A11.

Conversino, Mark J. 1997. "Sawdust Superpower: Perceptions of U.S. Casualty Tolerance in the Post-Gulf War Era." *Strategic Review* 25, no. 1 (Winter): 15–23.

Cooper, Scott A. 2003."The Politics of Air Strikes." In *Immaculate Warfare*, ed. Stephen D. Wrage, 71–84. Westport, CT: Praeger.

Copeland, Dale C. 2000. *The Origins of Major War*. Ithaca, NY: Cornell University Press.

Cordesman, Anthony H. 2001. *The Lessons and Non-Lessons of the Air and Missile Campaign in Kosovo*. Westport, CT: Praeger.

Cordesman, Anthony H. 2003. *The Iraq War: Strategy, Tactics, and Military Lessons*. Washington, DC: Center for Strategic and International Studies.

Cordesman, Anthony H. 2004. *The War After the War: Strategic Lessons of Iraq and Afghanistan*. Washington, DC: Center for Strategic and International Studies.

Corum, James S., and Wray R. Johnson. 2003. *Airpower in Small Wars: Fighting Guerrillas and Terrorists*. Lawrence: University Press of Kansas.

Cosmas, Graham A. 1971. *An Army for Empire: The United States Army in the Spanish-American War*. Columbia: University of Missouri Press.

Crane, Conrad C. 1993. *Bombs, Cities, and Civilians*. Lawrence: University Press of Kansas.

Crews, Robert D. 2008. "Introduction." In *The Taliban and the Crisis of Afghanistan*, ed. Robert D. Crews and Amin Tarzi, 1–62. Cambridge, MA: Harvard University Press.

Crumpton, Henry A. 2005. "Intelligence and War: Afghanistan, 2001–2002." In *Transforming U.S. Intelligence*, ed. Jennifer E. Sims and Burton Gerber, 162–75. Washington, DC: Georgetown University Press.

Daalder, Ivo H. 2000. *Getting to Dayton: The Making of America's Bosnia Policy*. Washington, DC: Brookings.

Daalder, Ivo H., and Michael E. O'Hanlon. 2000. *Winning Ugly: NATO's War to Save Kosovo*. Washington, DC: Brookings.

Dallaire, Roméo. 2003. *Shake Hands with the Devil: The Failure of Humanity in Rwanda*. Toronto: Random House Canada.

Davis, Richard G. 1993. *Carl A. Spaatz and the Air War in Europe*. Washington, DC: Smithsonian Institution Press.

Deady, Timothy K. 2005. "Lessons from a Successful Counterinsurgency: The Philippines, 1899–1902," *Parameters* 35, no. 1 (Spring): 53–68.

Desch, Michael C. 2002. "Democracy and Victory: Why Regime Type Hardly Matters." *International Security* 27, no. 2 (Fall): 5–47.

Desch, Michael C. 2008. *Power and Military Effectiveness: The Fallacy of Democratic Triumphalism*. Baltimore, MD: Johns Hopkins University Press.

deSutter, Paula A. 1997. *Denial and Jeopardy: Deterring Iranian Use of NBC Weapons*. Washington, DC: National Defense University Press.

Diamond, Larry. 2006. *Squandered Victory: The American Occupation and the Bungled Effort to Bring Democracy to Iraq*. New York: Henry Holt.

Djukić, Slavoljub. 2001. Milošević and Marković*: A Lust for Power*. Montreal: McGill-Queen's University Press.

Doder, Dusko, and Louise Branson. 1999. *Milosevic: Portrait of a Tyrant*. New York: Free Press.

Donnelly, Thomas. 2004. *Operation Iraqi Freedom: A Strategic Assessment*. Washington, DC: American Enterprise Institute.

Douhet, Giulio. 1942. *Command of the Air*. New York: Coward-McCann.

Downes, Alexander B. 2008. *Targeting Civilians in War*. Ithaca, NY: Cornell University Press.

Doyle, Michael. 1986. "Liberalism and World Politics." *American Political Science Review* 80, no. 4 (December): 1151–69.

Drew, Dennis M. 1998. "U.S. Airpower Theory and the Insurgent Challenge: A Short Journey to Confusion." *Journal of Military History* 62, no. 4 (October): 809–32.

Drew, Elizabeth. 1994. *On the Edge: The Clinton Presidency*. New York: Simon & Schuster.

"Dutch troops end Afghanistan deployment." 2010. BBC (August 1), http://www.bbc.co.uk/news/world-south-asia-10829837. Accessed August 14, 2011.

Echevarria, Antulio J., II. 2005. "Deconstructing the Theory of Fourth-Generation War." *Contemporary Security Policy* 26, no. 2 (August): 233–41.

Eikenberry, Karl W. 1996. "Take No Casualties." *Parameters* 26, no. 2 (Summer): 109–18.

Ervin, Clark Kent. 2006. *Open Target: Where America Is Vulnerable to Attack*. New York: Palgrave Macmillan.

Fallows, James. 2006. *Blind Into Baghdad: America's War in Iraq*. New York: Vintage.

Farrell, Theo. 2005. "World Culture and Military Power." *Security Studies* 14, no. 3 (July–September): 448–88.

Faust, Karl Irving. 1970. *Campaigning in the Philippines*. New York: Arno.

Fazey, Cindy. 2007. "Responding to the Opium Dilemma." In *Building a New Afghanistan*, ed. Robert I. Rotberg, 178–204. Cambridge, MA: World Peace Foundation.

Feaver, Peter D. 2003. *Armed Servants: Agency, Oversight, and Civil-Military Relations*. Cambridge, MA: Harvard University Press.

Feaver, Peter D., and Christopher Gelpi. 1999. "A Look at Casualty Aversion: How Many Deaths Is Acceptable? A Surprising Answer." *Washington Post* (November 7): B3.

Feaver, Peter D., and Christopher Gelpi. 2004. *Choosing Your Battles: American Civil-Military Relations and the Use of Force*. Princeton, NJ: Princeton University Press.

Feith, Douglas J. 2008. *War and Decision: Inside the Pentagon at the Dawn of the War on Terror*. New York: Harper.

Finel, Bernard I. 1998–1999. "Taking Offense at Offense-Defense Theory." *International Security* 23, no. 3 (Winter): 182–89.

Finnemore, Martha, and Kathryn Sikkink. 1998. "International Norm Dynamics and Political Change." *International Organization* 52, no. 4 (Autumn): 887–917.

FitzGerald, Frances. 1972. *Fire in the Lake: The Vietnamese and the Americans in Vietnam*. New York: Vintage.

Fitzsimmons, James R. 1995. "The Coming Military Revolution: Opportunities and Risks." *Parameters* 25, no. 2 (Summer): 30–36.

Flynn, Michael T., Matt Pottinger, and Paul D. Batchelor. 2010. *Fixing Intel: A Blueprint for Making Intelligence Relevant in Afghanistan*. Washington, DC: Center for a New American Security.

Foreman, Jonathan. 1999. "The Casualty Myth." *National Review* (May 3): 40–41.

Foulk, Vincent L. 2007. *The Battle for Fallujah: Occupation, Resistance and Stalemate in the War in Iraq*. Jefferson, NC: McFarland.

Franks, Tommy R. 2004. *American Soldier*. New York: ReganBooks.

Freedman, Lawrence. 1998. *The Revolution in Strategic Affairs*. Adelphi Paper 318. London: Oxford University Press.

Freedman, Lawrence, and Efraim Karsh. 1993. *The Gulf Conflict 1990–1991: Diplomacy and War in the New World Order*. Princeton, NJ: Princeton University Press.

Friedberg, Aaron L. 1987–1988. "The Assessment of Military Power: A Review Essay." *International Security* 12, no. 3 (Winter): 190–202.

Gacek, Christopher M. 1994. *The Logic of Force: The Dilemma of Limited War in American Foreign Policy*. New York: Columbia University Press.

Galland, Adolf. 1954. *The First and the Last: The Rise and Fall of the Luftwaffe, 1938–1945*. New York: Ballantine.

Galula, David. 1964. *Counterinsurgency Warfare: Theory and Practice*. New York: Frederick A. Praeger.

Gamson, William A. 1968. *Power and Discontent*. Homewood, IL: Dorsey.

Gansler, Jacques S. 2002. "Protecting Cyberspace." In *Transforming America's Military*, ed. Hans Binnendijk, 331–44. Washington, DC: National Defense University Press.

Gates, John Morgan. 1973. *Schoolbooks and Krags: The United States Army in the Philippines, 1898–1902*. Westport, CT: Greenwood.

Gelb, Leslie H. 1979. *The Irony of Vietnam: The System Worked*. With Richard K. Betts. Washington, DC: Brookings.

Gellman, Barton. 1998. "U.S. and China Nearly Came to Blows in 1996." *Washington Post* (June 21): A20.

Gellman, Barton. 1999. "How the U.S. and Allies Went to War." *Washington Post* (April 18): A1.

Gelpi, Christopher, Peter D. Feaver, and Jason Reifler. 2009. *Paying the Human Costs of War: American Public Opinion and Casualties in Military Conflicts*. Princeton, NJ: Princeton University Press.

Gentry, John A. 1998. "Military Force in an Age of National Cowardice." *Washington Quarterly* 21, no. 4 (Autumn): 179–91.

Gentry, John A. 2000. "Complex Civil-Military Operations: A U.S. Military-centric Perspective." *Naval War College Review* 52, no. 4 (Autumn): 57–76.

Gentry, John A. 2002/03. "Doomed to Fail: America's Blind Faith in Military Technology." *Parameters* 32, no. 4 (Winter): 88–103.

Gentry, John A. 2006. "Norms and Military Power: NATO's War Against Yugoslavia." *Security Studies* 15, no. 2 (April–June): 187–224.

Gentry, John A. 2008. "Intelligence Failure Reframed." *Political Science Quarterly* 123, no. 2 (Summer): 247–70.

Gentry, John A. 2010. "Intelligence Learning and Adaptation: Lessons from Counterinsurgency Wars." *Intelligence and National Security* 25, no. 1 (February): 50–75.

Gentry, John A. 2010. "Norms as Weapons of War." *Defense & Security Analysis* 26, no. 1 (March): 11–30.

Gentry, John A. 2011. "Casualty Management: Shaping Civil-Military Operational Environments." *Comparative Strategy* 30, no. 3 (July–August): 242–53.

Geyer, Georgie Anne. 1998. "Kosovo Countdown: Any Plans After the Bombs?" *Washington Times* (October 5).

Gilpin, Robert. 1981. *War and Change in World Politics.* Cambridge: Cambridge University Press.

Glaser, Charles L. 1994/1995. "Realists as Optimists: Cooperation as Self-Help." *International Security* 19, no. 3 (Winter): 50–90.

Glaser, Charles L., and Chaim Kaufman. 1998. "What Is the Offense-Defense Balance and Can We Measure It?" *International Security* 22, no. 4 (Spring): 44–82.

Glasser, Susan B., and Steve Coll. 2005. "The Web as Weapon." *Washington Post* (August 9): A1.

Gocić, Goran. 2000. "Symbolic Warfare: Nato Versus the Serbian Media." In *Degraded Capability: The Media and the Kosovo Crisis,* ed. Philip Hammond and Edward S. Herman. London: Pluto: 88–96.

Godwin, Paul H. B. 1997. "Military technology and doctrine in Chinese military planning: compensating for obsolescence." In *Military Capacity and the Risk of War: China, India, Pakistan and Iran,* ed. Eric Arnett, 39–60. Oxford: Oxford University Press.

Godwin, Paul H. B. 1999. "The PLA Faces the Twenty-First Century: Reflections on Technology, Strategy, Operations, and Doctrine." In *China's Military Faces the Future,* ed. James R. Lilley and David Shambaugh. New York: M. E. Sharpe: 39–63.

Goff, Peter, ed. 1999. *The Kosovo News and Propaganda War.* Vienna: International Press Institute.

Gordon, Michael R., and Bernard E. Trainor. 1995. *The Generals' War.* Boston: Little, Brown.

Gordon, Michael R., and Bernard E. Trainor. 2006. *Cobra II: The Inside Story of the Invasion and Occupation of Iraq.* New York: Pantheon.

Graham, Bradley. 2009. *By His Own Rules: The Ambitions, Successes, and Ultimate Failures of Donald Rumsfeld.* New York: PublicAffairs.

Gray, Colin S. 1981. "National Styles in Strategy: The American Example." *International Security* 6, no. 2 (Autumn): 21–47.

Grieco, Joseph M. 1988. "Anarchy and the Limits of Cooperation: A Realist Critique of the Newest Liberal Institutionalism." *International Organization* 42, no. 3 (Summer): 485–507.

Grissom, Adam. 2006. "The Future of Innovation Studies." *Journal of Strategic Studies* 29, no. 5 (October): 905–34.

Gudmundsson, Bruce I. 1989. *Stormtroop Tactics: Innovation in the German Army, 1914–1918.* Westport, CT: Praeger.

Gul, Imtiaz. 2010. *The Most Dangerous Place: Pakistan's Lawless Frontier.* New York: Viking.

Gunaratna, Rohan. 2003. *Inside Al Qaeda.* New York: Berkeley.

Haas, Ernst. 1953. "The Balance of Power: Prescription, Concept, or Propaganda." *World Politics* 5, no. 4 (July): 442–77.

Halberstam, David. 2001. *War in a Time of Peace.* New York: Touchstone.

Hamilton, Donald W. 1998. *The Art of Insurgency: American Military Policy and the Failure of Strategy in Southeast Asia.* Westport, CT: Praeger.

Hammes, T. X. 2010. *Private Contractors in Conflict Zones: The Good, the Bad, and the Strategic Impact.* Strategic Forum no. 260. Washington, DC: Institute for National Strategic Studies, National Defense University.

Hammes, Thomas X. 2005. "War Evolving into the Fourth Generation." *Contemporary Security Policy* 26, no. 2 (August): 189–221.

Hammes, Thomas X. 2006. *The Sling and the Stone.* St. Paul, MN: Zenith.

Handel, Michael I. 1996. *Masters of War: Classical Strategic Thought.* 2nd ed. London: Frank Cass.

Harris, Arthur. 1947 [2005]. *Bomber Offensive.* Barnsley, UK: Pen & Sword.

Harris, John F., and Bradley Graham. 1999. "Clinton Is Reassessing Sufficiency of Air War." *Washington Post* (June 3): A1.

Hashim, Ahmed S. 2006. *Insurgency and Counter-Insurgency in Iraq.* Ithaca, NY: Cornell University Press.

Hastings, Max. 1979. *Bomber Command.* New York: Dial.

Hendrickson, David C., and Robert W. Tucker. 2005. *Revisions in Need of Revising: What Went Wrong in the Iraq War.* Carlisle, PA: U.S. Army War College.

Herring, Eric, and Glen Rangwala. 2006. *Iraq in Fragments: The Occupation and Its Legacy.* Ithaca, NY: Cornell University Press.

Herring, George C. 1979. *America's Longest War: The United States and Vietnam, 1950–1975.* New York: John Wiley & Sons.

Herring, George C. 1982. "America's Strategy in Vietnam: The Post-War Debate." *Military Affairs* 46, no. 2 (April): 57–63.

Herrington, Stuart A. 1982. *Silence Was a Weapon: The Vietnam War in the Villages.* Novato, CA: Presidio.

Hersch, Seymour M. 2004. *Chain of Command: The Road from 9/11 to Abu Ghraib.* New York: HarperCollins.

Hoeffding, Oleg. 1966. *Bombing North Vietnam: An Appraisal of Economic and Political Effects.* RM-5213. Santa Monica, CA: RAND.

Hoffman, Bruce. 2003. *Al Qaeda, Trends in Terrorism and Future Potentialities: An Assessment.* P-8078. Santa Monica, CA: RAND.

Hoffman, Bruce. 2007. "Scarier Than Bin Laden." *Washington Post* (September 9): B1.

Hoffman, F. G. 1996. *Decisive Force: The New American Way of War.* Westport, CT: Praeger.

Hoffman, Frank G. 2005. "Small Wars Revisited: The United States and Nontraditional Wars." *Journal of Strategic Studies* 28, no. 6 (December): 913–40.

Hoffman, Frank G. 2007. *Conflict in the 21st Century: The Rise of Hybrid Wars.* Arlington, VA: Potomac Institute for Policy Studies.

Hoffman, Frank G. 2007. "Neoclassical Counterinsurgency?" *Parameters* 37, no. 2 (Summer): 71–87.

Holbrooke, Richard. 1998. *To End a War.* New York: Random House.

Horowitz, Michael, and Dan Reiter. 2001. "When Does Bombing Work?: Quantitative Empirical Tests, 1917–1999." *Journal of Conflict Resolution* 45, no. 2 (April): 147–73.

Hosmer, Stephen T. 2001. *Why Milosevic Decided to Settle When He Did*. Santa Monica, CA: RAND.

Howell, William G., and Jon C. Pevehouse. 2007. "When Congress Stops Wars: Partisan Politics and Presidential Power." *Foreign Affairs* 86, no. 5 (September/October): 95–107.

Human Rights Watch. 1999. *A Week of Terror in Drenica: Human Rights Violations in Kosovo* (February 1). New York: Human Rights Watch.

Human Rights Watch. 1999. *Yugoslav Government War Crimes in Racak* (January 29). New York: Human RIghts Watch.

Hyde, Charles K. 2000. "Casualty Aversion: Implications for Policy Makers and Senior Military Officers." *Aerospace Power Journal* 14, no. 2 (Summer): 17–27.

Ignatieff, Michael. 2000. *Virtual War*. New York: Metropolitan.

Ignatius, David. 2007. "Next Challenge in Iraq." *Washington Post* (October 21): B7.

Ikenberry, G. John, and Charles Kupchan. 1990. "Socialization and Hegemonic Power." *International Organization* 44, no. 3 (Summer): 283–315.

Independent International Commission on Kosovo. 2000. *Kosovo Report*. Oxford: Oxford University Press.

"In the line of duty: Canada's casualties." 2011. CBC News (June 26), http://www.cbc.ca/news/background/afghanistan/casualties/list.html. Accessed August 14, 2011.

Iqbal, Malik Zafir. 2010. "An Appraisal of the Afghanistan-Pakistan Strategy to Counter Terrorism," *Parameters* 40, no. 2 (Summer): 14–32.

Isby, David. 2010. *Afghanistan: Graveyard of Empires: A New History of the Borderland*. New York: Pegasus.

Isikoff, Michael, and David Corn. 2006. *Hubris: The Inside Story of Spin, Scandal, and the Selling of the War in Iraq*. New York: Crown.

Jalali, Ali A. 2007. "The Legacy of War and the Challenge of Peace Building." In *Building a New Afghanistan*, ed. Robert I. Rotberg, 22–55. Cambridge, MA: World Peace Foundation.

Jalali, Ali A. 2009. "Winning in Afghanistan." *Parameters* 39, no. 1 (Spring): 5–21.

Janis, Irving. 1972. *Victims of Groupthink: A Psychological Study of Foreign-Policy Decisions and Fiascos*. Boston: Houghton Mifflin.

Jenkins, Brian M. 1970. *The Unchangeable War*. RM-6278–2-ARPA. Santa Monica, CA: RAND.

Jervis, Robert. 1976. *Perception and Misperception in International Politics*. Princeton, NJ: Princeton University Press.

Jervis, Robert. 1979. "Cooperation Under the Security Dilemma." *World Politics* 30, no. 2 (January): 167–214.

Joes, Anthony James. 2000. *America and Guerrilla Warfare*. Lexington: University Press of Kentucky.

Johnson, Lyndon Baines. 1971. *The Vantage Point: Perspectives of the Presidency, 1963–1969*. New York: Holt, Rinehart and Winston.

Johnson, Thomas H., and M. Chris Mason. 2008. "No Sign until the Burst of Fire: Understanding the Pakistan-Afghanistan Frontier." *International Security* 32, no. 4 (Spring): 41–77.

Jones, Ronald H. 2005. *Terrorist Beheadings: Cultural and Strategic Implications*. Carlisle, PA: U.S. Army Strategic Studies Institute.

Jones, Seth. 2009. *In the Graveyard of Empires: America's War in Afghanistan*. New York: W. W. Norton.

Jones, Seth G. 2008. *Counterinsurgency in Afghanistan*. Santa Monica, CA: RAND.

Jones, Seth G. 2008. "The Rise of Afghanistan's Insurgency: State Failure and Jihad." *International Security* 32, no. 4 (Spring): 7–40.

Judah, Tim. 2002. *Kosovo: War and Revenge*. New Haven, CT: Yale University Press.

Kahl, Colin H. 2006. "How We Fight." *Foreign Affairs* 85, no. 6 (November/December): 83–101.

Kahl, Colin H. 2007. "In the Crossfire or the Crosshairs: Norms, Civilian Casualties, and U.S. Conduct in Iraq." *International Security* 32, no. 1 (Summer): 7–46.

Kaldor, Mary. 1981. *The Baroque Arsenal*. New York: Hill and Wang.

Karnow, Stanley. 1983. *In Our Image: America's Empire in the Philippines*. New York: Ballantine.

Karnow, Stanley. 1989. *Vietnam: A History*. New York: Viking.

Karzai, Hekmat. 2007. "Strengthening Security in Contemporary Afghanistan: Coping with the Taliban." In *Building a New Afghanistan*, ed. Robert I. Rotberg, 56–81. Cambridge, MA: World Peace Foundation.

Kattenburg, Paul M. 1980. *The Vietnam Trauma in American Foreign Policy, 1945–75*. New Brunswick, NJ: Transaction.

Katzenstein, Peter J., ed. 1996. *The Culture of National Security: Norms and Identity in World Politics*. New York: Columbia University Press.

Kay, Sean. 2004. "NATO, the Kosovo War and Neoliberal Theory." *Contemporary Security Policy* 25, no. 2 (August): 252–79.

Keaney, Thomas A., and Eliot A. Cohen. 1993. *Gulf War Air Power Survey (GWAPS) Summary Report*. Washington, DC: Government Printing Office.

Keck, Margaret E., and Kathryn Sikkink. 1998. *Activists Beyond Borders*. Ithaca, NY: Cornell University Press.

Keegan, John. 2004. *The Iraq War*. New York: Alfred A. Knopf.

Kepel, Gilles, and Jean-Pierre Milelli, ed. 2008. *Al Qaeda in Its Own Words*. Cambridge, MA: Harvard University Press.

Khong, Yuen Foong. 1992. *Analogies at War: Korea, Munich, Dien Bien Phu and the Vietnam Decisions of 1965*. Princeton, NJ: Princeton University Press.

Kier, Elizabeth. 1997. *Imagining War: French and British Military Doctrine Between the Wars*. Princeton, NJ: Princeton University Press.

Kilcullen, David J. 2005. "Countering Global Insurgency." *Journal of Strategic Studies* 28, no. 4 (August): 597–617.

Kilcullen, David. 2009. *The Accidental Guerrilla: Fighting Small Wars in the Midst of a Big One*. New York: Oxford University Press.

Kissinger, Henry A. 1969. "The Vietnam Negotiations." *Foreign Affairs* 47, no. 2 (January): 211–34.

Kissinger, Henry. 1994. *Diplomacy*. New York: Simon & Schuster.

Kitfield, James. 1995. *Prodigal Soldiers*. Washington, DC: Brassey's.

Kitfield, James. 2005. *War and Destiny*. Washington, DC: Potomac Books.

Klein, Alec. 2008. "The Complex Crux of Wireless Warfare." *Washington Post* (January 24): D1.

Klein, Yitzhak. 1991. "A Theory of Strategic Culture." *Comparative Strategy* 10, no. 1 (January–March): 3–23.

Knickmeyer, Ellen. 2005. "Surge in Deaths Reveals More Lethal Foe." *Washington Post* (August 7): A18.

Knights, Michael Andrew. 2005. *Cradle of Conflict: Iraq and the Birth of Modern U.S. Military Power*. Annapolis, MD: Naval Institute Press.

Knorr, Klaus. 1970. *Military Power and Potential*. Lexington, MA: D.C. Heath.

Knorr, Klaus. 1975. *The Power of Nations: The Political Economy of International Relations*. New York: Basic Books.

Knorr, Klaus. 1977. "Economic Interdependence and National Security in Historical Perspective." In *Economic Issues and National Security*, ed. Klaus Knorr and Frank N. Trager, 19-66. Lawrence: Regents Press of Kansas.

Kohn, Richard. 1994. "Out of Control: The Crisis in Civil-Military Relations." *National Interest* 35 (Spring): 3–17.

Kohn, Richard. 2001. "The Early Retirement of Gen Ronald R. Fogelman, Chief of Staff, United States Air Force." *Aerospace Power Journal* (Spring): 6–23.

Kolko, Gabriel. 1994. *Anatomy of a War: Vietnam, the United States, and the Modern Historical Experience*. New York: New Press.

Komer, Robert W. 1971. "Impact of Pacification on Insurgency in South Vietnam." *Journal of International Affairs* 25, no. 1 (Winter): 278–362.

Komer, Robert W. 1986. *Bureaucracy at War: U.S. Performance in the Vietnam Conflict*. Boulder, CO: Westview.

Krepinevich, Andrew F., Jr. 1986. *The Army and Vietnam*. Baltimore, MD: Johns Hopkins University Press.

Krepinevich, Andrew F., Jr. 2005. "How to Win in Iraq." *Foreign Affairs* 84, no. 5 (September/October): 87–104.

Kretchik, Walter E. 1997. "Force Protection Disparities." *Military Review* 77, no. 4 (July–August): 73–78.

Kull, Steven, and I. M. Destler. 1999. *Misreading the Public: The Myth of a New Isolationism*. Washington, DC: Brookings.

Kuperman, Alan J. 2006. "Suicidal Rebellions and the Moral Hazard of Humanitarian Intervention." In *Gambling on Humanitarian Intervention: Moral Hazard, Rebellion and Civil War*, ed. Timothy W. Crawford and Alan J. Kuperman, 1–25. New York: Routledge.

Kuperman, Alan J. 2008. "The Moral Hazard of Humanitarian Intervention: Lessons from the Balkans." *International Studies Quarterly* 52, no. 1 (March): 49–80.

Lambeth, Benjamin S. 2001. *NATO's Air War for Kosovo: A Strategic and Operational Assessment*. Santa Monica, CA: RAND.

Larson, Eric V. 1996. *The Historical Role of Casualties in Domestic Support for U.S. Military Operations*. MR-726-RC. Santa Monica, CA: RAND.

Larson, Eric V., and Bogdan Savych. 2007. *Misfortunes of War: Press and Public Reactions to Civilian Casualties in Wartime*. Santa Monica, CA: RAND.

Layne, Christopher. 1993. "The Unipolar Illusion: Why New Great Powers Will Arise." *International Security* 17, no. 4 (Spring): 5–51.

Layne, Christopher. 2001. "Collateral Damage in Yugoslavia." In *NATO's Empty Victory*, ed. Ted Galen Carpenter, 51–58. Washington, DC: Cato.

Layne, Christopher. 2006. "Impotent Power: Re-examining the Nature of America's Hegemonic Power." *National Interest* 85 (September/October): 41–48.

Layne, Christopher. 2006. *The Peace of Illusion: American Grand Strategy from 1940 to the Present*. Ithaca, NY: Cornell University Press.

Lebow, Richard Ned. 1981. *Between Peace and War: The Nature of International Crisis*. Baltimore, MD: Johns Hopkins University Press.

Legro, Jeffrey W. 1994. "Military Culture and Inadvertent Escalation in World War II." *International Security* 18, no. 4 (Spring): 108–42.

Legro, Jeffrey W. 1995. *Cooperation Under Fire*. Ithaca, NY: Cornell University Press.

Leites, Nathan. 1953. *A Study of Bolshevism*. Glencoe, IL: Free Press.

Levy, Jack S. 1983. *War in the Modern Great Power System, 1495–1975*. Lexington: University Press of Kentucky.

Levy, Jack S. 1984. "The Offense/Defense Balance of Military Technology: A Theoretical and Historical Analysis." *International Studies Quarterly* 28, no. 2 (June): 219–38.

Levy, Jack S. 1994. "Learning and Foreign Policy: Sweeping a Conceptual Minefield." *International Organization* 48, no. 2 (Spring): 279–312.

Lewy, Guenter. 1978. *America in Vietnam*. New York: Oxford University Press.

Liddell Hart, B. H. 1970. *History of the Second World War*. New York: Putnam.

Liddell Hart, B. H. 1991. *Strategy*. 2nd.rev. ed. New York: Meridian.

Lieber, Keir A. 2000. "Grasping the Technological Peace: The Offense-Defense Balance and International Security." *International Security* 25, no. 1 (Summer): 71–104.

Lieber, Keir A. 2005. *War and the Engineers: The Primacy of Politics over Technology*. Ithaca, NY: Cornell University Press.

Lieber, Keir A., and Gerard Alexander. 2005. "Waiting for Balancing: Why the World Is Not Pushing Back." *International Security* 30, no. 1 (Summer): 109–39.

Lind, William, Keith Nightengale, John Schmidt, Joseph Sutton, and Gary Wilson. 1989. "The Changing Face of War: Into the Fourth Generation." *Marine Corps Gazette* (October): 22–26.

Linn, Brian McAllister. 1989. *The U.S. Army and Counterinsurgency in the Philippine War, 1899–1902*. Chapel Hill: University of North Carolina Press.

Linn, Brian McAllister. 2000. *The Philippine War 1899–1902*. Lawrence: University Press of Kansas.

Linn, Brian McAllister. 2007. *The Echo of Battle: The Army's Way of War*. Cambridge, MA: Harvard University Press.

Locher, James R., III. 2002. *Victory on the Potomac: The Goldwater-Nichols Act Unifies the Pentagon*. College Station: Texas A&M University Press.

Lord, Carnes. 1995. "The Role of the United States in Small Wars." *Annals of the American Academy of Political and Social Science* 541, Small Wars (September): 89–100.

Luttwak, Edward N. 1984. *The Pentagon and the Art of War*. New York: Simon & Schuster.

Luttwak, Edward N. 1996. "A Post-Heroic Military Policy." *Foreign Affairs* 75, no. 4 (July–August): 33–44.

Luttwak, Edward N. 1999. "From Vietnam to *Desert Fox*: Civil-Military Relations in Modern Democracies." *Survival* 41, no. 1 (Spring): 99–112.

Lynn-Jones, Sean M. 1995. "Offense-Defense Theory and Its Critics." *Security Studies* 4, no. 4 (Summer): 660–91.

Lyon, Charlie. 2001. "Operation Allied Force: A Lesson on Strategy, Risk, and Tactical Execution." *Comparative Strategy* 20, no. 1 (January–March): 57–75.

Macdonald, Peter. 1993. *Giap: The Victor in Vietnam*. New York: W. W. Norton.

Mack, Andrew. 1975. "Why Big Nations Lose Small Wars: The Politics of Asymmetric Conflict." *World Politics* 27, no. 2 (January): 175–200.

Maclear, Michael. 1981. *The Ten Thousand Day War: Vietnam 1945–1975*. New York: St. Martin's.

Marquis, Susan L. 1997. *Unconventional Warfare: Rebuilding U.S. Special Operations Forces*. Washington, DC: Brookings.

Martel, William C. 2007. *Victory in War: Foundations of Modern Military Policy*. New York: Cambridge University Press.

Martin, Pierre, and Mark R. Brawley, ed. 2000. *Alliance Politics, Kosovo, and NATO's War: Allied Force or Forced Allies?* New York: Palgrave.

Matlary, Janne Haaland. 2006. *Values and Weapons: From Humanitarian Intervention to Regime Change?* New York: Palgrave Macmillan.

Matthews, William. 1995. "Shali's Warning: Casualties Are Inevitable." *Air Force Times* (October 16): 27.

Maze, Rick. 2011. "Amid global unrest, lawmakers ask: Will troops get enough dwell time?" *Air Force Times* (March 29): 8.

McCoy, Alfred W. 2007. "Torture in the Crucible of Counterinsurgency." In *Iraq and the Lessons of Vietnam*, ed. Lloyd G. Gardiner and Marilyn B. Young, 230–62. New York: New Press.

McMaster, H. R. 1997. *Dereliction of Duty: Lyndon Johnson, Robert McNamara, the Joint Chiefs of Staff, and the Lies That Led to Vietnam*. New York: HarperPerennial.

McNamara, Robert S. 1995. *In Retrospect: The Tragedy and Lessons of Vietnam*. New York: Vintage.

Mearsheimer, John J. 1983. *Conventional Deterrence*. Ithaca, NY: Cornell University Press.

Mearsheimer, John J. 2001. *The Tragedy of Great Power Politics*. New York: Norton.

Mearsheimer, John J., and Stephen M. Walt. 2007. *The Israel Lobby and U.S. Foreign Policy*. New York: Farrar, Strauss and Giroux.

Meckler, Laura, and Adam Entous. 2011. "U.S. Forces Kill Osama Bin Laden." *Wall Street Journal* (May 2): A1.

Meilinger, Phillip S. 2004. "The Origins of Effects-Based Operations." *Joint Force Quarterly* 35 (Autumn): 116–22.

Merom, Gil. 2003. *How Democracies Lose Small Wars*. Cambridge: Cambridge University Press.

Metz, Steven. 2001. "Strategic Asymmetry." *Military Review* 81, no. 3 (July–August): 25–28.

Metz, Steven. 2003–2004. "Insurgency and Counterinsurgency in Iraq." *Washington Quarterly* 21, no. 1 (Winter): 25–36.

Metz, Steven. 2007. *Rethinking Insurgency*. Carlisle, PA: U.S. Army War College Strategic Studies Institute.

Metz, Steven. 2008. "Three Years and You're Out." *National Interest* 93 (January/February): 54–57.

Michaels, Jim. 2007. "19,000 insurgents killed in Iraq since '03." *USA Today* (September 26), http://www.usatoday.com/news/world/iraq/2007-09-26-insurgents_N.htm. Accessed August 14, 2011.

Mierzejewski, Alfred C. 1988. *The Collapse of the German War Economy, 1944–1945: Allied Air Power and the German National Railway*. Chapel Hill: University of North Carolina Press.

Miles, Hugh. 2005. *Al-Jazeera: The Inside Story of the Arab News Channel That Is Challenging the West*. New York: Grove.

Military History Institute of Vietnam. 2002. *Victory in Vietnam: The Official History of the People's Army of Vietnam 1954–1975*, translated by Merle L. Probbenov. Lawrence: University Press of Kansas.

Miller, Stuart Creighton. 1982. *"Benevolent Assimilation": The American Conquest of the Philippines, 1899–1903*. New Haven, CT: Yale University Press.

Millett, Allan R. 1988. "The United States Armed Forces in the Second World War." In *Military Effectiveness, Volume III: The Second World War*, ed. Allan R. Millett and Williamson Murray, 45–89. Boston: Allen & Unwin.

Mockaitis, Thomas R. 2008. *Iraq and the Challenge of Counterinsurgency*. Westport, CT: Praeger Security International.

Momyer, William W. 2003. *Air Power in Three Wars*. Maxwell AFB, AL: Air University Press.

Moore, Alan. 1997. "Inside Looking Out." *Soldiers* (March): 16–17.

Morganthau, Hans. 1960. *Politics Among Nations*. 3rd ed. New York: Alfred A. Knopf.

Moyar, Mark. 1997. *Phoenix and the Birds of Prey: The CIA's Secret War to Destroy the Viet Cong*. Annapolis, MD: Naval Institute Press.

Moyar, Mark. 2006. *Triumph Forsaken: The Vietnam War, 1954–1965*. Cambridge: Cambridge University Press.

Mueller, John. 2005. "The Iraq Syndrome." *Foreign Affairs* 84, no. 6 (Nov./Dec.): 44–54.

Mueller, John E. 1980. "The Search for the 'Breaking Point' in Vietnam." *International Studies Quarterly* 24, no. 4 (December): 497–519.

Mueller, John E. 1994. *Policy and Opinion in the Gulf War*. Chicago: University of Chicago Press.

Murray, Williamson. 1985. *Luftwaffe*. Baltimore, MD: Nautical & Aviation Publishing.

Murray, Williamson. 1988. "British Military Effectiveness in the Second World War." In *Military Effectiveness, Volume III: The Second World War*, ed. Allan R. Millett and Williamson Murray, 90–135. Boston: Allen & Unwin.

Murray, Williamson. 1999. "Does Military Culture Matter?" *Orbis* (Winter): 27–42.

Murray, Williamson, and MacGregor Knox. 2001. "Conclusion: The Future behind Us." In *The Dynamics of Military Revolution*, ed. MacGregor Knox and Williamson Murray, 175–94. Cambridge: Cambridge University Press.

Murray, Williamson, and Allan R. Millet, ed. 1992. *Calculations: Net Assessment and the Coming of World War II*. New York: Free Press.

Murray, Williamson, and Robert H. Scales Jr. 2003. *The Iraq War: A Military History*. Cambridge, MA: Belknap.

Myers, Steven Lee. 1999. "U.S. Wields Defter Weapon Against Iraq: Concrete Bomb." *New York Times* (October 7): 3A.

Nagl, John A. 2005. *Learning to Eat Soup with a Knife: Counterinsurgency Lessons from Malaya and Vietnam*. Chicago: University of Chicago Press.

Nagl, John A. 2010. "A Better War in Afghanistan." *Joint Force Quarterly* 56 (1st Quarter): 32–39.

Nardulli, Bruce, R., Walter L. Perry, Bruce R. Pirnie, John Gordon IV, and John G. McGinn. 2002. *Disjointed War: Military Operations in Kosovo, 1999*. Santa Monica, CA: RAND.

National Commission on Terrorist Attacks on the United States. 2004. *The 9/11 Commission Report*. New York: W. W. Norton.

Neillands, Robin. 2001. *The Bomber War*. New York: Overlook.

Neumann, Ronald E. 2009. *The Other War: Winning and Losing in Afghanistan*. Washington, DC: Potomac.

Neustadt, Richard E., and Ernest R. May. 1986. *Thinking in Time: The Uses of History for Decision-Makers*. New York: Free Press.

O'Hanlon, Michael, and Mike Mochizuki. 2003. *Crisis on the Korean Peninsula: How to Deal with a Nuclear North Korea*. New York: McGraw-Hill.

O'Hanlon, Michael E. 2005. *Defense Strategies for the Post-Saddam Era*. Washington, DC: Brookings.

O'Hanlon, Michael E., and Hassina Sherjan. 2010. *Toughing It Out in Afghanistan*. Washington, DC: Brookings.

Oliker, Olga, Richard Kauzlerich, James Dobbins, Kurt W. Basseuner, Donald L. Sampler, John G. McGinn, Michael J. Dziedzic, et al. 2004. *Aid During Conflict: Interaction Between Military and Civilian Assistance Providers in Afghanistan, September 2001–June 2002*. Santa Monica, CA: RAND.

Olson, John E. 1999. "The Effect of Civilian Casualties on USAF Bombing Policy in Vietnam." *Air Power History* 46, no. 4 (Winter): 30–41.

Organski, A.F.K. 1968. *World Politics*. 2nd ed. New York: Alfred A. Knopf.

Organski, A.F.K., and Jacek Kugler. 1980. *The War Ledger*. Chicago: University of Chicago Press.

Osgood, Robert. 1957. *Limited War*. Chicago: University of Chicago Press.

Overy, Richard. 1995. *Why the Allies Won*. New York: W. W. Norton.

Owen, John M. 1994. "How Liberalism Produces Democratic Peace." *International Security* 19, no. 2 (Fall): 87–125.

Packer, George. 2005. *The Assassin's Gate: America in Iraq*. New York: Farrar, Strauss and Giroux.

Paley, Amit R., and Joshua Partlow. 2008. "Evolution of a U.S. General in Iraq." *Washington Post* (February 16): A1.

Palmer, Bruce, Jr. 1984. *The 25-Year War: America's Military Role in Vietnam*. Lexington: University Press of Kentucky.

Pape, Robert. A., Jr. 1990. "Coercive Air Power in the Vietnam War." *International Security* 15, no. 2 (Autumn): 103–46.

Pape, Robert. A. 1996. *Bombing to Win: Air Power and Coercion in War*. Ithaca, NY: Cornell University Press.

Pape, Robert. A. 1997/1998. "The Air Force Strikes Back: A Reply to Barry Watts and John Warden." *Security Studies* 7, no. 2 (Winter): 191–214.

Pape, Robert A. 2005. *Dying to Win: The Strategic Logic of Suicide Bombing*. New York: Random House.

Pape, Robert A. 2005. "Soft Balancing against the United States." *International Security* 30, no. 1 (Summer): 7–45.

Partlow, Joshua. 2010. "Karzai calls on U.S. to lighten troop presence." *Washington Post* (November 14): A1.

Partlow, Joshua. 2011. "Oft-targeted Kandahar police chief Is killed." *Washington Post* (April 16): A1.

Paul, T. V. 1994. *Asymmetric Conflicts: War Initiation By Weaker Powers*. Cambridge: Cambridge University Press.

Paul, T. V. 2005. "Soft Balancing in the Age of U.S. Primacy." *International Security* 30, no. 1 (Summer): 46–71.

Pentagon Papers, Senator Gravel edition. 1971. *The Defense Department History of United States Decisionmaking on Vietnam*. Boston: Beacon.

Perret, Geoffrey. 1993. *Winged Victory: The Army Air Forces in World War II*. New York: Random House.

Peters, Gretchen. 2009. *Seeds of Terror: How Heroin in Bankrolling the Taliban and al Qaeda*. New York: Thomas Dunne.

Petraeus, David H. 1986. "Lessons of History and Lessons of Vietnam." *Parameters* 16, no. 3 (Autumn): 43–53.

Phillips, David L. 2005. *Losing Iraq: Inside the Postwar Reconstruction Fiasco*. New York: Westview.

Pike, Douglas. 1966. *Viet Cong*. Cambridge, MA: MIT Press.

Pike, Douglas. 1969. *War, Peace, and the Viet Cong*. Cambridge, MA: MIT Press.

Pike, Douglas. 1986. *PAVN: People's Army of Vietnam*. Novato, CA: Presidio.

Pillsbury, Michael. 1999. "Chinese Views of Future Warfare." In *China's Military Faces the Future*, ed. James R. Lilley and David Shambaugh, 64–84. New York: M. E. Sharpe.

Pollack, Kenneth M. 2002. *Arabs at War: Military Effectiveness, 1948–1991*. Lincoln: University of Nebraska Press.

Posen, Barry R. 1984. *Sources of Military Doctrine*. Ithaca, NY: Cornell University Press.

Posen, Barry R. 1993. "Nationalism, the Mass Army, and Military Power." *International Security* 18, no. 2 (Autumn): 80–124.

Posen, Barry R. 1993. "The Security Dilemma and Ethnic Conflict." *Survival* 35, no. 1 (Spring): 27–47.

Posen, Barry R. 1996. "Military Responses to Refugee Disasters." *International Security* 21, no. 4 (Summer): 72–111.

Posen, Barry R. 2000. "The War for Kosovo." *International Security* 24, no. 4 (Spring): 39–84.

Powell, Colin. 1992/93. "U.S. Forces: Challenges Ahead," *Foreign Affairs* 72, no. 5 (Winter): 32–45.

Powell, Colin. 1995. *My American Journey*. New York: Random House.

Prados, John. 1995. *The Hidden History of the Vietnam War*. Chicago: Ivan R. Dee.

Press, Daryl G. 2005. *Calculating Credibility: How Leaders Assess Military Threats*. Ithaca, NY: Cornell University Press.

Price, Richard H. 1998. "Reversing the Gun Sights: Transnational Civil Society Targets Land Mines." *International Organization* 52, no. 3 (Summer): 613–44.

Priest, Dana. 1999. "A Decisive Battle That Never Was." *Washington Post* (September 19): A1.

Priest, Dana. 1999. "France Balked at NATO Targets." *Washington Post* (September 20): A1.

Qaio, Liang, and Wang Xiangsui. 2002. *Unrestricted Warfare*. Panama City: Pan American Publishing.

Quester, George. 1977. *Offense and Defense in the International System*. New York: John Wiley.

Race, Jeffrey. 1972. *War Comes to Long An: Revolutionary Conflict in a Vietnamese Province*. Berkeley: University of California Press.

Rashid, Ahmed. 2010. *Taliban: Militant Islam, Oil and Fundamentalism in Central Asia*. New Haven, CT: Yale University Press.

Record, Jeffrey. 1998. *The Wrong War: Why We Lost in Vietnam*. Annapolis, MD: Naval Institute Press.

Record, Jeffrey. 2000. "Force-Protection Fetishism: Sources, Consequences, and (?) Solutions." *Aerospace Power Journal* (Summer): 4–11.

Record, Jeffrey. 2004. *Dark Victory: America's Second War Against Iraq*. Annapolis, MD: Naval Institute Press.

Record, Jeffrey. 2005/2006. "Why the Strong Lose Wars." *Parameters* 35, no. 4 (Winter): 16–31.

Record, Jeffrey. 2007. *Beating Goliath: Why Insurgencies Win*. Washington, DC: Potomac.

Reiter, Dan. 1996. *Crucible of Beliefs: Learning, Alliances, and World Wars*. Ithaca, NY: Cornell University Press.

Reiter, Dan, and Curtis Meek. 1999. "Determinants of Military Strategy, 1903–1994: A Quantitative Empirical Test." *International Studies Quarterly* 43, no. 2 (June): 363–87.

Reiter, Dan, and Allan C. Stam III. 1998. "Democracy and Battlefield Military Effectiveness." *Journal of Conflict Resolution* 42, no. 3 (June): 259–77.

Reiter, Dan, and Allan C. Stam. 2002. *Democracies at War*. Princeton, NJ: Princeton University Press.

Reveron, Derek S. 2004. "U.S. European Command: General Wesley Clark and the War for Kosovo." In *America's Viceroys: The Military and U.S. Foreign Policy*, ed. Derek S. Reveron, 95–126. New York: Palgrave.

Ricks, Thomas E. 2006. *Fiasco: The American Military Adventure in Iraq*. New York: Penguin.

Ricks, Thomas E. 2009. *The Gamble: General David Petraeus and the American Military Adventure in Iraq, 2006–2008*. New York: Penguin.

Riehm, Peter J. A. 1997. "The USS *Harlan County* Affair." *Military Review* 77, no. 4 (July–August): 31–36.

Rogers, John. 2008. "I Love It. But I Have to Leave It." *Washington Post* (March 16): B1.

Rogers, Simon, and Ami Sedghi. 2011. "Afghanistan civilian casualties: year by year, month by month." *The Guardian* (UK) (March 10), http://www.guardian.co.uk/news/datablog/2010/aug/10/afghanistan-civilian-casualties-statistics, accessed (September 24): 2011.

Rose, Gideon. 1998. "Neoclassical Realism and Theories of Foreign Policy." *World Politics* 51, no. 1 (October): 144–72.

Rose, Gideon. 2010. *How Wars End: Why We Always Fight the Last Battle*. New York: Simon & Schuster.

Rose, Michael. 1998. *Fighting for Peace*. London: Harvill.

Rosen, Stephen Peter. 1996. *Societies and Military Power*. Ithaca, NY: Cornell University Press.

Rosen. Steven. 1982. "Vietnam and the American Theory of Limited War." *International Security* 7, no. 2 (Fall): 82–113.

Rothstein, Hy S. 2006. *Afghanistan and the Troubled Future of Unconventional Warfare*. Annapolis, MD: Naval Institute Press.

Rumsfeld, Donald. 2011. *Known and Unknown: A Memoir*. New York: Sentinel.

Rumsfeld, Donald H. 2002. "Transforming the Military." *Foreign Affairs* 81, no. 3 (May–June): 20–32.

Russett, Bruce. 1993. *Grasping the Democratic Peace: Principles for a Post-Cold War World*. Princeton, NJ: Princeton University Press.

Sanchez, Ricardo S. 2008. *Wiser in Battle: A Soldier's Story*. New York: HarperCollins.

Sapolsky, Harvey M., and Jeremy Shapiro. 1996. "Casualties, Technology, and America's Future Wars." *Parameters* 26, no. 2 (Summer): 119–27.

Scarborough, Rowan. 2004. *Rumsfeld's War*. Washington, DC: Regnery.

Schaffer, Ronald. 1985. *Wings of Judgment: American Bombing in World War II*. New York: Oxford University Press.

Schelling, Thomas C. 1963. *The Strategy of Conflict*. Cambridge, MA: Harvard University Press.

Schelling, Thomas C. 1966. *Arms and Influence*. New Haven, CT: Yale University Press.

Schirmer, Daniel B. 1972. *Republic or Empire: American Resistance to the Philippine War*. Cambridge, MA: Schenkman.

Schooner, Steven L. 2008. "Why Contractor Fatalities Matter." *Parameters* 38, no. 3 (Autumn 2008): 78–91.

Schoonover, Thomas. 2003. *Uncle Sam's War of 1898 and the Origins of Globalization*. Lexington: University Press of Kentucky.

Schroeder, Paul. 1994. "Historical Reality vs. Neo-realist Theory." *International Security* 19, no. 1 (Summer): 108–48.

Schroen, Gary C. 2005. *First In: An Insider's Account of How the CIA Spearheaded the War on Terror in Afghanistan*. New York: Ballantine.

Schwarz, Benjamin C. 1994. *Casualties, Public Opinion, and U.S. Military Intervention: Implications for Regional Deterrence*. MR-431-A/AF. Santa Monica, CA: RAND.

Schwarzkopf, H. Norman. 1992. *It Doesn't Take a Hero*. New York: Bantam.

Schweller, Randall L. 1992. Domestic Structure and Preventive War: Are Democracies More Pacific?" *World Politics* 44, no. 2 (January): 235–69.

Schweller, Randall L. 2004. "Unanswered Threats: A Neoclassical Realist Theory of Underbalancing." *International Security* 29, no. 2 (Fall): 159–201.

Schweller, Randall L. 2006. *Unanswered Threats: Political Constraints on the Balance of Power*. Princeton, NJ: Princeton University Press.

Sciolino, Elaine, and Steven Lee Myers. 2001. "Bush Says 'Time Is Running Out'; U.S. Plans to Act Largely Alone." *New York Times* (October 7): 1A.

Sebti, Bassam. 2007. "What I Risked as an Iraqi Journalist." *Washington Post* (October 21): B1.

Sell, Louis. 2002. *Slobodan Milosevic and the Destruction of Yugoslavia*. Durham, NC: Duke University Press.

Semple, Michael. 2009. *Reconciliation in Afghanistan*. Washington, DC: U.S. Institute of Peace Press.

Shafer, D. Michael. 1988. *Deadly Paradigms: The Failure of U.S. Counterinsurgency Policy*. Princeton, NJ: Princeton University Press.

Shambaugh, David. 2002. *Modernizing China's Military: Progress, Problems, Prospects*. Berkeley: University of California Press.

Sharp, Ulysses S. Grant. 1978. *Strategy for Defeat: Vietnam in Retrospect*. San Rafael, CA: Presidio.

Sheehan, Neil. 1988. *A Bright Shining Lie: John Paul Vann and America in Vietnam*. New York: Random House.

Shimshoni, Jonathan. 1990/1991. "Technology, Military Advantage, and World War I: A Case of Military Entrepreneurship." *International Security* 15, no. 3 (Winter): 187–215.

Short, Michael. 1999. "Lessons Learned from Military Operations and Relief Efforts in Kosovo." Testimony at Hearings of U.S. Senate Armed Services Committee, October 21.

Silbey, David J. 2007. *A War of Frontier and Empire: The Philippine-American War of 1899–1902*. New York: Hill and Wang.

Singer, J. David, and Melvin Small. 1966. "The Composition and Status Ordering of the International System, 1815–1940." *World Politics* 18, no. 2 (Winter): 236–82.

Singer, J. David. 1990. "Reconstructing the Correlates of War Data Set on Material Capabilities of States, 1816–1985." In *Measuring the Correlates of War*, ed. J. David Singer and Paul F. Diehl, 53–71. Ann Arbor: University of Michigan Press.

Slater, Jerome. 2002. "Ideology vs. the National Interest: Bush, Sharon, and U.S. Policy in the Israeli-Palestinian Conflict." *Security Studies* 12, no. 1 (Autumn): 164–206.

Slim, Hugo. 2008. *Killing Civilians: Method, Madness, and Morality in War*. New York: Columbia University Press.

Smith, R. Jeffrey. 1999. "A G.I.'s Home Is His Fortress." *Washington Post* (October 5): A1.

Smith, Rupert. 2007. *The Utility of Force: The Art of War in the Modern World*. New York: Alfred A. Knopf.

Snyder, Glenn H. 1984. "The Security Dilemma in Alliance Politics." *World Politics* 36, no. 4 (July): 461–95.

Snyder, Jack. 1991. *Myths of Empire: Domestic Politics and International Ambition*. Ithaca, NY: Cornell University Press.

Snyder, Jack L. 1984. *The Ideology of the Offensive: Military Decision Making and the Disasters of 1914*. Ithaca, NY: Cornell University Press.

Soderberg, Nancy. 2005. *The Superpower Myth: The Use and Misuse of American Might*. Hoboken, NJ: John Wiley & Sons.

Sorley, Lewis. 1999. *A Better War: The Unexamined Victories and Final Tragedy of America's Last Years in Vietnam*. New York: Harcourt Brace.

Sorley, Lewis. 2004. *Vietnam Chronicles: The Abrams Tapes 1968–1972*. Lubbock: Texas Tech University Press.

Spector, Ronald H. 1993. *After Tet: The Bloodiest Year in Vietnam*. New York: Free Press.

Speer, Albert. 1970. *Inside the Third Reich*. New York: Macmillan.

Sremac, Danielle S. 1999. *War of Words: Washington Tackles the Yugoslav Crisis*. Westport, CT: Praeger.

Stam, Allan C. 1996. *Win, Lose, or Draw: Domestic Politics and the Crucible of War*. Ann Arbor: University of Michigan Press.

Stigler, Andrew L. 2002/2003. "A Clear Victory for Air Power: NATO's Empty Threat to Invade Kosovo." *International Security* 27, no. 3 (Winter): 124–57.

Strachan, Hew. 2003. *The First World War*. New York: Penguin.

Strickland, Paul C. 2000. "USAF Aerospace-Power Doctrine: Decisive or Coercive?" *Aerospace Power Journal* 14, no. 3 (Fall): 13–25.

Sullivan, Patricia L. 2007. "War Aims and War Outcomes: Why Powerful States Lose Limited Wars." *Journal of Conflict Resolution* 51, no. 3 (June): 496–524.

Summers, Harry G., Jr. 1982. *On Strategy: A Critical Analysis of the Vietnam War*. New York: Presidio.

Sun-Tzu. 1993. *The Art of Warfare*, translated by Roger Ames. New York: Ballantine.

Suskind, Ron. 2006. *The One Percent Doctrine: Deep Inside America's Pursuit of Its Enemies Since 9/11*. New York: Simon & Schuster.

Talentino, Andrea Kathryn. 2005. *Military Intervention after the Cold War: Evolution of Theory and Practice*. Athens: Ohio University Press.

Tannenwald, Nina. 1999. "The Nuclear Taboo: The United States and the Normative Basis of Nuclear Non-Use." *International Organization* 53, no. 3 (Summer): 433–68.

Tanner, Stephen. 2002. *Afghanistan: A Military History from Alexander the Great to the Fall of the Taliban*. Cambridge, MA: Da Capo.

Taylor, Frederick. 2004. *Dresden*. New York: HarperCollins.

Tenet, George. 2007. *At the Center of the Storm: My Years at the CIA*. New York: HarperCollins.

Thomas, Ward. 2001. *The Ethics of Destruction*. Ithaca, NY: Cornell University Press.

Thompson, Robert. 1966. *Defeating Communist Insurgency: The Lessons of Malaya and Vietnam*. New York: Praeger.

Thompson, Robert. 1970. *No Exit from Vietnam*. New York: McKay.

Thompson, Robert. 1974. *Peace Is Not at Hand*. New York: McKay.

Thompson, Robert. 1989. *Make for the Hills: Memories of Far Eastern Wars*. London: Leo Cooper.

Thurow, Roger. 1994. "Serbs Bet that the West Won't Risk the Thing They Fear: Ground Troops." *Wall Street Journal* (April 21): A10.

Tilford, Earl H., Jr. 1991. *Setup: What the Air Force Did in Vietnam and Why*. Maxwell AFB, AL: Air University Press.

Tilford, Earl H., Jr. 1999–2000. "Operation Allied Force and the Role of Air Power." *Parameters* 29, no. 4 (Winter): 24–38.

Tocqueville, Alexis de. 1954. *Democracy in America*. New York: Vantage.

Tooze, Adam. 2006. *The Wages of Destruction: The Making and Breaking of the Nazi Economy*. New York: Viking.

Tran, Van Dong. 1978. *Our Endless War: Inside Vietnam*. Novato, CA: Presidio.

"'Troops in Contact': Air Strikes and Civilian Deaths in Afghanistan." Human Rights Watch (September 8), http://www.hrw.org/en/reports/2008/09/08/troops-contact-0. Accessed February 6, 2011.

Truong Nhu Tang. 1985. *A Vietcong Memoir*. New York: Vintage.

Turner, Kathleen J. 1985. *Lyndon Johnson's Dual War: Vietnam and the Press*. Chicago: University of Chicago Press.

Turner, Stansfield. 2005. *Burn Before Reading: Presidents, CIA Directors, and Secret Intelligence*. New York: Hyperion.

Tyson, Ann Scott. 2008. "FBI Agent: Saddam Didn't Expect Invasion." *Washington Post* (January 26): A13.

Ucko, David H. 2009. *The New Counterinsurgency Era: Transforming the U.S. Military for Modern Wars*. Washington, DC: Georgetown University Press.

UK Ministry of Defence, "Operations in Afghanistan: British Fatalities." http://www.mod.uk/DefenceInternet/FactSheets/OperationsFactsheets/OperationsInAfghanistanBritishFatalities.htm. Accessed August 14, 2011.

Ullman, Harlan, and James Wade Jr. 1996. *Shock and Awe: Achieving Rapid Dominance.* Washington, DC: Government Printing Office.

UN Assistance Mission in Afghanistan. 2011. "Shifting Tactics Drive Record High Afghan Civilian Death Toll in First Half of 2011." http://unama.unmissions.org/Default.aspx?tabid=1741&ctl=Details&mid=1882&ItemID=14443. Accessed August 14, 2011.

U.S. Army. 1993. *Operations.* Field Manual 100–5. Washington, DC: U.S. Army.

U.S. Army. 2006. *Counterinsurgency.* Field Manual 3–24 (and jointly as Marine Corps Warfighting Publication 3–33.5). Washington, DC: U.S. Army.

U.S. Department of Defense. 2005. *Annual Report to the President and to the Congress.* Washington, DC: Government Printing Office.

U.S. Department of Defense. 2011. *Department of Defense Dictionary of Military and Associated Terms.* Joint Publication 1–02. Washington DC: Government Printing Office.

U.S. Department of Defense. 2011. "Operation Enduring Freedom U.S. Casualty Status." http://www.defense.gov/news/casualty.pdf. Accessed April 22, 2011.

U.S. Department of Defense, Gordon England, Deputy Secretary. 2005. "Military Support for Stability, Security, Transition and Reconstruction (SSTR) Operations." Department of Defense Directive 3000.5, November 28.

U.S. General Accounting Office. 2001. *Kosovo Air Operations: Need to Maintain Alliance Cohesion Resulted in Doctrinal Departures.* GAO-01–784. Washington, DC: Government Printing Office

Valentine, Douglas. 1990. *The Phoenix Program.* New York: William Morrow.

Van Creveld, Martin. 1991. *The Transformation of War.* New York: Free Press, 1991.

Van Evera, Stephen. 1998. "Offense, Defense, and the Causes of War." *International Security* 22, no. 4 (Spring): 5–43.

Van Evera, Stephen. 1999. *Causes of War: Power and the Roots of Conflict.* Ithaca, NY: Cornell University Press.

Vien, Cao Van. 1983. *The Final Collapse.* Washington, DC: Center of Military History, U.S. Army.

Walt, Stephen M. 1987. *The Origins of Alliances.* Ithaca, NY: Cornell University Press.

Waltz, Kenneth N. 1979. *Theory of International Politics.* Boston: Addison-Wesley.

Waltz, Kenneth N. 2000. "Structural Realism after the Cold War." *International Security* 25, no. 1 (Summer): 5–41.

Warden, John A., III. 1992. "Employing Air Power in the Twenty-First Century." In *The Future of Air Power in the Aftermath of the Gulf War,* ed. Richard H. Schultz and Robert L. Pfalzgraff, 57–69. Maxwell AFB, AL: Air University Press.

Warden, John A., III. 1997/98. "Success in Modern War: A Response to Robert Pape's *Bombing to Win.*" *Security Studies* 7, no. 2 (Winter): 172–90.

Warden, John A., III. 1998. *The Air Campaign: Planning for Combat.* Washington, DC: National Defense University Press.

Warner, Michael. 2010. "'US Intelligence in Vietnam': The Official Version(s)." *Intelligence and National Security* 25, no. 5 (October): 611–37.

Watts, Barry D. 1984. *The Foundations of U.S. Air Doctrine: The Problem of Friction in War*. Maxwell AFB, AL: Air University Press.

Watts, Barry D. 1996. *Clausewitzian Friction and Future War*. McNair Papers 52. Washington, DC: Institute for National Strategic Studies, National Defense University.

Watts, Barry D. 1997/1998. "Ignoring Reality: Problems of Theory and Evidence in Security Studies." *Security Studies* 7, no. 2 (Winter): 115–71.

Weigley, Russell F. 1973. *The American Way of War: A History of United States Military Strategy and Policy*. Bloomington: Indiana University Press.

Weinberger, Caspar W. 1984. "The Uses of Military Power." Speech to the National Press Club, Washington, DC, November 28.

Weiner, Tim. 2007. *Legacy of Ashes: The History of the CIA*. New York: Doubleday.

Welch, Richard E., Jr. 1980. *Response to Imperialism: The United States and the Philippine-American War, 1899–1902*. Chapel Hill: University of North Carolina Press.

Welle, Joshua W. 2010. "Civil-Military Integration in Afghanistan: Creating Unity of Command." *Joint Force Quarterly* 56 (1st Quarter): 54–59.

Wendt, Diego. 1990. "Using a Sledgehammer to Kill a Gnat: The Air Force's Failure to Comprehend Insurgent Doctrine During Operation Rolling Thunder." *Air Power Journal* 4, no. 2 (Summer): 52–64.

West, Bing. 2005. *No True Glory: A Frontline Account of the Battle for Fallujah*. New York: Bantam Dell.

Westmoreland, William C. 1976. *A Soldier Reports*. Garden City, NJ: Doubleday.

White, Josh. 2008. "U.S. Boosts Its Use of Airstrikes in Iraq." *Washington Post* (January 17): A1.

White House Office of the Press Secretary. 1998. "Address by the President to the Nation on Iraq Air Strike." (December 16): 3.

Whitlock, Craig. 2008. "U.S. Network Falters in Mideast Mission." *Washington Post* (June 23): A1.

Whitlock, Craig. 2008. "Al-Qaeda's Growing Online Offensive." *Washington Post* (June 24): A1.

Whitlock, Craig. 2011. "U.S. terrorist hunter had long prepared for call." *Washington Post* (May 5): A1.

Wiebes, Cees. 2003. *Intelligence and the War in Bosnia, 1992–1995*. Hamburg: LIT Verlag Münster.

Williamson, Oliver E. 1985. *The Economic Institutions of Capitalism*. New York: Free Press.

Wintour, Patrick, and Richard Norton-Taylor. 2010. "British troops to withdraw from Sangin in Afghanistan." *The Guardian* (July 6), http://www.guardian.co.uk/uk/2010/jul/06/british-troops-sangin-withdrawal-afghanistan. Accessed August 14, 2011.

Wirtz, James J. 1991. "Intelligence to Please? The Order of Battle Controversy during the Vietnam War." *Political Science Quarterly* 106, no. 2 (Summer): 239–63.

Wirtz, James J. 2004. "The Balance of Power Paradox." In *Balance of Power: Theory and Practice in the 21st Century*, ed. T. V. Paul, James J. Wirtz, and Michel Fortmann, 127–49. Stanford, CA: Stanford University Press.

Wohlforth, William C. 1999. "The Stability of a Unipolar World." *International Security* 24, no. 1 (Summer): 5–41.

Woodward, Bob. 1991. *The Commanders*. New York: Simon & Schuster.

Woodward, Bob. 2002. *Bush at War*. New York: Simon & Schuster.
Woodward, Bob. 2004. *Plan of Attack*. New York: Simon & Schuster.
Woodward, Bob. 2006. *State of Denial*. New York: Simon & Schuster.
Woodward, Bob. 2008. *The War Within: A Secret White House History 2006–2008*. New York: Simon & Schuster.
Woodward, Bob. 2010. *Obama's Wars*. New York: Simon & Schuster.
Wright, Quincy. 1964. *A Study of War*. Chicago: University of Chicago Press.
Wright, Tom, and Quais Tohid. 2010. "Drones Push Taliban from a Pakistani Haven." *Washington Post* (November 5): A10.
Younossi, Obaid, Peter Dahl Thruelsen, Jonathan Vaccaro, Jerry M. Sollinger, and Brian Grady. 2009. *The Long March: Building an Afghan National Army*. Santa Monica, CA: RAND.

Index

About the Author

JOHN A. GENTRY is an adjunct professor at the National Intelligence University, Washington, DC. He is a retired U.S. Army Reserve officer with assignments mainly in special operations and intelligence analysis, including active duty with the 1st Special Forces Group in East Asia and a tour as a reservist with NATO's Peace Implementation Force (IFOR) in Bosnia in 1996. He spent 12 years as an intelligence analyst at the Central Intelligence Agency, where he worked primarily on Warsaw Pact issues and was for two years a senior analyst on the staff of the National Intelligence Officer for Warning. Gentry writes mainly on military, political, and intelligence topics and is author of *Lost Promise: How CIA Analysis Misserves the Nation* (1993) and articles in a variety of academic and policy journals. He received a PhD in political science from the George Washington University.